PRAISE FOR

AIR MANAGEMENT FOR THE FIRE SERVICE

"If we take what the authors of this book are telling us seriously, we greatly minimize our chances of *not going home*."

—*Billy Goldfeder,*
Deputy Fire Chief, Loveland-Symmes (OH) Fire Dept.

"This is a landmark book and a defining moment for our profession. Generations who come after may take this work as a given, but clearly those of us who watched the journey and witnessed the struggle know that Casey, Mike, Phil, and Steve have set the fire service down a safer road."

—*Chief Bobby Halton, ret.*
Editor in Chief, Fire Engineering
Education Director, FDIC

"For years, it has been common practice to wear SCBAs that allow fireground personnel to operate in hazardous environments. Unfortunately, it has also been common practice to use SCBAs until the low air warning alarm sounds before evaluating how much air is left to allow a safe exit, even when egress from the contaminated environment may not be accomplished in a timely manner. With fireground injuries and deaths on the rise, it is time for a change in philosophy regarding the use of SCBAs. This book is a must read for every firefighter who must wear SCBAs into contaminated atmosphere, and every officer responsible for personnel who wear SCBAs."

—John Mittendorf
Battalion Chief (ret., Los Angeles Fire Department)

"Today's fires produce some of the most toxic, explosive, and chemically-complex smoke that firefighters have ever faced. One breath—*one*—can alter a life forever or simply end it. The authors of this book understand today's smoke and have created many solutions to help manage the threat. It's one thing to identify problems—and yet another to create solutions. Mike, Casey, Steve, and Phil have thoroughly identified the problem *and* created solutions that can help you make a difference. Don't just read the book—embrace the concepts, practice the application, and help *make it safe* at your next fire."

Dave Dodson, Battalion Chief (ret.)
Author, The Art of Reading Smoke

AIR MANAGEMENT FOR THE FIRE SERVICE

AIR MANAGEMENT FOR THE FIRE SERVICE

Mike Gagliano

Casey Phillips

Phillip Jose

Steve Bernocco

Copyright © 2008 by
PennWell Corporation
1421 South Sheridan Road
Tulsa, Oklahoma 74112-6600 USA

800.752.9764
+1.918.831.9421
sales@pennwell.com
www.pennwellbooks.com
www.pennwell.com

Director: Mary McGee
Managing Editor: Steve Hill
Production Manager: Sheila Brock
Production Editor: Tony Quinn
Cover Designer: Clark Bell
Book Designer: Wes Rowell
Layout Artist: Susan E. Ormston Thompson

Photo credits: Front cover: Glen Ellman (Fortworthfire.com), Stefan Svensson and Peter Lundgren. Back cover: John Hanley Photo, Lt. Tim Dungan (Seattle Fire Department)

Library of Congress Cataloging-in-Publication Data

Air management for the fire service / Mike Gagliano ... [et al.].
   p. cm.
 Includes bibliographical references and index.
 ISBN-13: 978-1-59370-129-1
 1. Fire extinction--Equipment and supplies. 2. Fire prevention--Equipment and supplies. I. Gagliano, Mike.
 TH9310.5.A47 2008
 628.9'25--dc22

2007041131

Printed in the United States of America

1 2 3 4 5   12 11 10 09 08

DEDICATION

STEVE BERNOCCO

To my wife, Bonnie, and my children, Rossella, Stefano, and Antonio, whose understanding and support are limitless.

To all the officers and firefighters in the SFD I have had the honor to work with over the years.

And finally, to Tom Brennan, who took the time to listen to four guys from Seattle. Tom, you taught us so much. Thanks.

CASEY PHILLIPS

To my wife Lynn, without the love and support you've given me I would not be the person I am.

To all of the officers and firefighters that trained, mentored, molded and nurtured me.

Thank you!

PHILLIP JOSE

To my wife Trish, my daughter Maria, and my son Ian: The grace, love, and support you provide humbles me.

To the Madhatters: Take care of each other as you took care of me.

To my brothers and sisters, all of you, thank you.

MIKE GAGLIANO

To my lovely bride Anne: You are my wife, soul mate and best friend. I thank God every day for the privilege of sharing life with you.

To Mike and Rick: You amazed me when you were born and filled my days with joy as you grew. I am proud of the godly men you've become and am honored to be your father.

To Mom, Dad and Melissa: The family every kid wishes they had. Thanks for loving me with both your words and your deeds.

To my adopted family: The firefighters who have been and continue to be such a huge and wonderful part of my life.

And most of all I thank God, for all of the above and for your grace and love. The greatest day of my life was when I bowed my knee to Jesus Christ, my Lord and Savior.

ACKNOWLEDGMENTS

The Seattle Guys would like to thank the following:

Chief Alan Brunacini—America's Fire Chief. You are a great friend and mentor.

Bobby Halton—The moral and spiritual voice of today's fire service. We love you brother.

Diane Feldman—Chief Brennan's star pupil, and our trusted friend.

John Mittendorf—You set the standard for what every firefighter wishes to be both personally and professionally.

Dave Dodson—Thanks for your encouragement and example of excellence.

Billy Goldfeder—Thanks for your tireless efforts to see that everyone goes home.

To all our assistant instructors who help us carry our message to the troops. You were chosen because you are the best the fire service has to offer: Joel Andrus, Rick Verlinda, Tim Frank, Brian Maier, Jay Robinson, Joaquin Hubbard, John Bowen, Mike Wagner, John Cameron, Vance Anderson, and Brian O'Brien. And to those who will follow.

To all the outstanding contributors to our book whose photos and writing humbled and thrilled us, thanks.

Special thanks to:

Rick Verlinda—Your commitment to training inspires and makes a difference every day;

Brian Schaeffer and all great firefighters in Spokane, Spokane Valley, and Yakima;

Warren Merritt and our top notch neighbors in Bellevue;

Brian Hendrickson for all the encouragement and assistance with this book;

Dan Rossos and the courageous NFPA 1404 committee—Thanks for making the tough decisions;

Jim Bowie and our wonderful friends in South Carolina;

Sarah Larson and the Minnesota firefighters;

Ray Brown and all the Utah firefighters;

Forest Reeder and the great Illinois fire instructors;

Jim Crawford and the Pittsburgh Guys—The Royal Order of the Pink Toe/Woof, Woof, Woof;

Mike Dugan/Mike Ciampo and the New York Guys—No wait, it gets better!

Jim McCormack and the Indy Crew—Rock on brothers;

Mary Jane Dittmar, Julie Simmons, Steve Hill, Jerry Naylis, Francie Halcomb, Tony Quinn, and all the great Pennwell folks;

Harvey Eisner—Thanks for giving us our first shot.

Anne Gagliano—Thanks for making this book so much better.

We are also grateful to the following and so many more:

Chris Yob, Ron MacDougall, Mike Milam, Jesse Youngs, Raul Angulo, Scott Krushak, Gregory Dean, Tim Sendelbach, Skip Coleman, Rob Schnepp, John Lewis, Rob Moran, Gary Morris, David Bernzweig, Nicky Giordano and the logistics crew, Ed Olmstead and the AV crew, Glen Ellman, Bart Bradberry, Dave Kraski, the families of Bret Tarver and Mark Noble, and all the committed firefighters out there who are taking the message of Air Management to firefighters across the nation.

And finally, we acknowledge the men and women of the Seattle Fire Department. Our list of mentors and what their contributions to our lives have meant would create a whole new book. We truly stand on the shoulders of so many wonderful firefighters who have served the citizens of Seattle with honor and excellence. It is our privilege to work among the best in the business and to proudly proclaim: We are Seattle Firefighters.

—Thanks from Phil, Casey, Mike, and Steve—The Seattle Guys.

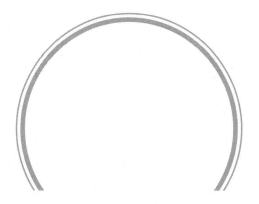

CONTENTS

PART 1 — THE NEED

PART 2 — THE MANDATE

PART 3 — THE SOLUTION

FOREWORD

Several years ago while I was walking through the halls of FDIC, I wandered into a classroom where four firefighters from Seattle were preaching a gospel called Air Management. While we always knew our air consumption was a serious issue, until now no one had figured out how to address it in the American fire service. Running out of air was like a crazy aunt in the basement; everyone knew it was happening and no one knew what exactly to do about it. The steadily rising numbers of firefighters dying from asphyxia, after running out of air during firefights, had heightened our interest in air management. We had identified part of the issue in significantly larger structures, more complicated floor plans and very difficult ventilation profiles in the now common, big box-type commercial structures. This did not explain, however, the rise we saw in residential firefighter asphyxiation.

Several attempts to answer the question of air management were being presented. Things such as personal air consumption rates were identified as a good starting point, as were theories regarding the disorientation sequence and enclosed structures. However, not until the Point of No Return and the Rule Of Air Management (ROAM) did anyone present a workable theory that blended the strategic, the tactical, and the individual firefighter. When Casey, Steve, Phil, and Mike began talking we knew they had found something. Air management is critical to the care and well-being of today's firefighters. Air management is not just how to monitor and regulate your air reserves, it is how to survive modern fires.

One of the most encompassing problems that faces today's firefighters has an eerie connection with a defining movie of my generation: *The Graduate*. In that movie, Dustin Hoffman is given one word of advice about the future: it was "plastics." If Hoffman had been a 1960s firefighter, the advice could not have been any more prophetic. Polymers or plastics have changed the nature of fire behavior. The heat release rates associated with polymers are connected with the threat of flashover, which in the 1970s was an anomaly; now it is almost an accepted stage in fire behavior models.

The toxicity of fire gasses, again related to polymers, is now an impending epidemic for our nation's fire service. The dense, blinding, carbon monoxide-filled black smoke we once associated only with tire fires now comes spewing from residential fires. Tragically, we know from dozens of cases of disoriented and lost firefighters that it will cause you to lose your point of reference almost instantaneously. This hideous smoke, this Black Death, or, as Mike so aptly named it, this "Breath from Hell," if it doesn't get you today, has undoubtedly laid its seeds of cancer for tomorrow. The American fire service is preparing itself to somehow provide comfort and support for the hundreds of thousands of cancer cases which are now being discovered in the firefighters of 1970 to 2000. Firefighters learned quickly to fear the black smoke. However, many firefighters had no idea that the light colored, seemly innocuous smoke also carried its painful suffering and threat of death for them down the road.

The Rule of Air Management is more than a nice thing to know, more than a rule to follow on the fire ground. It is a matter of life and death—yours and those you respond with. Many years ago we taught young firefighters to get low in a fire when they ran out of air, to suck the carpet. I taught this and am sorry I ever created such a possibility in any firefighters mind. If you run out of air in a working fire today, you are in mortal danger.

There is no good air at the floor anymore, no effective filtering methods, no matter what others may say to the contrary. Respectfully fearing smoke does not make you a coward; it makes you an effective firefighter. The use of cheaters is the lowest common denominator defining what is acceptable; you do no one any good when you are incapacitated and cheaters will get you there in a hurry. When a firefighter gets away with something, or seemingly so, that does not make it safe or acceptable. Most importantly, past success does not guarantee future success. The chemical mix could be different, the levels of toxins slightly greater, and that is all it may take for tragedy to occur.

The toxicity of smoke rich in hydrogen cyanide, a narcotic gas, may be responsible for firefighters fighting off rescuers—what we have traditionally called the rescue myth. The cardio toxicity of cyanide may also account for firefighters with little to no atherosclerotic disease dying after being exposed to smoke heavy with hydrogen cyanide. The reality that more needs to happen on the fireground to protect firefighters from the off-gassed products of combustion is being championed by my four friends from Seattle. They have in the pages that follow given us our history, defined our challenge, given us a program and provided us a roadmap for safer fireground operations. It is something that works no matter the size or makeup of your department.

This is a landmark book and a defining moment for our profession. Generations who come after may take this work as a given, but clearly those of us who watched the journey and witnessed the struggle know that Casey, Mike, Phil, and Steve have set the fire service down a safer road. Equally important is their example of providing a product, not for their glory or to advance themselves, but simply to help others. This is truly the definition of what it means to be an American firefighter. As you read this book, remember well the price that has so honorably been paid learning these lessons. Honor your department, honor your profession, be a thinking firefighter and live to serve. Be your brother's keeper.

—*Chief Bobby Halton, ret.*
Editor in Chief, Fire Engineering
Education Director, FDIC

INTRODUCTION

WHY AIR MANAGEMENT IS IMPORTANT

The air that firefighters bring with them on their backs to structure fires is critical to their health and safety. As firefighter deaths due to asphyxiation continue to rise, and as close calls caused by running out of air in structure fires multiply, the fire service is beginning to understand the importance of having an effective air management policy. This book will give you both the tools and the training material to make air management a part of your team's, your crew's, and your department's standard operating procedure.

Before the advent and use of the Self-Contained Breathing Apparatus (SCBA) in the fire service, aggressive firefighters attacking structure fires were forced to breathe in the smoke generated from these fires. This is where we get those once time-honored terms "Smoke-Eaters" and "Leather Lungs," terms that accurately and tragically describe the firefighters of this pre-SCBA era.

These pre-SCBA firefighters had no respiratory protection from the products of combustion, and were exposed to all types of poisons and carcinogens—hydrogen cyanide, formaldehyde, carbon monoxide, and arsenic—poisons and carcinogens that these firefighters knew nothing about.

The long-term health effects for these pre-SCBA firefighters were not good, and to this day, cancers and other diseases continue to take their toll on our older and retired brother and sister firefighters.

Not too long ago, most of the materials found in single-family dwellings or apartment houses were made of Class A materials—wood, cotton, wool, and paper made up the majority of the common combustibles. Class A materials produce various poisons and toxins when burned. However, the smoke from Class A materials is not nearly as toxic as the smoke of the burning contents found in today's buildings.

Today, many of the contents in residential and commercial structures are made of plastics or other synthetics. Examples include: carpets, tables, chairs, beds, workstations, wall coverings, fixtures, appliances, and computers. Plastics, which are really Class B solids, generate extremely toxic gases when exposed to heat, and burn much hotter than the fires of years past. There have been numerous studies quantifying the deadly smoke produced from these plastics and synthetics. Today, most firefighters know that the SCBA is their best protection from the IDLH (Immediately Dangerous to Life and Health) atmospheres of modern-day fires.

Because today's smoke is so toxic, firefighters cannot afford to run out of air in a structure fire. This is why air management is so important and why we have written this book. A firefighter who takes his or her mask off in a smoke-filled structure may get themselves killed, and will certainly expose themselves to toxins and cancer-causing gases that will injure them down the line.

Simply put, firefighters must do a better job of managing the air they bring with them on their backs.

It is our belief that firefighters should manage their air the way SCUBA divers do. SCUBA divers are constantly aware of both their air consumption and the amount of air they have left in their tanks.

Firefighters entering today's fires must know how much air they have in their SCBA before they enter the hazardous atmosphere, manage that air along the way, and leave the hazardous atmosphere, before they breathe into their emergency air reserve. The time to exit is before the low-air warning alarm activates. We believe that today's firefighters must consider themselves "smoke divers," and must have a comprehensive, workable air management model to follow.

In this book, we will present and discuss relevant case studies where firefighters were injured or died in structures as a result of a failure to manage the air in their SCBA. We use these case studies with the utmost respect, so that we might learn from those who have come before us. We will also introduce the concepts of the Rule Of Air Management and the Point of No Return. We believe these concepts will help firefighters manage their air, which makes for a safer and more efficient fireground.

Included in this book are questions for discussion at the end of each chapter, and training tips that will help you train your crews, your stations, or your departments. Our goal is to provide you with the model and the tools to implement an effective, easy-to-use air management program for your organization.

Finally, we believe that firefighter safety is everyone's responsibility—the individual member's, the company officer's, the incident commander's, and the fire chief's. "Everyone goes home," shouldn't be just an empty phrase, but a belief system that must become part of our culture. And because more than 60 percent of all firefighter deaths in structure fires are due to asphyxiation, or running out of air, we believe it is our duty to provide the fire service with an air management solution that works.

It is important for every firefighter to realize a very simple reality: "Your air is *your* responsibility!"

—*Steve, Mike, Casey, and Phil—The Seattle Guys.*

ALL ROADS LEAD TO ROAM

This book provides a comprehensive look at the Need, the Mandate and the Solution for effective Air Management for firefighters. At the heart of our training is a very simple concept: The Rule of Air Management (ROAM). You will see this term used throughout the pages of our text. And while it is fully developed in chapter 10, it is important to know the definition of ROAM as it is a foundational component of every chapter.

The Rule of Air Management states:

Know how much air you have in your SCBA and manage that air so that you leave the hazardous environment before your low-air alarm activates.

The journey you are about to take, as you read this book, will dramatically impact the way you operate on the fireground. The ROAM will be a significant part of that. We sincerely hope it becomes the new firefighter skill that defines how you approach Air Management from this day forward.

PART 1

THE NEED

HISTORY AS TEACHER

INTRODUCTION

The fire service has a proud and honorable sense of tradition. Traditions are present in everything we do each day at the fire station. Look at the process of reporting for duty. Virtually every paid, volunteer, or combination department has a defined procedure for documenting who is at the firehouse, what time they arrived, and what time they left. Departments also hold a meeting at the beginning of the shift. For paid and combination departments, this is the roll call. For volunteer departments, drill nights start with a formal meeting about the events of the day. In most departments, the roll call or the shift meeting has changed little since the procedure was first written. The same relationship between past and current practice is evident in many of our fireground procedures. History defines much of what we do. Each department must combine respect for its traditions with a willingness to adapt based on the demands of the modern fire service.

HISTORY IN THE SEATTLE FIRE DEPARTMENT

The Seattle Fire Department provides an example of tradition in the way apparatus are cleaned each day. For those unfamiliar with the City of Seattle, on occasion it rains; when the author of this text entered the Seattle Fire Department, the accepted practice on those occasions when it rained was to clean the apparatus on return to the station after each alarm. In addition, the underside of the rig, including the wheel wells, was cleaned in the morning prior to the change of shifts. With increasing run volumes, building-inspection rounds, and training requirements, this practice fell into disfavor with line firefighters over the next decade.

The Seattle Fire Department looked at the reason behind the original practice of cleaning the rig on returning to quarters. It was discovered that the original purpose of postresponse cleaning was to remove horse dung from the rig before it dried. This tradition had clung to the fire department long after the reason was put to pasture. The practice of washing the rigs after each response became too burdensome to continue. Some apparatus were regularly responding to more than 10 runs per day. In addition, other duties were a higher priority than keeping the apparatus clean. Over the course of a decade, the practice was changed. The apparatus now receive daily cleaning during rainy weather and as necessary during periods of dry weather.

This short diversion down memory lane illustrates that while the fire service has a great and proud history, clinging to tradition is not always good for an organization or the profession. This chapter demonstrates that many of the ways in which self-contained breathing apparatus (SCBA) is used in the modern fire department are the result of history and tradition. Examples are given to show how this history reflects the story of the dung on the apparatus wheels—to demonstrate that our practice should perhaps be changed to reflect a progressive approach to the use of one of our most important pieces of safety equipment.

THE MANDATORY MASK

The time was the early 1980s. I was a relatively new (less than five years' service) member of a mid-sized career fire department (approximately 325 members). The department had members with less than one year's service and a handful of members with over 35 years' service, dating back to just after World War II. For many of the more senior members, the majority of their career had been during the era when SCBA was used only as a last resort and a firefighter was measured by the amount of smoke he could eat.

Early in my career, I responded to a fire in a garden apartment. On arrival, we observed heavy smoke pushing out from the first and second floors of the building from one of the middle units. A woman was hanging out of a second-story window, clutching a baby in her hands. There was heavy smoke pushing out from the window all around her.

The captain came right out of the cab and attempted to make the stairs and pull the woman to safety. Unfortunately, he made a critical mistake. He failed to mask up. Just as quickly as he tried to make entry, he was back out at the apparatus gasping for air. The smoke was just too much.

I passed the captain on my way into the building. Having my mask on made it relatively easy to get to the woman, grab the child, and get back to safety. As I exited the building, the captain was still beside the apparatus, trying to catch his breath and sucking on the oxygen mask.

This event galvanized my belief in the value of SCBA. The other reason why I believe in the value of SCBA is that an alarming number of firefighters seem to die all too early after retirement.

After several years with the department, I became part of the local union safety committee. One of the details I looked at was the number of on-the-job injuries. In a single six-month period, our department had approximately 60 smoke inhalation injuries. How could this happen if we had SCBA? The answer was simple. There was no policy requiring members to actually use their SCBA. And if it wasn't required—plus made a firefighter look less macho—then it just wasn't done.

When this was brought to the attention of the fire chief and his administration, a mandatory-mask rule was soon established. At first, there was resistance. However, that resistance quickly faded once the reasoning behind the new rule became evident: As a result of the mandatory-mask rule, in the six-month period following its establishment, there were only six smoke inhalation injuries—a drop of 90%.

Author's name withheld by request

THE INTRODUCTION OF SCBA

Firefighters of old had no choice in the matter of whether or not to breathe smoke. The firefighters before the turn of the 20th century had only one option—learning to eat smoke. To put the fire out, each firefighter had to be able to function in a smoky environment. Even during these times, firefighters attempted to protect themselves from smoke by sucking in air "filtered" by their mustaches, by staying low, and by aggressively ventilating structures. During this era, the ability to be a smoke eater became embedded as a value of a good firefighter. The most effective personnel were able to choke down the smoke, advance on the fire, and put it out. Toughness in the breathing of smoke was the mark of someone to be respected, even revered.

Experimental versions of SCBA were being developed and tested in the United States and Europe in the late 19th century (fig. 1–1). One of the first of these was the Apparatus Aldini, which consisted of a thick mask of asbestos that covered the head.[1] This and many other inventions were adapted through trial-and-error testing until the design of a compressed-gas cylinder carried on a backpack, which has become the method preferred throughout the world today. The patent for this method was granted to A. Lacour in 1863 for his Improved Respiring Apparatus.[2]

Over the next century, product development led to consistent improvement of SCBA. Competition between manufacturers, government regulations, firefighting successes and failures, firefighter fatalities, and improvements in technology all contributed to the ongoing development of SCBA. Consequently, modern SCBA bear little resemblance to the first products provided to firefighters. Lightweight material, improvements in compressed-gas technology, gauges, facepiece designs, heads-up displays, and many other improvements give us a tool that is now indispensable on the fireground.

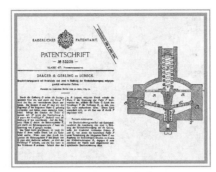

Fig. 1–1. An early version of the SCBA regulator
(Facsimile courtesy Draeger Safety)

Yet, as reliable as modern respiratory protection is, there are still firefighters entering environments that are immediately dangerous to life and health (IDLH environments) without using SCBA. Incredibly, the mentality that forgoing the use of the SCBA is an acceptable, and even desirable, practice persists. Recently, a career lieutenant in a big-city department died after entering a fire building without his SCBA. In response to this fatality, the National Institute of Occupational Health and Safety (NIOSH) conducted a line-of-duty death investigation. The following recommendation is from the NIOSH Fire Fighter Fatality Investigation:[3]

> *Require . . . that all fire fighters wear their SCBAs whenever there is a chance they might be exposed to a toxic or oxygen-deficient atmosphere, including during the initial assessment.*

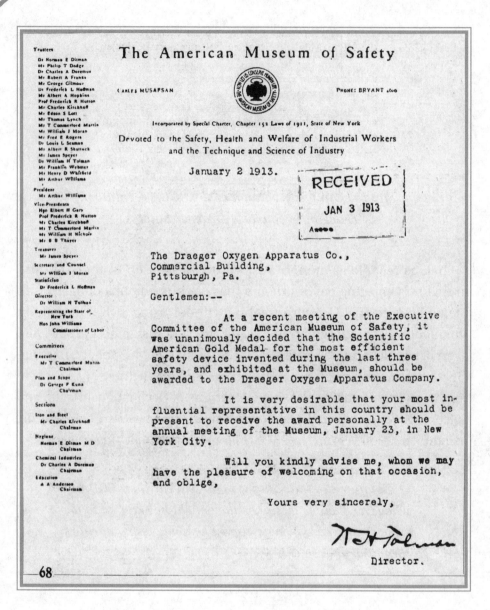

Fig. 1–2. Letter documenting the American Museum of Safety Gold Medal, awarded to the Draeger Oxygen Apparatus Company in 1913 for the most efficient safety device invented (Facsimile courtesy Draeger Safety)

The culture of the fire service reflects more than a century of development and use of SCBA (fig. 1–2). After all this time, it seems impossible to believe that this is the best that we can do. Firefighters are excellent problem solvers. We rely on our experience and our equipment to achieve the result of saving lives and property. Is it possible that the cultural mind-set that allows firefighters—especially company officers—to refuse to abide by the very simple rule to wear your mask is too difficult to overcome? Where did this mind-set originate, and how can we get past it?

THE INTERMITTENT-USE MIND-SET

Early SCBA were definitely the result of necessity's being the mother of invention. Firefighters understood that exposure to smoke limited success in extinguishing fires as rapidly as possible. The first breathing apparatus had a very limited service time. Pressure and air volume were limited owing to the materials available for providing compressed breathing air. Firefighters—by necessity, not choice—developed a habit of using SCBA only when necessary. This practice, combined with the glorification of the smoke eater, provided fertile ground for the intermittent-use mind-set that encourages firefighters to go without their masks. The decade-by-decade process of improving SCBA without changing the intermittent-use philosophy prolonged the reverence of the smoke eater; consequently, generations of firefighters were unnecessarily exposed to products of combustion (fig. 1–3). During this time, the products of combustion also changed from primarily Class A materials to primarily Class B materials, which produce a larger quantity and more dangerous quality of smoke (see chap. 5).

Intermittent use, while a necessity when SCBA were new, became ingrained as a regular practice. The original users of SCBA might have been more than happy to use the device all the time; however, the design limitations did not allow it, and this was translated into an intermittent-use philosophy.

Fig. 1–3. Rescue of a firefighter from the Fire Department of New York, demonstrating the need for quality SCBA, as well as the rapid intervention team, at the turn of the century (Photograph courtesy Draeger Safety)

SCBA, once introduced, produced a change in the fire service. New firefighters looked to veteran firefighters for guidance on proper procedure. These veterans, cognizant of the limitation of the first SCBA models, had learned to operate in the intermittent-use mode.

Well-respected veterans were the ones who could eat the most smoke before using their SCBA or were able to function without using the SCBA at all. New firefighters embraced this mind-set because they wanted to be like the tough, competent firefighters they admired. These new firefighters grew to become veterans; hence, the pattern continues to this day, even though changes in SCBA technology (fig. 1–4) have removed the intermittent-use requirement.

Fig. 1–4. Old versions of the SCBA bear only a mild resemblance to the modern, high-tech equipment mandated by National Fire Protection Association standards today (Photograph courtesy Draeger Safety)

THE ENIGMA OF SMOKE

Webster's defines smoke as "vaporous matter arising from something burning." This simplistic definition fails to address two major considerations with regard to smoke—and, most importantly, with regard to the fire service.

The first consideration, derived from Fire Science 101, is that the vaporous matter referred to in Webster's definition is a direct result from "what" is burning. Every firefighter is well aware that several years ago, the primary burning materials were wood, wool, paper, and cotton—conventional materials. Today, the primary materials are petrochemical products— plastics and/or synthetic materials. The obvious result of this dichotomy is twofold: faster and hotter fires; and flashover, not backdrafts! Remember, the National Fire Protection Association (NFPA) currently underscores the importance of flashover on fireground operations; their statistics indicate the modern fire service suffers from three primary fireground problems, one of which is flashover.

The second consideration is the content of modern smoke. Although the smoke of structure fires has never been known for its medicinal effects on the human body, the content of smoke derived from modern fires has dramatically changed. Particulates and gases are now present that were not found in smoke several years ago, and as a result, smoke has become significantly more dangerous to the human body. This can and will have a direct effect on your safety and longevity, depending on how much emphasis you put on reading smoke and not breathing it!

The second perception (firefighter safety) became important to me three years ago as I sat in a doctor's office and was notified that I had tested positive for prostrate cancer. Although I was fortunate in that the cancer was easily cured, I distinctly remember the conversation that I had with the doctor. When the doctor inquired about the nature of my vocation, I informed him that I had retired from the fire service after 30 years with the Los Angeles City Fire Department. The doctor's exact words were "What do you guys do at fires, breath smoke?" As I nodded my head in affirmation, he added, "It's no wonder I see a lot of you guys in here."

After my visit with that doctor, I had numerous opportunities to reflect on my experience with fireground operations. Although I was always taught the importance of safety—and, in particular, wearing SCBA inside buildings—the need for SCBA outside buildings, especially in vertical ventilation operation, did not receive the same emphasis. I distinctly remember going to the roof for ventilation, stepping from the ladder to the roof, removing my SCBA and placing it on the roof near the ladder so that I would not forget it when returning to the ground. Interestingly, this was common practice during the 1960s and 1970s. Unfortunately, it remains a common practice in some departments today. You may be thinking, "I don't remove my SCBA on the roof," but there is a significant difference between wearing your SCBA and using it!

A weak defense (a very weak defense) for not using SCBA on the roof during ventilation operations is that SCBA can hamper visibility, flexibility, and maneuverability, which are all beneficial to roof ventilation operations. However, the primary requirement for any fireground operation should be firefighter safety—present and future! If there is any doubt about the importance of using (as opposed to wearing) SCBA during roof ventilation operations, take the opportunity to watch the video Phoenix Ladder 27, available from the Phoenix Fire Department. Without the use of SCBA, the Phoenix Fire Department would have lost three firefighters in a routine roof ventilation operation. Instead, the three firefighters walked away from the incident.

In conclusion, staying out of smoke is important from both a present and a future perspective. Unfortunately, while it is easy to focus only on the present benefits of using SCBA, don't forget the future benefits, which may be the most important. If you fail to remember this, some doctor visits may not be pleasant.

—*John Mittendorf (Battalion Chief, retired, Los Angeles City Fire Department)*

Following the lead of the veteran on your crew is usually a good practice. Most of the lessons that we learn from senior members are well developed and form the basis for competent firegound actions. However, the difference with respect to SCBA is that the practice of intermittent use has outlasted any operational deficiencies in the equipment. The continued practice of intermittently using SCBA has provided fertile ground for resistance to the statutory requirements to use SCBA in all IDLH or possible IDLH environments. New firefighters were taught the "way we always do it"—without fully understanding the reasons behind the method or understanding that changes in the technology no longer required that SCBA be used in the old way.

The practice of intermittently using SCBA has been carried over to today's fire service with deadly results. Just as the practice of cleaning dung off the apparatus has become unnecessary, so too has the practice of intermittently using SCBA. Modern firefighters are exposed to a more toxic environment than our predecessors were. (These environmental factors are thoroughly elaborated in chap. 5.) No longer are we significantly limited by the capabilities of SCBA; no longer can we claim ignorance to the effects of long-term exposure to products of combustion; no longer can we claim that SCBA isn't up to the task; and no longer can we claim that the way to operate should be the same as it was a decade or a century ago. The practice of intermittently using SCBA is no longer acceptable.

Each fire department must make it clear throughout the organization that SCBA use is not optional (fig. 1–5). To eradicate the practice of intermittent use, we must change the behavior of the firefighter. This change must be supported at all levels in the organization. Strategic-level change comes with policy development and implementation. Tactical-level change comes with increased training and emphasis. Task-level change comes with good leadership and good followership on the fireground (fig. 1–5).

SCBA AND THE PERSONAL ALERT SAFETY SYSTEM

No discussion of modern SCBA would be complete without a discussion of the personal alert safety system (PASS). The PASS device provides an alert signal whenever a firefighter gets into trouble. The device monitors the movement of the firefighter and has two modes for activation of the alarm: manual and automatic. Manual activation of the alarm requires that a firefighter who is experiencing a problem move a switch from "auto" to "on"; the switch requires the sliding, pressing, or twisting of a large, easy-to-operate button. In automatic mode, the PASS device activates an alarm when a firefighter does not move for a short period, normally 30 seconds.

Fig. 1–5. An entire company of the Pittsburgh Fire Bureau wearing their new SCBA early 1900s. The move toward use of SCBA has been incremental, requiring leadership to implement a change in behavior. (Photograph courtesy Draeger Safety)

The original PASS devices required that the firefighter activate the device separately from the SCBA (fig. 1–6). PASS devices quickly developed a reputation for false alarms. As the units degraded, through normal wear and tear, even movements of a hardworking firefighter did not prevent the device from activating and issuing an alarm. Because of the rise of false-alarm activations, firefighters developed the practice of deliberately not activating PASS devices prior to entry into an IDLH environment.

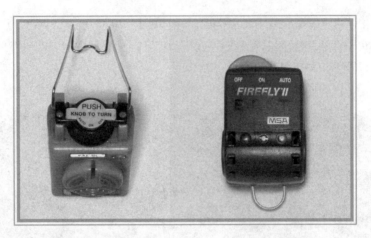

Fig. 1–6. Two versions of original PASS
devices, requiring user activation. The "on"
and "auto" modes can be seen clearly.

It seems unlikely that firefighters would allow a piece of equipment that did not work properly to just sit on their SCBA. There is no other piece of equipment that firefighters would leave on the rig if it were not working properly. In all other cases of faulty equipment, firefighters would have it serviced. Unfortunately, with the PASS device, initial resistance to its use combined with the unreliability of the first commercially available units created a situation in which firefighters felt comfortable not using the device at all. This practice has contributed to firefighter fatalities in several instances. In at least one instance, two firefighters were lost and the only one rescued had an activated PASS device.[4]

The current National Fire Protection Association (NFPA) standard requires that the PASS device be integrated into the SCBA (fig. 1–7). When the SCBA is used, the PASS device automatically activates. This is an example of engineering overcoming resistance. Compliance with the requirement to use the PASS device was achieved without requiring that firefighters change their behavior.

Fig. 1–7. Modern, integrated PASS device, which activates automatically when air is supplied to the SCBA regulator (Top photograph courtesy Draeger Safety)

TRAINING AND LEARNING FROM HISTORY

History has provided the lessons for how, why, and when to use the SCBA. Many of these lessons are provided through the NIOSH Fire Fighter Fatality Investigation and Prevention Program. Throughout this book, reports from NIOSH will be used as a reference. These lessons were provided at a brutal cost to the families and fire departments involved in the fatality incident. One lesson that will receive the continued focus of this book is the importance of air management and following the *rule of air management* (ROAM), which is covered thoroughly in chapter 10.

Training is a critical function of any fire department. Moving a fire department's pattern of SCBA use out of the past, despite resistance, and into the present requires training. Step 1 is providing NFPA-compliant SCBAs with integrated PASS devices. Step 2 is mandating the use of SCBA during any activity in IDLH or possible IDLH environments. Step 3 is ensuring that firefighters manage the air in the SCBA, consistent with ROAM (see chap. 10).

TRAINING WITH SCBA AND ROAM

Training in the use of SCBA and the ROAM is mandated by NFPA 1404, which requires training in an individual air management program.[5] This prescribed training program includes three components, outlined in NFPA 1404, A.5.1.4(2), as follows:

> *This program will develop the ability of an individual to manage his or her air consumption as part of a team during a work period. This can require team members to rotate positions of heavy work to light work so air consumption is equalized among team members. The individual air management program should include the following directives:*
>
> > *(1) Exit from an IDLH atmosphere should be before consumption of reserve air supply begins.*
> >
> > *(2) Low air alarm is notification the individual is consuming their reserve air supply.*
> >
> > *(3) Activation of the reserve air alarm is an immediate action item for the individual and the team.*

Training based on the NFPA standard and following the ROAM will provide an increased margin for error and, thus, an increase in overall scene safety. Operating in accordance with NFPA 1404 and following the ROAM will preclude the low-air alarm as expected background noise on the fireground. Following the adoption of this standard, this alarm will be restored to its original purpose as an indication of an actual emergency. Changing the low-air warning from a normal fireground sound into an emergency alarm will dramatically improve the way the fire service operates.

FINAL THOUGHTS

This change will not occur without resistance. Resistance to this improvement in firefighter safety will be overcome with time, training, and effort. Leadership will be needed at all levels in the organization to implement this change. Looking back on other dramatic changes to our operations, we can clearly see the roll of history as teacher. It will be no different when an organization changes to meet the requirements of NFPA 1404 and the ROAM.

A history lesson is provided by the fatality of Bret Tarver in the Southwest Supermarket fire in Phoenix on March 14, 2001 (see chap. 7).[6] This lesson will be repeated in the future. Left for readers to decide is how many lessons history must provide before they are willing to learn. History can be a brutal instructor. Better to learn the lesson when others have paid the price than to wait until you, your crew, or your department must pay.

MAKING IT HAPPEN

1. To start implementing air management in your fire department, go to Manageyourair.com and download SMART Drill 60-1 for free.

2. Take a look at your department procedures and history to find areas where you can make the case for change.

3. Recognize whether your department has an intermittent-use philosophy as it applies to the SCBA and, if so, determine why this mind-set has been allowed.

4. Ask around for stories about when a senior firefighter supposedly had to breathe some nasty smoke. Correlate this to SCBA use in the modern era and move the discussion to appropriate SCBA policy.

5. Ask the veterans in your department about equipment that has malfunctioned on an incident, the outcome, and how they prevented the situation from recurring.

STUDY QUESTIONS

1. What methods did firefighters use to "filter" their breathing air before the invention of the SCBA?

2. When and by whom was the prototypical SCBA, consisting of a compressed-gas cylinder carried on a backpack, patented?

3. How did the first PASS devices contribute to the false-alarm mentality?

4. How did firefighters react to the PASS device's tendency to false alarm?

5. What national organization requires the integration of the PASS device into new models of SCBA?

6. What is the most common reaction to change in the fire service?

7. What NFPA standard dictates the requirements for training firefighters to use the SCBA?

8. What three components of an individual air management program are outlined in the NFPA standard?

9. How does the low-air alarm contribute to the false-alarm mentality in firefighters?

10. What firefighter died in 2001, resulting in the recommendation from NIOSH to train firefighters to manage their air?

NOTES

[1] Hashagen, P. 1998. The American fire service: 1648–1998. *Firehouse.* Special Section in the September, 1998 edition.

[2] Ibid.

[3] NIOSH Firefighter Fatality Investigation 2004-05. Residential basement fire claims the life of career lieutenant—Pennsylvania.

[4] NIOSH Firefighter Fatality Investigation 1997-16. One firefighter dies of smoke inhalation, one overcome by smoke while fighting an attic fire—New York.

[5] NFPA 1404—Standard for Fire Service Respiratory Protection Training, 2007 edition.

[6] NIOSH Firefighter Fatality Investigation 2001-13. Supermarket fire claims the life of one career firefighter and critically injures another career firefighter—Arizona.

CHAPTER 2

TIME IN A BOTTLE?

INTRODUCTION

Since the beginning of drill school, the cylinders we use in our SCBA have been described the same way—as *30-minute, 45-minute,* or *60-minute* cylinders. This terminology is standard among firefighters from coast to coast. It is also completely wrong.

Most firefighters know that they will not get 30 minutes from a 30-minute cylinder. So why describe them that way? Does it make sense to term a pump that gives 1,500 gallons per minute (GPM) as a 3,000 GPM pump? Would you allow a 90-foot aerial to be labeled at 100 feet? Inaccurate and misleading information has no place on the emergency scene. Yet that is exactly how we label our SCBA cylinders.

Misidentification of the cylinder has led to problems for firefighters that must be addressed if we are to operate safely on the modern fireground. A change needs to be made in how cylinders are described. In addition, a review of work cycles, crew rotations, and the role of rehabilitation during the emergency operation is appropriate.

COMMON TERMINOLOGY

The discussion should begin with terminology. Common practice in the fire service has been to refer to cylinders in terms of expected duration of use, expressed in minutes. The result is the 30-minute, 45-minute, and 60-minute references with which we are all familiar. For the purposes of this book, a new methodology will be adopted to discuss cylinder size and capacity in terms of volume. We will refer to the 30-minute cylinder as a *1,200-liter (1200L) cylinder,* the 45-minute cylinder as an *1,800-liter (1800L) cylinder,* and the 60-minute cylinder as a *2,400-liter (2400L) cylinder.* This change in terminology also eliminates the need to consider whether the cylinder is high or low pressure.

Furthermore, there are distinctions among the various sizes of cylinders as they relate to emergency operations. A partial list of considerations includes cylinder size, work cycles, fluid loss, body core temperature, and crew rotations, as well as how these factors influence rehabilitation policies.

THE 1200L CYLINDER

The 1200L cylinder has been the standard in the U.S. fire service since the era of modern SCBA began. This cylinder has provided excellent and reliable service for decades. It has seen the fire service through the change from low-pressure (2,216 pounds per square inch [psi]) to high-pressure (4,500 psi) cylinders. The 1200L cylinder has also been the standard through material changes from steel, to hoop wound, to the modern carbon fiber composite cylinder used today.

Firefighter's personal experience has demonstrated that the 1200L cylinder will not provide a working firefighter with 30 minutes of air.

The 1200L cylinder will last an average firefighter actively engaged in suppression activities about 15–18 minutes. Acknowledgment of this time frame for working in the hazard area, on air, has several advantages.

STRATEGY CHANGES AND THE 1200L CYLINDER

The work cycle of about 15–18 minutes—as defined by the air volume of the 1200L cylinder, without using air management techniques—provides several innate advantages to fireground operations. The first advantage is that it provides for initial crew rotation out of the hazard area approximately 15 minutes after arrival at the incident. This is consistent with recommendations that incident commanders (ICs) should consider a change in strategy after 15 minutes. If the IC has crews deployed in the offensive mode for 15 minutes, then consideration must be given to whether the strategy is still effective. The following key questions need to be asked:

- Is the situation getting better or worse?
- What impact is the fire having on the building construction?
- Has the fire extended to the structure?
- Is the fire still in the original compartment?
- Are sufficient resources on the scene, or responding, to support an ongoing offensive operation?
- What is really going on inside?
- Is it time to conduct a personnel accountability report (PAR) or roll call?

When the initial crews rotate out of the hazard area approximately 15 minutes into the incident, the IC can get a face-to-face report of conditions inside the structure. The IC can then use this information from personnel who have been in the building, along with other information gathered outside the structure and by radio, as part of the ongoing size-up.

Second, most firefighters' bodies have become acclimated, through repetition, to the 15–18 minute work cycle using the 1200L cylinders. SCBA have the low-air alarm set to activate at 25% of the rated capacity. Firefighters use 900 liters of air for entry and work in the hazard area. Firefighters then begin to exit once the low-air alarm activates. This practice leaves 300 liters of air for exit from the hazard area. This has been going on for so long that most firefighters have become accustomed to the work cycle—to the point that varying from it may have adverse impacts.

Firefighters work through a 1200L cylinder before rotating out of the hazard area to check in with the IC. While out of the hazard area, the team will re-supply their air, hydrate, and prepare for reassignment to the hazard area. This practice allows firefighters to get a modicum of rest and fluid intake, and it allows the incident command system (ICS) to maintain accountability of personnel. Once the crew has completed this cycle, most departments will reassign them to the action with a second cylinder. Firefighters then perform a second consecutive work cycle, again using 900 liters for entry and work and 300 liters for exit after activation of the low-air alarm, before they rotate out of the hazard area and are reassigned to rehabilitation. This system is commonly referred to as the *two-cylinder rule*. National publications support—and many departments use—a practice allowing firefighting teams to follow the two-cylinder rule before mandatory rotation to rehabilitation.[1] While this is a good standard practice, ICs and company officers should always be aware of—and prepared to deal with—the possibility that symptoms of dehydration and elevated body core temperatures may be present after only one rotation in the hazard area.

Most progressive departments will mandate a trip to rehabilitation after a firefighter has used two full cylinders. If you are not already doing so, your department should seriously consider implementing such a policy. The firefighters have been working hard, in full personal protective equipment (PPE), for at least 30–45 minutes by this time. They need a break to get fluids on board and dissipate body heat. The need for rehabilitation after two cylinders is even more important with the advent of modern PPE, because of its greater encapsulation of the wearer.

WORK RATE AND AIR CONSUMPTION

When firefighters work hard in full PPE, the air consumption rate increases throughout the work cycle. The highest rate of air consumption will be achieved at the end of the work cycle. While firefighters work in the hazard area, they are under an increasing burden of physical stressors. These include elevated body core temperature, reduced hydration levels, and overall physical fatigue from the workload. These factors affect all firefighters at differing rates, depending on a variety of factors that may or may not be under their control.

CONTROLLABLE FACTORS AFFECTING AIR CONSUMPTION. Two controllable factors that affect the rate of air consumption are fitness and training (fig. 2–1). It is generally true that the more fit a firefighter is, the longer he or she will be able to tolerate a specific workload and the lower the overall fatigue experienced. Increased fitness will reduce the rate of air consumption based on the impact that the workload has on the firefighter. In addition, fit people can acclimatize to working in elevated temperatures more quickly.[2]

Fig. 2–1. Firefighters being trained in vertical ventilation.
Firefighters who receive regular training in basic skills achieve
better results and experience less stress than untrained firefighters.

Training can enhance control of the rate of air consumption by
increasing the competence and confidence of firefighters in performing the
tasks to which they are assigned. Well-trained firefighters complete their
assigned tasks with less work expended, in less time, and more efficiently
than poorly trained firefighters. Increased training means increased
effectiveness (fig. 2–2). Moreover, an increase in the effectiveness of the
operation has a positive impact on the firefighter by reducing stress,
workload, heat generated, and time to complete the task.

Fig. 2–2. Effective vertical ventilation of a cockloft fire, which resulted in rapid fire extinguishment. Well-trained crews have a positive impact on the fire scene.

UNCONTROLLABLE FACTORS AFFECTING AIR CONSUMPTION. Uncontrollable factors that affect the rate of air consumption for a team of firefighters include the weather, the operational environment, and the task assigned. Weather can have an impact on a crew at both the hot and cold extremes. The operational environment, while controllable to an extent, is not under the control of the individual crew assigned to an area of operation. Certainly, every effort should be made to provide effective and timely ventilation, but this will not always occur before crews are operating in IDLH environments.

Operational environments often include temperatures in excess of ambient, increasing the thermal load on firefighters. Firefighters experience increases in respiratory drive and air volume used as a result of the increased thermal load. In addition, the task assigned to the crew will be dependent on the needs of the incident and the decisions of the IC. The assigned task will need to be carried out to the best of the crew's ability. Only in rare circumstances involving extreme risk to firefighters and with no benefit to the operation will company officers have the discretion and responsibility to decline operational assignments.

Additional factors that affect the rate of air consumption are the emotional stress of operating in an IDLH environment, the potential for harm to ourselves, the possibility or reality of victims, the difficultly of keeping track of position in the building, and the dynamics of our crew These, as well as other details, can have a dramatic impact on situational awareness. Paradoxically, these factors are both controllable and uncontrollable in nature. Humans have an instinctive fight-or-flight response when placed in mortal danger, regardless of whether the danger is real or perceived. Emotional stress will be added when there are reports of trapped victims in the fire building. Firefighters will also experience varying levels of stress, depending on their individual experience level and whether they have operated at similar incidents in the past.

Taken together, these factors result in an increasing respiratory demand over the course of the firefighter's time in the operational environment. Respiratory demands may increase from a resting rate of near 40 liters/minute to a required flow rate of over 300 liters/minute.[3] This increasing demand significantly limits exit time if firefighters are relying on the low-air alarm. If firefighters delay making the decision to exit the hazard area until activation of the low-air alarm, then they have only 300 liters of air remaining until the cylinder will be empty and they are forced to breathe products of combustion. While the instructional materials provided with SCBA and cylinders indicate that 300 liters will provide up to six minutes of air, the reality is that a hardworking, physically stressed firefighter can easily blow through 300 liters of air in under three minutes.

Three minutes will barely cover the reflex time for the rapid intervention team (RIT) to launch and make entry into the hazard area. Three minutes does not provide adequate time for rescue of a firefighter should anything unexpected happen during egress from the structure.

THE 1800L CYLINDER

The 1800L cylinder is quickly becoming the cylinder of choice for fire operations. The 1800L cylinder offers a good balance of increased air capacity and acceptable weight profile. Firefighters using the 1800L cylinder also have an increased volume of air in the emergency reserve; furthermore, as will be shown later, when using 1800L cylinders and following ROAM, firefighters can maintain their work/rest intervals very close to the operational periods of firefighters using the 1200L cylinder not following the ROAM.

The 1800L cylinder allows fire departments to implement an aggressive air management philosophy while maintaining the standard 1200L cylinder work-cycle time (table 2–1; see also chap. 11). Firefighters beginning their operation with a full 1800L cylinder and following the ROAM have 1,350 liters of air for entry, work, and exit. Firefighters beginning their operation with a full 1200L cylinder and not following the ROAM have 1,200 liters of air for entry, work, and exit. With 1,350 liters of air for entry, work, and exit from an IDLH environment, the 1800L cylinder increases the air available to the firefighter by only 150 liters relative to the standard 1200L cylinder. Work-cycle times for firefighters using 1800L cylinders and following ROAM will closely reflect those of firefighters using 1200L cylinders and no air management plan. Work-cycle time will increase only 90–120 seconds while using an 1800L cylinder, and this increase is more than compensated by the increased margin of error that the ROAM provides.

Table 2–1. ROAM table. Work-cycle times are compared for the 1200L, 1800L, and 2400L cylinders.

Cylinder Size and Air Management	Number of Liters	75% in Liters	50% in Liters	25% in Liters (RESERVE)	SFD Allowed Operational AIR	Increase over 1200L Cylinder	Increase Time at 100 L/min.	Margin for Error at 100 L/min.
30 no ROAM	1200	900	600	300	1200 liters	none	none	none
45 no ROAM	1800	1350	900	450	1800 liters	increase of 50%	6 mins.	1:30 mins.
45 ROAM	1800	1350	900	450	1350 liters	increase of 12.5%	1:30 mins.	4:30 mins.
60 no ROAM	2400	1800	1200	600	2400 liters	increase of 100%	12 mins.	3:00 mins.
60 ROAM	2400	1800	1200	600	1800 liters	increase of 50%	6 mins.	6:00 mins.

One advantage of using an 1800L cylinder in combination with the ROAM is a much-needed increase in the margin for error (table 2–1). Fire departments that institute an aggressive air management policy including the ROAM will provide their firefighters with a margin of error of 450 liters of air by maintaining the 25% of air in the cylinder after the low-air alarm activates for the emergency reserve. An air management policy that includes the ROAM requires that firefighters exit the IDLH environment prior to depletion of the emergency reserve air (25% of the rated capacity).[4]

Fire departments that use 1200L cylinders without the ROAM provide firefighters with zero margin for error. Since firefighters can remain in the hazard area until the low-air alarm activates, they use 75% of their air (900 liters) for entry and work in the IDLH environment and 25% of their air (300 liters) for exit from the IDLH environment. There is no air set aside for an emergency reserve, as indicated in NFPA 1404.[5] Using 1800L cylinders and following the ROAM, firefighters have 75% of their air (1,350 liters) for entry, work, and exit from the IDLH environment. This leaves 25% of their air (450 liters) as an emergency reserve.

THE LUNAR DOCTRINE

Fire departments should have a standardized method for a downed firefighter to declare a mayday. There are many methods that fire departments can use to report information. Acronyms like *LUNAR* and *ESCAPE* are mnemonics that help firefighters to remember the sequence of critical information that the IC and the RIT need to launch an efficient and effective rescue operation.

LUNAR stands for
- Location
- Unit
- Name
- Assignment/air supply
- Resources needed

New communication systems and fire-alarm centers have the ability to identify a firefighter with the key of a microphone on a portable radio. At the beginning of the shift, members sign on to the unit's computer-aided dispatch (CAD) roster, which correlates the assignment of a portable radio to each member in the company by name and employee number. Whenever a transmission is made from that portable radio, the dispatcher immediately knows to whom the radio is assigned.

Not all fire departments have this capability, however. With that in mind, we chose nevertheless to include LUNAR in the National Fire Academy Mayday curriculum for reasons that will shortly become clear.

When the word *Mayday* is heard over the fireground radio, it is understood that the RIT is needed to enter the IDLH environment to rescue a firefighter. With that one word, the IC can begin the process of determining who called for help and where they are supposed to be by
- Radioing CAD for identification
- Requesting a PAR
- Cross-referencing the accountability system

After the word *Mayday* is communicated, if the only words the downed firefighter can get out conveys their location—for example, "division 2"—then at least we know that he or she is on the second floor. The IC should know what companies were assigned to the second floor. When giving the location, we teach the firefighters to give their quadrant on the floor using *alpha*, *bravo*, *charlie*, and *delta*.

If a downed firefighter reports his or her *unit* as Engine 33, for example, then the IC has an even better idea of the downed firefighter's location. By referring to the tactical worksheet, the IC should know the assignment and the location of Engine 33's crew on the fireground.

By giving his or her *name*, the downed firefighter lets the IC and the RIT team know exactly who to look for. Having the name is critical if more than one firefighter is in trouble. There are real-world examples of RITs that have assisted a lost firefighter out of the building thinking they had rescued the firefighter who called the Mayday when, sadly, the firefighter who called the Mayday was still in the building.

Many fire departments avoid using names on the radio unless necessary. However, when there is more than one Mayday called at an incident, it becomes absolutely necessary. It is too difficult to remember who is *Rescue 5, position A* or *Engine 18, position 3*, for example. At the incident in Phoenix that resulted in Bret Tarver's death, there were 12 separate Maydays! You may have firefighters working trades or working overtime from different platoons, multiple alarm companies coming from the other battalions, or mutual aid companies coming from other cities. They will not know what the riding assignments are on Truck 6 or who's riding positions 1, 2, 3, or 4 on Engine 8.

Turnout coats (i.e., PPE) with the firefighter's name stenciled on the back with reflective fluorescent lettering are becoming the norm. The name on the back of a turnout coat may be the only sure confirmation that you have the right person. The IC and the RIT need that name.

Assignment lets the IC know what the firefighter was doing—for example, search and rescue or checking for extension. This information narrows down the search and prompts the IC to consider whether the task is critical to the firefighter rescue or whether it should be reassigned to another company. Just because a firefighter calls a Mayday does not mean the fire goes out. The fire still needs to be controlled. While some assignments will continue to be necessary to facilitate the firefighter rescue, others may be unnecessary in light of the Mayday and the changes it represents to the incident priorities.

Air is what a trapped firefighter needs the most. Once a firefighter is out of air and starts to breathe in smoke, that firefighter will quickly succumb to the effects of carbon monoxide and superheated gases. Emergency air can be transfilled into the SCBA bottle, buying more time for the RIT and the firefighter. Air management for firefighters inside an IDLH environment is comparable to scuba divers' monitoring of their air while underwater. The amount of air left can be reported in psi or expressed as a percentage. For example, the firefighter might state "My air is at 1,020 psi" or "My air is at 50%." This gives the RIT and the IC an idea of how much time they have before that firefighter runs out of air.

Finally, *resources needed* informs the IC and the RIT as to what has actually happened (i.e., what the problem is) and what equipment is necessary for the entry team. For example, they might hear, "I'm stuck in some wires. I need wire cutters," or "I'm pinned under a beam. Bring in pry bars and a chain saw."

When this information is put together, it is communicated like this:

> Mayday! Mayday! Mayday! Division 2, alpha/bravo corner, Engine 33, Stanley, search and rescue. Air at 50%. I'm stuck under a ceiling collapse. Bring pry bars and a chainsaw. Fire in the attic.

This may seem simple, but under stress, getting it correct is not easy. Firefighters must practice often if they expect to be 100% competent. The IC and the RIT also need to train their listening skills to get the information correct the first time.

Whatever acronym or information communications system your fire department chooses to use for Mayday, the training requirement remains the same: practice, practice, practice.

—Dr. Burton Clark (National Fire Academy, EFO, CFO)
and Raul A. Angulo (Captain, Seattle Fire Department)

HOW LONG CAN THE AIR LAST?

Firefighters trained in emergency air conservation techniques have demonstrated the ability to make 300 liters of air last for more than 45 minutes in training scenarios. Great training programs, such as the Smoke Divers program at the Connecticut Fire Academy, stress familiarization with SCBA and give students the opportunity to practice control of respiratory rate during simulated firefighting tasks.[6] The skills needed to maximize the length of time that a firefighter can survive on his or her last 25% can be improved only through training. With regular and effective training, firefighters in Mayday situations will be able to make decisions about how best to ensure their survivability. Many training programs give firefighters the skills they need to breach walls, floors, or other building components in order to move through a building. A primary consideration for any firefighter who is lost or trapped in a fire building is the amount of air they have left.

Firefighters who find themselves in a Mayday situation must be aware that the first step is to call for help. Firefighters often delay the call for help, thereby reducing their survivability, because of their belief that they can get themselves out of trouble. The reality is that well-trained firefighters will call the Mayday the moment they realize they are in trouble.[7] Calling the Mayday is the most important step once the situation has gone bad. Calling the Mayday should be practiced regularly to ensure that the first time a firefighter experiences a Mayday situation is not the first time he or she transmits a Mayday. The National Fire Academy Mayday curriculum recommends using the *LUNAR* format for transmitting the Mayday.

Once a firefighter has transmitted the Mayday, all decisions should be viewed in the context of three major factors:

- Air
    - How much air do I have left, and how long will it last at my current breathing rate?
    - What is the surrounding air like, and will it be breathable?
    - What actions can I take to minimize my air use, and how will those affect my survivability?
    - What options do I have?
    - Should I attempt to exit or remain where I am?

- Tools
    - What tools did I bring with me?
    - How can these tools be used?
    - If I take action, what will be the benefits?
    - If I take action, what will be the cost in air?

- Communication
    - Can I direct the RIT to my location accurately?
    - How can I improve the tenability of this space?
    - What can the IC do to improve my survivability?
    - Is the RIT established and ready to deploy right now?
    - How much training do they have?
    - How confident am I that they will be able to find me?

Decision making for a person in the Mayday situation is a critical function. There is no comprehensive list of all the possible variables that may exist on a given fireground at the time of the Mayday. Firefighters must not only train regularly on calling the Mayday but also diligently prepare their minds for the decision making necessary to survive the situation. Providing multiple opportunities for firefighters to experience a Mayday and incorporating decision-making components into the drill is very effective. Such training can also be expensive and time consuming. While there is no substitute for hands-on, reality-based training, it is nevertheless possible to train based on the factors that can impact the firefighter in a Mayday situation. This training can be done in a tabletop or beanery environment as a decision-making exercise outside the hazard area of a hands-on drill.

REALISTIC RIT EXPECTATIONS

In training scenarios, under perfect conditions, RITs in the Seattle Fire Department have demonstrated the ability to consistently locate and provide air to downed firefighters within 10–15 minutes. How long will your reserve last if you don't create some margin for error? Should you hunker down and wait? Should you attempt self-extrication? Should you do some combination of these activities? The answers are situational. Too often this type of thinking is not included in firefighter self-rescue curricula.

Once you start breathing the products of combustion, you are far beyond the *point of no return* and will be lucky to survive. This does not mean that such situations are not survivable, nor should any training program even suggest the idea of giving up. Firefighters can survive—and have survived—such experiences. This book provides methods of avoiding out-of-air situations in the first place. Effective training programs must include prevention techniques such as the ROAM, decision making for Mayday situations, and survival techniques for out-of-air situations.

THE 2400L CYLINDER

The 2400L cylinder packs a wallop in terms of both air supply and weight (table 2–2). While many hazardous materials (hazmat) and confined-space entry teams have been using the 2400L cylinder for years, it is a poor choice for regular fireground use. While it may seem appropriate to offer our personnel the largest air supply possible, there is a diminishing return with the added weight and size of the 2400L cylinder. Even with an air management policy in place, the 2400L cylinder provides 1,800 liters of air for entry, work, and exit. While this represents a 100% increase over the 1200L cylinder, there are a number of associated drawbacks.

Table 2–2. Cylinder size and weight (Courtesy Mine Safety Appliances)

Cylinder Type	NIOSH Service Life Rating	Pressure	Weight (Empty)
L-30	30-minute	2216 psig	7 lb. 9 oz.
L-30+	30-minute	3000 psig	10 lb. 2 oz.
Fiberglass Hoop-Wound	30-minute	2216 psig	13 lb. 0 oz.
Aluminum	30-minute	2216 psig	18 lb. 0 oz.
H-30	30-minute	4500 psig	7 lb. 3 oz.
H-45	45-minute	4500 psig	10 lb. 9 oz.
H-45 Low Profile	45-minute	4500 psig	10 lb. 6 oz.
H-60	60-minute	4500 psig	12 lb. 11 oz.

Studies have shown that weight carried on the back and supported by the SCBA harness can have a negative impact on the firefighter in several ways (fig. 2–3).[8] First, with an increase in weight, a firefighter will adopt a more hunched posture, to maintain the same center of gravity. Second, the additional weight is transferred through the SCBA harness to the chest and back in the area of the support structure for the lungs. This increases the effort required in order to breathe and results in the breathing muscles becoming tired more quickly. These effects were noted in a British study:[2]

Fig. 2–3. Visual differences between the smaller, 1800L, and larger, 2400L cylinders. The 2400L cylinder places an increased load on operational firefighters.

Such alterations to normal gait are counteracted by eccentric and isometric contraction of various muscle groups that include the hamstrings, the muscles of the lower back and the abdominal wall, and various muscles in the shoulders and neck. Many of these same muscle groups also act as accessory muscles of respiration during times of high ventilatory demand. Additionally, isometric contraction of the shoulders, upper chest and upper-limbs, which has been shown to restrict blood flow may impact negatively on respiratory muscle function. Finally, all of these factors may be exacerbated by thermal stress, and may result in sub-optimal firefighting performance.

The increased effort required in order to breathe under the load of a 2400L cylinder further manifests itself in reduced ability to work rapidly. The decrease in the ability to perform work and the physical and mental stress involved impair the ability to make good decisions. Importantly, each decision made in this environment can be classified as a *critical decision point.* Critical decision points are times when decisions must be made rapidly and under pressure, in *threatened loss-of-life* situations.[9]

PHYSIOLOGICAL IMPACTS OF THE 2400L CYLINDER

There are three pitfalls associated with larger cylinder sizes. First, increased muscular activity results in increased heat production and increased thermal stress on the firefighter. Second, this creates an even greater respiratory drive, resulting in a cycle of negative impacts on the firefighter wearing the 2400L cylinder. Third, the potential exists for the firefighter to experience increased body core temperatures that may reach a dangerous level within a single 2400L-cylinder work cycle; studies indicate that high body core temperatures are reached relatively quickly in hot training environments.[10]

Departments that opt for the 2400L cylinder as their primary firefighting tool create a situation in which firefighters will be physically exhausted after a single rotation. This has always been true of firefighters performing hazmat and confined-space operations. Each entry team is supported by a backup team. If multiple entry teams are required, then the first entry team is rotated out and a fresh crew takes up the entry duties. It would be unacceptable in a hazmat or confined-space entry to have an entry team come out, switch out their 2400L cylinders, and head right back in. Why would this be an acceptable practice for structural firefighting?

Alternative use of the 2400L cylinder: Rapid intervention

One possible use on the structural fireground is to have the RIT in 2400L cylinders. While this is a controversial proposal, the Seattle Fire Department has implemented a system that uses the 2400L cylinder for rapid intervention operations. The RIT also bring with them a rescue air kit (RAK), containing a 2400L cylinder connected to both a RIT fitting and a complete facepiece assembly. This allows the RIT to provide air to downed firefighters regardless of whether their SCBA are intact.

Seattle is blessed with enough resources to provide a RIT function on all firegrounds. In addition, Seattle has been fortunate to have creative and passionate advocacy for the rapid intervention/RIT program from Captain Brian Shearer and Safety Battalion Chief Rick Verlinda. Seattle uses 12–14 firefighters designated as a *rapid intervention group* (RIG) within the ICS (fig. 2–4). Seattle RIGs receive additional training and equipment and follow standardized operational guidelines (SOGs) for arrival at an incident and preparation for a potential Mayday. Policy requires that ICs back up an activated RIG with another, similarly equipped RIG. In this way, Seattle ensures that the backup team has the same level of protection as the initial RIG entry team and provides the initial RIG entry team with the ability to be reassigned to rehabilitation after using a single 2400L cylinder.

Fig. 2–4. Seattle Fire Department RIG. Good training allowed this RIG to get all the necessary equipment to the assigned staging location quickly. Fortunately, their service was not required at this major warehouse fire. (Photograph courtesy Seattle Fire Fighters Union, IAFF Local 27)

CONCLUSION: THE SEATTLE WAY

The Seattle Fire Department has adopted the 1800L cylinder in combination with an aggressive air management philosophy for fireground operations. In addition, the Seattle Fire Department, in consultation with the Washington State Department of Labor and Industries, is considering a comprehensive rehabilitation policy in recognition of the fact that increases in work-cycle times require increases in rest-cycle (rehabilitation) times. Recommended procedures detail five ways to earn an assignment to rehabilitation:

1. You (or a member of your team) have used two 1800L cylinders while practicing good air management and exiting prior to activation of the low-air alarm. Assignment to rehabilitation is integrated into the regular two-cylinder work/rest interval.

2. You (or a member of your team) have used one 1800L cylinder without practicing good air management and have remained in the IDLH environment until after the activation of the low-pressure alarm. Since you have chosen a long work cycle, the department will provide a long rest cycle and will assign you to rehabilitation. (In addition, since activation of a low-pressure warning is an immediate-action item, as established by NFPA 1404, you must communicate with the IC or another member of the ICS immediately.)

3. You have used one 2400L cylinder in any capacity. Since you have had a long work cycle, the department will ensure that you have an appropriate rest cycle and will assign you to rehabilitation.

4. You (or a member of your team) have used two 1800L cylinders previously during this incident and have used one 1800L cylinder during this rotation. Assignment to rehabilitation is in recognition of the cumulative effect of multiple work cycles during an incident or a work shift.

5. The IC assigns you to rehabilitation for any reason.

MAKING IT HAPPEN

1. Conduct training exercises using various workloads and identify which firefighters on your team use the most air. Use this information when assigning tasks in the IDLH environment.

2. Conduct training exercises to identify which size of cylinder will provide your department with the best combination of effective work/rest intervals and an adequate margin for safety.

3. Work with your department wellness/fitness representatives and safety committees to identify standardized work/rest/rehabilitation intervals and make them policy.

4. Ensure that all department personnel understand the role that the Mayday plays on the fireground and can effectively respond to a Mayday call.

5. Ensure that all department personnel understand how work rates, breathing rates, and cylinder size affect the length of the survivability window for a downed firefighter.

6. Ensure that rapid intervention teams (e.g., RITs and RIGs) understand their role on the fireground, including their responsibility to avoid becoming part of the problem.

STUDY QUESTIONS

1. Should firefighting SCBA cylinders be rated by time or volume?

2. If a firefighter is using a 1200L (30-minute) cylinder and works in the hazard area until the low-air alarm activates, what is the margin for error on the way out of the structure?

3. A firefighter in trouble should call a Mayday using what USFA format?

4. What three major factors should the firefighter in a Mayday situation consider when making decisions after the Mayday is called?

5. What is the difference in work-cycle air between using a 1200L (30-minute) cylinder without following the ROAM and using an 1800L (45-minute) cylinder and following the ROAM?

6. In a training situation, how long can a firefighter expect 300L of air in a 1200L cylinder to last after activation of the low-air alarm?

7. What negative impacts can be caused by carrying the increased weight of the 2400L (60-minute) cylinder?

8. How do the impacts of the increased weight of the 2400L cylinder affect respiratory demand?

9. What impact does the increased load of the 2400L cylinder have on thermal stress?

10. What cylinder provides a good balance between work cycle, rest cycle, and trips to rehabilitation, while providing a good margin for error when used in conjunction with the ROAM?

NOTES

1 Klaen, B., and R. Sanders. 2000. *Structural Firefighting,* Quincy, MA: National Fire Protection Association.

2 Parsons, K. C. 1993. *Human Thermal Environments.* London: Taylor and Francis.

3 S.E.A. Group. 2002. "Will your respirator let you breathe?" http://www.sea.com.au/docs/articles/nswfb_firedrill.pdf.

4 NFPA 1404—Standard for Fire Service Respiratory Protection Training. 2007 edition.

5 Ibid.

6 Smoke divers. 2006. Windsor Locks, CT: Commission on Fire Prevention and Control, Connecticut Fire Academy, June Fire School.

7 Effective Fire Service Training. 2006. The 39's incident: Seattle Fire near miss experience. From the presentation "How Firefighters Die and the Point of No Return." Fire Department Instructors Conference, Indianapolis, IN.

8 Operational Physiological Capabilities of Firefighters: Literature Review and Research Recommendations. Office of the Deputy Prime Minister. http://www.odpm.gov.uk.

9 Klein, G., R. Calderwood, and A. Clinton-cirocco. 1988. *Rapid Decision Making on the Fire Ground,* U.S. Army Research Institute for the Behavioral and Social Sciences, 5001 Eisenhower Avenue, Alexandria, VA.

10 Operational Physiological Capabilities of Firefighters: Literature Review and Research Recommendations. Office of the Deputy Prime Minister. http://www.odpm.gov.uk.

CHAPTER 3

SAVED BY THE BELL?

INTRODUCTION

The low-air alarm on SCBA, also called the *bell* or *vibra-alert,* has traditionally signaled firefighters to exit the IDLH environment. Until that activates, firefighters don't give much thought to how much air they have, what their rate of air use is, or how much air they'll need in order to exit safely. Many times they don't even realize that their own bell is the one ringing. Multiple low-air alarms are commonly heard sounding within a structure fire—and rarely are they perceived as a concern. The problem with the practice of using the bell as the primary indication of when to leave is straightforward: It is not meant to serve that purpose, and using it as such dramatically increases your chances of becoming a close call or fatality.

The low-air alarm is a warning that you have begun using the final 25% of your air supply, or the *emergency reserve.* That air was never intended for normal fireground/emergency operations. While it has been utilized in that capacity erroneously, that practice needs to stop.

The emergency reserve is aptly named, as it is meant for just that: emergencies, of the kind that sometimes befall even the best-trained firefighters when thrust into the dynamic and deadly situations that firefighters call work. The emergency reserve is intended for the sudden collapse, entrapment, or disorientation that overtakes an individual or a crew. It is the air that provides the time for rescue teams and RITs to achieve access, provide essential air and water, and extricate firefighters in trouble. To put it bluntly, it is *your air*.

WHAT'S WITH THE BELL, ANYWAY?

The telltale vibration or ringing of the low-air alarm is an indication that 75% of the SCBA air has been depleted. In general terms, it is a notification that you have approximately one to seven minutes before the bottle is empty, depending on your rate of air use at the time of alarm activation. The low-air alarm indicates that firefighters are using the air of their emergency reserve (fig. 3–1).

Fig. 3–1. Low-air alarm and emergency reserve. The first 75% of the gauge is air intended for normal operations. The final 25%, usually colored red, is the emergency reserve and is intended only for firefighter emergencies. (Photograph by Lt. Kyle White, Seattle Fire Department)

It's easy to see why the fire service fell into the practice of using the emergency reserve for normal operations. Historically, most firefighters refused to wear the SCBA when they were first introduced, and this began the process. Taking valuable time to put on your mask was viewed as less than aggressive since every second in the firefight counts.

Moving forward in time, SCBA was ultimately accepted as a necessary component of the fireground, but the established thought pattern persisted. Firefighters desired to spend as much time as possible getting the job done, and that meant using every last breath of air in the bottle before exiting. This included the emergency reserve, and as a consequence, the chorus of multiple bells going off in the structure was perceived as routine and no cause for concern. Most firefighters are probably familiar with the steady pace of the alarm that slowly fades as air begins to run out. Many have probably experienced the feeling that the facepiece is collapsing toward them as air is depleted. Few ever forget the feeling of unexpectedly running out of air and having the mask suck onto your face.

The original reason why there was little concern for running out of air, even as SCBA gained acceptance, was because the smoke was wood based and did not readily incapacitate firefighters. However, it is now known that even *that* smoke contained deadly chemicals that are wreaking havoc on the firefighters of that time in the form of respiratory diseases and cancers. The modern smoke environment is understood by most sober firefighters to be a poisonous and deadly killer that can immediately render them unconscious and leave effects that will be felt many years after exposure.

All of these factors, combined with a reluctance to change, have brought the U.S. fire service to its current practice of not exiting the IDLH atmosphere until after the low-air alarm activates. This practice is killing firefighters every year.

ARE WE REALLY SMOKE DIVERS?

Interestingly, the dive industry is way ahead of the fire service in this area. Most divers quickly learn that dive masters will not tolerate divers who do not hold their reserve air sacred. The reason for this is simple: The underwater environment is deadly when breathable air is depleted. While this is common sense in the world of underwater diving, since our lungs were not designed to breathe water, why is it so foreign to the fire service? Our lungs were not designed to breathe smoke either (fig. 3–2).

*Fig. 3–2. Seattle Fire Department divers,
pictured in a routine drill on air management
and dive safety (Photograph by Lt. Tim Dungan,
Seattle Fire Department)*

Firefighters are basically *smoke divers.* They insert themselves into a hazardous atmosphere that will not sustain unprotected breathing and thus makes portable air a necessity. As in the dive world, that air is broken up into two components: air for normal operations and the emergency reserve. It is imperative that we initiate a change in behavior mandating that reserve air be treated in the same manner as it is by divers.

One classroom participant who listened to the dive analogy offered this enlightening observation:

> *I see the role the emergency reserve plays for firefighters as far more critical than for divers. When I dive, even if I run out of air, I always know which way is up. I at least know the direction of fresh air. Running out of air in a smoke-filled structure is a whole different matter. If I'm lost or disoriented, the way to fresh air is not immediately evident and may be difficult or impossible to find. Firefighters are much more likely than divers to have difficulty getting to fresh air when in trouble.*

This point should not be lost on firefighters. The dive industry takes the emergency reserve seriously enough to have started the practice of carrying an extra cylinder, called a *pony bottle,* to ensure safety. Firefighters are already carrying enough weight into the firefight that such an option is not realistic for our emergency operations. We need to commit, as an industry, to reclaiming the emergency reserve for its intended use.

WHEN AN ALARM BECOMES JUST ANOTHER NOISE

It doesn't take the rookie firefighter long to notice that the veterans don't pay much attention to low-air alarms. The buzz of the vibra-alert or the incessant ringing of the bells is usually viewed as a nuisance, rather than an indication of trouble. Many times, individuals whose alarms have activated aren't aware that it is their alarm that is sounding. While this is common practice, it certainly isn't smart.

Firefighters have anywhere from one to seven minutes of air left once their warning bell sounds before they will be breathing in whatever the hazardous atmosphere has to offer. In other words, they are minutes from inhaling hydrogen cyanide, phosgene, carbon monoxide, formaldehyde, and any of a host of other poisons that will have both short- and long-term effects on their health. Along with the added noise caused by the ringing bell comes the urgency naturally associated with a dwindling air supply. Most would readily admit that the urgency picks up dramatically as the alert starts to slow, indicating the coming depletion of the bottle.

A firefighter whose bell is ringing is minutes away from inhaling hydrogen cyanide, phosgene, carbon monoxide, formaldehyde, and any of a host of other poisons that will have both short- and long-term effects on their health.

As with the alarm from the original PASS device that was ignored owing to constant false alarms, a low-air alarm gives no indication whether the warning truly represents an emergency. The false-alarm mentality created by acceptance of these alarms going off routinely sets us up for disaster when fire-alarm activations elicit similarly complacent responses. Most firefighters have a few occupancies in their districts that are notorious for multiple false alarms. It is easy to be lulled into complacency because these alarms prove to be false time after time after time. The false-alarm mentality that this creates causes firefighters to be unprepared when the alarm turns out to be the real thing (fig. 3–3).

*Fig. 3–3. Firefighters assisting a member whose
low-air alarm has activated (Photograph by
Lt. Tim Dungan, Seattle Fire Department)*

Our philosophy is simple: *Any alarm activation in an IDLH environment
should be an immediate-action item.* It should be immediately investigated
by nearby crews, and a determination should be made as to whether the
individual is in trouble or has stayed in too long and is exiting. This view
was unanimously agreed on by the NFPA 1404 committee, who added it to
thecurrent edition of to the respiratory standard.

NO GAIN, LOTS OF PAIN

Most of our actions on the fireground are preceded by a rigorous
risk-benefit analysis. We evaluate the structure to determine if entry is
feasible based on structural stability, degree of fire involvement, and

numerous other factors. Taking the time to ascertain exactly what we would gain by sacrificing our emergency reserve proves to be interesting. The balance of benefit to risk is not what you might expect.

When asked how much additional work they can accomplish by working to the end of their bottles, classes usually answer around two to three minutes of work. That seems fair. In essence, all the heated arguments about staying in as long as possible and not being dragged out until we're ready are centered around these two to three minutes of actual work time, pulling ceilings or advancing additional lines. This is the price we place on violating our emergency reserve. Is it a reasonable one? When weighed against the consequences, it becomes readily apparent that firefighters play a fool's game when they continue operating in the same old way (fig.3–4).

Fig. 3–4. Violating the emergency reserve. During strenuous operations, how much actual work gets done when violating the emergency reserve? (Photograph by Lt. Tim Dungan, Seattle Fire Department)

HOW LONG CAN IT LAST?

As is evident from the preceding discussion, utilizing the emergency reserve leads to unnecessary risk and tragedy. The opposite is true when the reserve is used as it was intended. Giving your teams the 25% margin for error will eliminate many of the situations that have caused firefighter close calls and fatalities in structures.

As most of us know, there will always be times when a safe exit becomes impossible. A sudden collapse, a wrong turn, or an unexpected rescue can put us in a situation where the reserve may need to be used. These unforeseen events are all accounted for in the risk-benefit analysis and have nothing to do with the inappropriate routine use of the reserve at emergencies.

What should be obvious is that proper use of the emergency reserve puts the firefighter in a much better position to survive the unexpected and avoid becoming part of the problem. Recent drills conducted by the author's and other departments have produced dramatic results. By recreating situations in which firefighters have found themselves trapped or lost and in need of RIT deployment, the emergency reserve was proven to be an effective tool to achieve RIT success (fig. 3–5).

The trapped firefighters in figure 3–5 made the appropriate request for help and then went into air conservation mode to make the remaining air last as long as possible. In tests, firefighters were able to make their emergency reserve last as long as 45 minutes before the bottle was empty. Utilizing skip-breathing (see chap. 17) and other methods of breath control, the average firefighter was able to make the reserve last 25–30 minutes. In classes taught by the authors, departments have reported times of more than one hour in drills.

Fig. 3–5. The emergency reserve as survival tool. The emergency reserve becomes critical when something goes wrong. (Photograph by Lt. Tim Dungan, Seattle Fire Department)

Imagine your chance of a successful RIT operation if you had a 20-minute window in which to get air to the firefighter needing rescue. Expand that to 30 minutes or more; now how good would your chances be?

Think of the improved situational awareness and decision-making ability a lost/trapped firefighter would have without the added stress of immediately needing air. It is indisputable that having the emergency reserve at your disposal will make a dramatic impact on your survivability when things go wrong.

CREW ROTATION

Chapter 9 provides a thorough discussion of how crew rotation can be enhanced through good incident command system (ICS) procedures combined with air management. It is worth mentioning here as well, though, since ICs need a reasonable turnaround time to get relief crews into position. Firefighters who manage their air and give good reports to the IC will put themselves in a position to assign replacement crews and will not be in catch-up mode. This is in sharp contrast to the current actions of most fire departments, who rely on a series of low-air alarm activations for the impetus to provide relief.

Work cycles, as discussed in chapter 2, are also applicable here. Maintaining an appropriate level of reserve air will also facilitate the consistent application of good work cycles during emergencies.

IT'S YOUR AIR

In the final analysis, all of this reduces to one simple fact that every firefighter should know: The emergency reserve is *your air*. The first 75% of the bottle is air that is meant for the emergency to which you've responded. It is for the rescue, the suppression, the ventilation, and the overhaul. It is for the task you are assigned, the department you serve, and the citizens you protect. That air is rightly meant to be used in the aggressive accomplishment of your job. The emergency reserve, however, is not.

Your emergency reserve is the air that is meant to return you to your family when something goes wrong. It is for the rescue teams that are risking their lives to come after you and safely extricate you from danger. It is designed to give your children and loved ones every possible chance to avoid the grief of a firefighter funeral. It is your air—and it is your family's air (fig. 3–6).

Fig. 3–6. Your family's air. The emergency reserve is intended to get you home at the end of the day. (Photograph by Lt. Tim Dungan, Seattle Fire Department)

FOR WHOM THE BELL TOLLS

Most firefighters, knowingly or unknowingly, have elected to roll the dice on their lives and those of their crew. That's the bet that is casually made across the fire service when operations continue until the bell sounds. As has already been discussed, once the emergency reserve is being used, the firefighter is minutes away from being forced to remove his or her facepiece and breathe whatever air surrounds him or her. Often, that will be the fresh air that beckons from outside the IDLH atmosphere. The unarguable reality, however, is that many times the mask must be pulled off while still in the midst of the smoke. And it is here that the gamble exacts its toll: If you are waiting until your bell rings to exit the structure, you are betting your life that nothing will go wrong on your way out (fig. 3–7).

Fig. 3–7. A firefighter funeral. The fire service never forgets its fallen. Still, we must work harder to keep them from falling. (Photograph courtesy Seattle Fire Fighters Union, IAFF Local 27)

One to seven minutes of air is all that is keeping you from smoke inhalation, carbon monoxide or cyanide poisoning, disorientation, panic, asphyxiation, and/or death. It is a very slim margin on which to bet the health and safety of yourself and your crew. Yet that very gamble is routinely taken throughout the fire service.

When gamblers bet their money in any game of chance—for instance, at a casino—they know going in that there is a risk involved. That risk should be acceptable. A $20 bet at a blackjack table should be made knowing that the odds are ultimately against the player. The bright lights and huge profits of such establishments are not consonant with handing out big winnings to customers. The gambler knows the odds but lives with them because the risks are acceptable—or at least they should be.

How acceptable, then, is the notion of betting the safety of your crew on the emergency reserve when such a gamble is not necessary in the first place? A well-run, professional operation should have replacement crews ready to take over. If those additional reserve companies are not available, that only furthers the need to exit with appropriate reserve of air intact. How organized and successful will the RIT response be if those in charge of the operation cannot effectively handle basic crew rotation or don't have the resources to begin with? Once again, you are betting your life and the lives of your crew on the answer.

AIR MANAGEMENT IN BELOW-GRADE FIRE SUPPRESSION OPERATIONS

Fire suppression in below-grade areas is the most labor-intensive and dangerous fireground operation with which firefighters will ever be confronted. The demanding challenges posed by these incidents—such as limited egress, additional search considerations, numerous fire extension probabilities, and increased difficulty in accessing the area of origin—significantly increase the level of risk placed on responding companies.

A closer look at specific operational hardships associated with basement fires reveals a common denominator that encompasses all below-grade fire suppression operations. Simply stated, fires occurring in below-grade areas require that firefighters operate on the interior of a structure for longer periods of time. As a result, we precariously push our personal physical capabilities and the limitations of our SCBA to the point of no return.

Unfortunately, this distressing fact resonates throughout the fire service. Nationwide efforts to initiate mandatory-mask rules and increase training in SCBA use have failed to ensure that firefighters recognize the detrimental impact that improper SCBA use will have on their own safety and the safety of their crew. The tendency of firefighters to exhibit behavior patterns on the fireground that consistently cause an escalation of individual and crew risks is problematic. These dangers are only compounded when fire suppression operations are conducted in basements and other below-grade areas.

The most basic principle of air management is a personal and constant recognition of the amount of air remaining in your SCBA. This should be coupled with the critical requirement to effectively manage that air, to maintain a sufficient level of reserve to escape from hazardous environments. Failure to respect these principles is a significant weakness of many firefighters. This failure, together with dangerous, ego-based decision making by fire suppression crews, contributes overwhelmingly to fireground deaths, injuries, and close calls.

The risks facing fire suppression crews at below-grade fire incidents and the importance of adopting the ROAM as a department guideline are evident in several examples. Two fire incidents that occurred in Englewood, New Jersey, were directly linked to ineffective utilization of fireground air management techniques.

Michael Marino, an Englewood firefighter (now a lieutenant), became disoriented while conducting fire suppression operations in a 25 ft by 25 ft windowless commercial basement containing several rows of rack storage of common combustible materials. The following excerpt by Lieutenant Marino was published, shortly after the incident, as a supplement to an article in *Fire Engineering* magazine:

I had been a firefighter with the City of Englewood for a little more than six years. During this time I have responded on all types of alarms, ranging from the mundane everyday service call to a 4th alarm defensive firefight. However, there is one seemingly routine fire that I will never forget. It is this story I would like to share with you.

At 0843 hours on October 19, 1998 while members of Group 2 were conducting our daily routine work, an alarm was received for a reported fire in the basement of a commercial building in the city's southern industrial area. I was assigned the position of truck chauffeur for the day. Our department places a heavy emphasis on truck company operations. Procedures for structure related responses dictate the truck be first due with the first and second due engines following.

This gives the truck the opportunity to always have the front of the building and enhances our operations. While enroute to the scene, dispatch informed us that they were receiving calls reporting a working fire.

Upon arrival, units found smoke pouring from the front door of a one-story, block-long commercial building. The on duty captain transmitted a second alarm as we rolled into the street. As the other members of the truck company entered the structure they were met by heavy smoke coming from the basement door. The irons man forced entry and the team began their descent. Outside, engine companies were laying supply and attack lines in the street. I tried to position the tower ladder for ladder operations but was unsuccessful due to power lines blocking access to the front of the building. The interior truck company team was reporting a delay in finding the seat of the fire due to the presence of rack storage in the area. After conducting my own size up of the building I realized that we had a fire in a windowless basement and that my duties on the exterior would be limited. I decided to enter the basement and assist the other truck company members in finding the seat of the fire. While entering the basement I propped open the door with my halligan tool. I felt that this would serve as our only vent hole and I did not want the door to close and lock on us while we were in the basement. At the base of the stairs, I was met by a heavy-smoke condition with some heat. I could hear the sprinkler system operating on the fire but the question was where? The Engine Company was beginning their advance down the stairs; I grabbed a wall and began my search. Visibility was zero and the rack storage was creating a major orientation problem for me. During my search I came off the wall numerous times. Being the last member of the Truck Company to enter the room I was surprised when I found the fire in the rear 3/4 corner of the basement. I guess that I was lucky enough to turn the right way during my search. I found the fire burning in boxes and other material stored on the racks in this area. I radioed the Engine Company on the location of the fire and they began to aggressively move in and extinguish the fire.

I don't know exactly how long I was operating in the basement, maybe 10 to 15 minutes but right about this time my low air alarm activated. I knew that it had taken me some time to find the fire and it was going to take me about the same time to get out. This was not a good situation to be in. Deep in the basement with my low air alarm sounding. I turned around and attempted to exit by following the same path I had taken on the way in. I began to think that if my low air alarm is activated now I would not have enough time to get myself out of the area. I oriented myself with the wall and began to move toward the exit. I think I got about 20 feet into my retreat when I hit a dead end corner of the basement. Now I'm totally disoriented and lost in the room. I reach for my radio and I'm about to request a "Mayday," but this ego thing goes off in the back of my head and I make the decision not to call for help. I figure that I will be able to get myself out without any assistance. I decide to find the attack line and follow it out of the basement. Now I'm crawling fast and furious over debris on the floor, desperately trying to find the hose line. I can hear the

Engine Company operating on the fire, but I don't know where they are in the room. Bingo! It happens I come upon the hose line and begin to follow it out. I'm making good progress until it happens, I find myself at the nozzle! I had followed the hose line the wrong way! Apparently the engine company members had also run out of air and were in the process of removing themselves from the area. Now panic is quickly setting in. I know that I am in serious trouble, my mind is racing and it's making me forget all of the skills that I have grasped throughout my career. I had always believed that I could get myself out of any situation that I might be involved in; how wrong I was! Disorientation and panic will cause you to do things you don't normally do. I turn around on the hose and again begin to follow it out. I get to a point where the hose is kinked in a figure eight, I stand up to try to get myself through the maze and my air cuts out. Luckily for me I was able to get close enough to the stairs, and as I was standing and searching for the way out the second alarm companies who were making their way into the basement bumped into me. I screamed through my mask for them to direct me to the stairs; they do, and I begin to run up to the exit. But as my luck would have it, someone removed the halligan that I had used to prop open the basement door on my way in. As a result, the door had closed and locked leaving me stranded at the top of the stairs. I did not have a tool with me so I gave the door a couple of stiff shoulders and I was able to get myself out of the basement.

Once I was out on the street, and at the rehab area, I began to feel shaken by what just happened to me. Many different thoughts were racing through my head. The panic, the smoke, and the seriousness of the situation I was just in. I know that we all make mistakes, but in this business stupid decisions will cost you your life. I also know that I possess excellent Firefighting skills and feel confident in my arena. I have had many great teachers throughout my career, who gave me the opportunity to learn and become a better firefighter. However, I can tell you that once you become disoriented and panicked, all of those skills and experience can quickly disappear. At the very first hint of danger I should have radioed for help. There was a FAST team in the street and I didn't call them. My PASS alarm was on, but because I was moving, it never activated. I should have activated it manually. Since my first day on the job, I have always carried a 50-ft. piece of personal search rope in my pants pocket. I should have used it. I normally pay better attention to the amount of air I have used. I didn't monitor it like I usually do. All of these items may sound trivial to some. However, not utilizing them when you have to may cost you your life. I learned a lot from this experience, but what really sticks in my mind is that this story could have had a tragic ending all for what started out as a typical common fire. Obviously, the old quote "no fire is routine" certainly stands true in this case. I was asked to write about my experiences by Chief Moran. At first I was a little apprehensive, but he convinced me that it would be a great lesson learned to share with my fellow firefighters. It certainly has allowed me to improve my margin of safety on the fireground. I hope that it will do the same for all of you who take an interest.

The second incident occurred on December 12, 2003, and involved a working basement fire in a 30 ft by 30 ft two-story private dwelling—what we would typically describe as a *bread-and-butter* operation. However, this fire turned out to be anything but routine. On arrival at the incident, the on-duty platoon of two officers and eight firefighters found a heavy fire condition in the rear, or C side, of the structure.

Companies quickly went to work on the fire. After advancing the initial attack line through the first floor and down into the basement, the low-air alarm of one of the first due engine company firefighters activated. This firefighter made the decision to remain on the line with two other members as they pushed toward the seat of the fire. After remaining on the attack line, the firefighter eventually made the decision to remove himself from the hazardous area to the exterior of the structure. He notified the other members and began to ascend the interior basement stairs.

At the top of the stairs, the firefighter ran out of air. He immediately issued a Mayday over the radio, to communicate his situation to the IC and other members operating at the incident. The firefighter began searching for the hoseline to provide direction to the exterior. During this time, he came upon two other firefighters who had been moving toward the area. In the confusing and chaotic environment of the first-floor hallway, the three members became disoriented and lost contact with the hose. In an attempt to regain orientation and locate an exit, the three members ended up in a small bedroom in the first-floor B/C corner of the structure. The intensity of the situation increased as the low-air alarms of the other two firefighters activated at this time.

On entering the bedroom, the engine company firefighter who called the Mayday became severely compromised by the products of combustion and went unconscious in a corner of the room. Unaware of this member's grave condition, the other two firefighters continued to search for a means of egress. Fortunately, the distressed firefighters discovered a rear window and were able to remove themselves from the hazardous environment with the assistance of members on the exterior. Once outside, they notified the rear sector officer that another firefighter was still trapped within the dwelling. Three firefighters immediately made entrance into the room via the window and, after searching the area, came across the downed firefighter trapped under a dresser in a corner of the bedroom. He was instantaneously removed to the exterior, where efforts to resuscitate him were successful.

As chief of the Englewood Fire Department, I do not have a problem admitting that several key operational errors were made during the incident. Importantly, however, many of the factors contributing to the outcome—such as low staffing levels and lack of crew integrity—have been glossed over in order to focus on the major issue of SCBA air management in below-grade fire incidents. Readers can read the entire report on this incident at the Web site of the New Jersey Division of Fire Safety.

A critique of the air management decisions made at the two incidents reveals that the basic elements of the ROAM were not followed. This alone provides a powerful statement in support of the mandatory implementation of the air management concepts discussed in this book and the in support of the adoption of NFPA 1404. In common-sense terms, firefighters who conduct a function and perform an air check prior to donning their SCBA, effectively monitor the level of available air throughout the incident, and ensure that they have a reserve capable of safely allowing them to evacuate a hazardous environment will vastly improve their chances of surviving the multitude of air management–related emergencies occurring on the fireground.

—Robert Moran (Fire Chief, City of Englewood [NJ] Fire Department), John Lewis (Firefighter, City of Passaic [NJ] Fire Department), and Michael Marino (Lieutenant, City of Englewood Fire Department)

MAKING IT HAPPEN

1. Conduct drills in which a heavy workload is used until activation of the low-air alarm. Once the alarm is activated, see how long the firefighters can continue working at that level before running out of air.

2. Run ICS drills with companies making a notification on the interior when they have reached 50% of their air. Use this as a cue for the IC to begin initial movement of replacement crews from staging. Gauge effectiveness of crew rotation by comparing this model with your current practice.

STUDY QUESTIONS

1. What percentage of the SCBA bottle is left when the low-air alarm activates?

2. In the dive world, air is broken up into two components. What are they?

3. What response is evoked by the ringing of low-air alarm in most firefighters?

4. Alarms going off all the time on the fireground creates what type of mentality?

5. How does implementing the ROAM affect RIT operations?

6. Lack of air management causes ICs to be in what mode in regard to crew rotation?

7. What is the name for the final 25% of an SCBA bottle?

8. In the author's words, whose air is that emergency reserve?

HOW FIREFIGHTERS ARE DYING ON THE FIREGROUND

INTRODUCTION

This chapter details the most likely causes of firefighter deaths. The primary focus is on the five ways firefighters are dying in structure fires:

- Smoke
- Thermal insult
- Structural collapse
- Getting lost
- Running out of air

Two key points to emphasize are that the fire service has not come up with new ways to die and that air management can make a dramatically positive impact on the situations that continue to kill firefighters.

A HISTORICAL PERSPECTIVE ON FIREFIGHTER DEATHS

History allows us the luxury of hindsight. By heeding its lessons, we can wisely avoid repeating past mistakes.

The U.S. fire service is ignoring history when it comes to firefighter fatalities. For nearly 30 years, an average of 105 firefighters have died on the job each year (fig. 4–1).

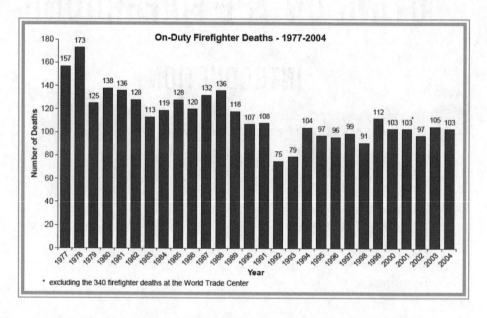

Fig. 4–1. The harsh reality of firefighter losses over the past several years (Source: U.S. Fire Administration, annual reports)

The data of figure 4–1 suggest several difficult questions. How are these firefighters dying each year? What lessons can we learn from these fatalities? And what can we do to decrease these numbers in the future?

First, how are firefighters in the U.S. fire service dying each year? The greatest number of firefighters are dying from heart attacks. The truth is that many firefighters are not screened adequately for heart health. Comprehensive wellness/fitness programs either have not been adopted by fire districts, departments, cities, and townships or have not been given the attention they deserve.

Vehicle accidents, when responding to or returning from an emergency (in a fire apparatus or in a private vehicle), are responsible for the next greatest percentage of firefighter fatalities each year. Week after week, we see news stories in which a fire apparatus or a private vehicle responding to an emergency becomes involved in a terrible accident in which a firefighter perishes. On Gordon Graham and Billy Goldfeder's Web site, Firefighterclosecalls.com, you can read about all the firefighters who die in vehicle accidents each month. It is amazing how many fire apparatus get into accidents, injuring or killing firefighters and civilians Many of these fatal accidents could be avoided by simply having the drivers slow down and wear their seatbelts.

What is most unfortunate about the majority of these vehicle accidents is that they could be avoided by the adoption by of an emergency vehicle accident prevention (EVAP) program. Ultimately, we must ask, are EVAP, apparatus operator, and training programs working? Evidently not.

Last, firefighters in the U.S. fire service are dying in structures. These fatalities deserve closer scrutiny.

FIREFIGHTERS DYING IN STRUCTURE FIRES

While the total number of firefighter fatalities has followed a downward trend over the past 20 years, the number of firefighter deaths per fire incident actually has returned to the levels seen in the 1980s. Figure 4–2 compares the total number of firefighter fatalities each year that are associated with responses to fires and the total number of fire incidents reported by the NFPA through 2003. Despite a slight decrease in the early 1990s, the level of firefighter fatalities has returned to the levels experienced in the 1980s. In particular, there have been marked increases in the rate of deaths due to traumatic injuries while operating inside structures. In the late 1970s, traumatic deaths inside structures occurred at a rate of 1.8 deaths per 100,000 structure fires; by the late 1990s, this had risen to almost 3 deaths per 100,000 structure fires; and even more recently, it has climbed to nearly 4 deaths per 100,000 structure fires (fig. 4–2).[1]

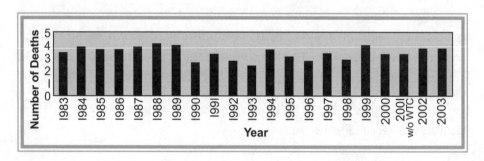

Fig. 4–2. Firefighter fatalities from the 1980s to 2003
(Source: U.S. Fire Administration, 2004 annual report)

The following are the major causes of traumatic deaths inside structures: smoke inhalation due to getting lost and running out of air, burns resulting from flashover or backdraft, and crushing injuries due to structural collapse (fig. 4–3). Interestingly, of the 87 firefighters who perished due to smoke inhalation while operating inside structures between 1990 and 2000, the majority became lost inside the structure and ran out of air in their SCBA.[2]

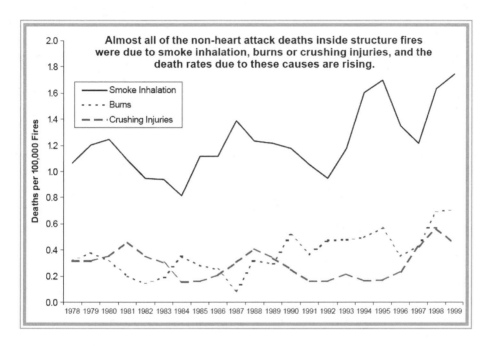

Fig. 4–3. Statistics showing that the majority of firefighter fatalities during this time period were from smoke inhalation (Source: U.S. Fire Service Fatalities in Structure Fires, 1977–2000)

The fireground is a very unforgiving learning environment. We know from statistics that the number of structure fires in the U.S. fire service is decreasing while the rate of firefighter fatalities in structures is not. Hence, even though there are fewer fires each year, the same number of firefighters are dying inside structures today as 30 years ago. This is counterintuitive; with the decreasing number of structure fires, there should be a corresponding decrease in the number of firefighter fatalities (fig. 4–4). Why is this happening? What is going wrong in the U.S. fire service?

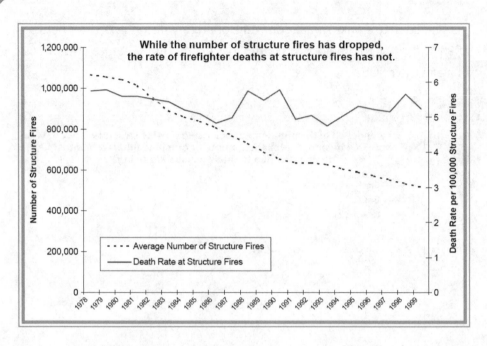

Fig. 4–4. Statistics showing that the number of structure fires in the United States is decreasing while the rate of firefighter fatalities in structures is not (Source: U.S. Fire Service Fatalities in Structure Fires, 1977–2000)

There are several reasons for the steady rate of firefighter fatalities in structure fires, including

- Lack of experience and training
- Poor leadership at the strategic, tactical, and task levels
- Lack of air management

The first and most important aspect of firefighter safety that has been compromised in interior firefighting is that firefighters are not monitoring their SCBA air supplies in fires. Therefore, they are running out of air and succumbing to the deadly chemicals found in smoke, which are, primarily, hydrogen cyanide and carbon monoxide (see chap. 5).

AIR SUPPLY AT STRUCTURE FIRES

Firefighters must constantly monitor their air supply at a structure fire. They must know how much air they have before they enter the structure, monitor that air while working inside the structure, and leave the IDLH atmosphere before their low-air alarm activates (see chap. 3). The time to exit the structure is before the low-air alarm activates. A comprehensive air management program is essential to the health and safety of firefighters.

Another reason for the constant rate of firefighter fatalities in structure fires is that because of the sharp decrease in structure fires nationally, firefighters can no longer count on gaining valuable fireground experience solely through on-the-job structure fires. Fire officers and firefighters are not getting the experience necessary to understand fire progression, fire behavior, and what happens to a building under fire conditions at today's infrequent structure fires. Therefore, fire departments across the nation have an obligation to provide frequent, high quality, realistic structure-fire training to ensure that firefighters, company officers, and chief officers learn as much as possible about structural firefighting.

In the modern context of weapons of mass destruction (WMD), many fire departments have replaced basic training in structural firefighting with specialized training in special operations. These fire departments have been applying for and receiving grants from the federal government to provide WMD training and equipment. This WMD money is seductive and has provided many fire departments around the nation with much-needed equipment—for example, new SCBA, air monitors, and thermal imaging cameras. While these improvements are good, fire departments that focus on WMD training at the expense of basic structural fire attack are doing their members a great disservice.

The reality is that all cities, towns, and municipalities are guaranteed to experience a structure fire sometime in their future. Consequently, which type of training is more valuable—special operations, in preparation for an event that may never happen, or structural firefighting, in preparation for fires that we know will occur and in which firefighters are dying every year?

HOW FIREFIGHTERS DIE IN STRUCTURES

SMOKE

The fires of the past are certainly not the fires of today. Yesterday's fires were primarily composed of natural products (Class A materials: wood, cotton, and wool), which produce smoke that contains carbon monoxide, nitrogen dioxide, and polycyclic aromatic hydrocarbons (PAHs). Breathing this smoke presents an enormous risk to the short- and long-term health of firefighters. Generations of past firefighters have suffered from cancers and pulmonary diseases due to this toxic smoke.

Today's fires consume contents that are made mostly of synthetics (Class B materials: plastics and other hydrocarbon-based materials), which produce some of the most dangerous and fatal chemicals known to humankind—chemicals like hydrogen cyanide, polyvinyl chloride, formaldehyde, hydrogen chloride, and carbon monoxide. Synthetic building contents have a smoke production level that may be as much as 500 times that of a similar fire burning wood and cotton contents.[3]

The Phoenix Fire Department recently completed a study that revealed that a typical room-and-contents fire in a modern structure produces levels of carbon monoxide above 20,000 parts per million.[4] This is enough carbon monoxide to kill a firefighter in only one or two breaths of this noxious smoke (table 4–1).

Table 4–1. Findings from a recently completed study by the Phoenix Fire Department. The study revealed that a typical room-and-contents fire in a modern structure produces levels of carbon monoxide above 20,000 parts per million. (Source: International Fire Service Training Association. 1995. Essentials of Fire Fighting. 4th ed. Stillwater, OK: Fire Protection Publications, Oklahoma State University)

Carbon Monoxide (CO)(ppm)	Carbon Monoxide in air (percent)	Symptoms
100	0.01	no symptoms – no damage
200	0.02	mild headache; few other symptoms
400	0.04	headache after 1 to 2 hours
800	0.08	headache after 45 minutes; nausea, collapse, and unconsciousness after 2 hours
1,000	0.1	dangerous; unconscious after 1 hour
1,600	0.16	headache, dizziness, nausea after 20 minutes
3,200	0.32	headache, dizziness, nausea after 5 to 10 minutes; unconsciousness after 30 minutes
6,400	0.64	headache, dizziness, nausea after 1 to 2 minutes; unconsciousness after 10 to 15 minutes
12,800	1.26	immediate unconsciousness, danger of death in 1 to 3 minutes

RUNNING OUT OF AIR IN TODAY'S SMOKE

Firefighters cannot afford to run out of air in the toxic smoke environment of today's structural fires. The evidence is conclusive: if you run out of air and breathe in smoke in a fire, you may not survive; and if you do, you have a high probability of developing some type of cancer in the future.

Lieutenant Chris Yob of the Seattle Fire Department experienced it firsthand, just before he jumped from a three-story building at an apartment fire in the Ballard district of Seattle, barely escaping with his life. As he was desperately fighting the effects of carbon monoxide (and probably hydrogen cyanide) poisoning from SCBA filter-breathing, he attempted to place his low-pressure hose out of a window of the fire building to obtain a breath of fresh air. The eddying smoke came back around, and he got a full breath of smoke that he described as tasting like "liquid, molten plastic." He said it was the worst thing he ever breathed, and he knew he couldn't stand another breath. In his dazed state, he attempted to jump to what he thought was an adjoining roof, only to fall three stories into a light well between the fire building and an adjacent structure. Luckily, Lieutenant Yob survived his fall and lived to tell his story.[5]

The only protection for firefighters in the toxic smoke environment of today's fires is the air they carry on their backs. Like scuba divers, firefighters must manage their air effectively and leave enough reserve air in case of unforeseen occurences while inside a structure fire. Firefighters must manage their air so that they leave the IDLH environment before the low-air alarm activates. This leaves an adequate emergency reserve and removes the noise of the low-air alarm from the fireground.

THERMAL INSULT

The energy crisis that occurred in the mid-1970s provided the impetus for a significant change in building construction techniques. Prior to the 1970s, buildings were not built with energy efficiency in mind; in other words, they were not necessarily well insulated and therefore could breathe. Firefighters responding to fires in these buildings often found that the fires had self-vented prior to their arrival.

Fast-forward to the modern, energy-efficient building. Modern buildings have been subjected to stringent building and energy codes that have evolved for the purpose of energy conservation. These modern buildings have double-wall construction, double- or triple-paned insulated windows, and high insulation ratings (i.e., high R values) in attics, walls, ceilings, and floors. This quest for energy efficiency has translated into "tighter" buildings, which do not breathe like older buildings. Instead, these modern buildings hold heat and superheated fire gases longer, creating a dangerous situation—namely flashover—for responding firefighters (figs. 4–5 through 4–8).

Fig. 4–5. This series of photos demonstrates the rapid fire progression of today's fires, as these Toronto firefighters are forces to escape from second-floor windows. (Photograph by John Hanley)

Fig. 4–6. (Photograph by John Hanley)

Fig. 4–7. (Photograph by John Hanley)

*Fig. 4–8. Toronto firefighters forced to escape from a
second-floor window owing to impending flashover
(Photograph by John Hanley)*

FLASHOVER

Flashover is the sudden ignition of combustible gases and smoke in an involved area, resulting in a rapid and intense increase in temperature. This sudden ignition is usually caused by thermal radiation feedback from the ceilings and upper walls of a room, which have been heated by the fire. With the sudden ignition of combustible fire gases and smoke and the rapid rise in heat, all of the exposed contents of the room simultaneously ignite, and the fire room flashes over (table 4–2).

Table 4–2. Interior fire attack—a very dangerous environment (Source: U.S. Fire Administration, annual reports)

Year	Number of Firefighter Deaths while Engaged in Fire Attack (Source-USFA Annual Reports)
2005	11
2004	16
2003	11
2002	13
2001	13
2000	13
1999	16
1998	18
1997	21
1996	9
1995	18
1994	7

Since modern buildings are better insulated and can hold thermal radiation longer, firefighters are much more likely to encounter flashover today than they were years ago. Moreover, thermal insult, or flashover, is killing firefighters (fig. 4–9).[6]

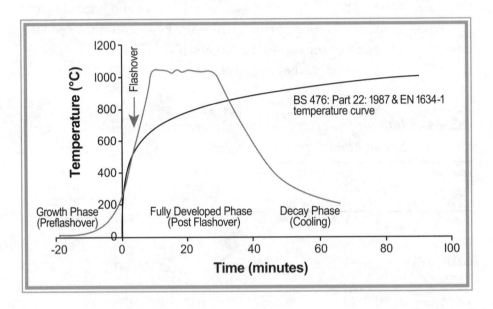

*Fig. 4–9. Graph showing that firefighters can be faced
with flashover conditions in as little as five minutes. The
time-temperature curve is plotted against a real fire curve, showing
the point of flashover. (Source: Chiltern International Fire)*

CASE STUDY 1: THREE FIREFIGHTERS DIE FROM THERMAL INSULT IN A SINGLE EVENT — KEOKUK, IOWA.[7] On December 22, 1999, firefighters from a career fire department responded to a structural fire at a residence. The fire was started when a stove turned on shortly after 0800 hours, igniting the materials on the stovetop.

A 49-year-old shift commander (victim 1), a 39-year-old engine operator (victim 2), and a 29-year-old engine operator (victim 3) lost their lives while performing search-and-rescue operations for three children who were trapped inside the burning structure (fig. 4–10).

Fig. 4–10. Side A, point of entry, with the front porch and heavy smoke from the second floor visible. (Source: NIOSH Fire Fighter Fatality Investigation 2000-04)

The crew of Aerial Truck 2 witnessed a woman and child trapped on the porch roof, and they were informed that three children were trapped inside the house. It is believed that the three victims were hit with a thermal blast of heat before the hand line burned through. The three victims failed to exit as 12 additional firefighters arrived on the scene through a callback method and began fire suppression and search-and-rescue operations. Victim 2 was located, removed, and transported to a nearby hospital, where he was pronounced dead. Victims 1 and 3 were later found and pronounced dead on the scene (figs. 4–11 and 4–12).

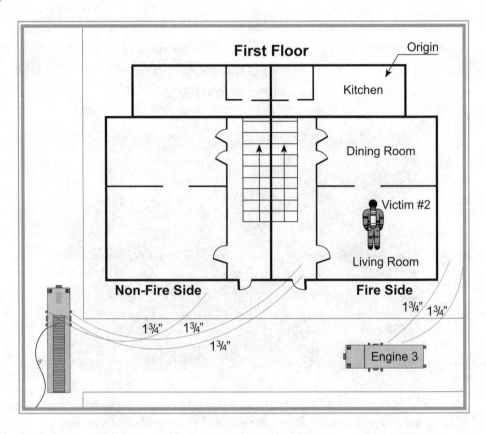

*Fig. 4–11. Diagram showing the location of
victim 2 on the first floor after flashover and
rapid fire extension (Source: NIOSH Fire Fighter
Fatality Investigation 2000-04)*

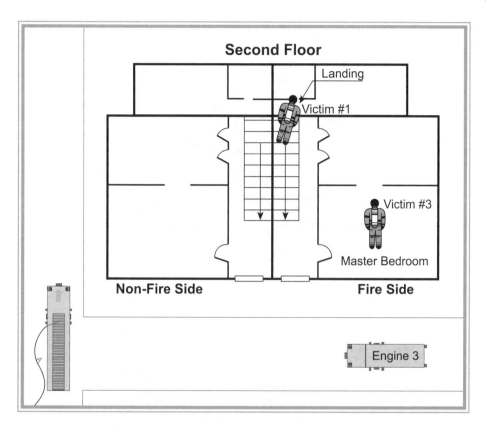

*Fig. 4–12. Diagram showing the location of
victims 1 and 3 on the second floor after
flashover and rapid fire extension (Source: NIOSH
Fire Fighter Fatality Investigation 2000-04)*

The incident occurred in a balloon-frame structure built in the 1870s as a single-family residence. In the 1970s, the structure was remodeled and divided into three separate apartment units. The apartment in which the incident occurred was a single unit that occupied both floors of the structure (fig. 4–13).

Fig. 4–13. The C side of the balloon-frame structure, built in the 1870s as a single-family residence, in which the incident occurred. The fire originated in the kitchen, at the lower left corner of the photograph. (Source: NIOSH Fire Fighter Fatality Investigation 2000-04)

The pathologist's reports list the causes of death of the victims as follows:

- Victim 1: Smoke inhalation and sudden exposure to intense heat
- Victim 2: Smoke inhalation and sudden exposure to an extremely hot environment
- Victim 3: Sudden exposure to intense heat

In simple terms, these three firefighters burned to death.

Significantly, these fatalities occurred in a very short time frame. Detailed investigation revealed that the time line from ignition to flashover and rapid fire spread was less than 10 minutes.

CASE STUDY 2: TWO FIREFIGHTERS DIE BY THERMAL INSULT.[8]
On April 18, 2005, a 23-year-old firefighter and a 39-year-old assistant lieutenant died after a smoke explosion at a town house complex. Both victims were on the first fire apparatus to arrive on scene and were advised that there were children inside the structure, on the second floor of the fire unit. The victims made entry into the structure and proceeded to the second floor with a charged hoseline. Within minutes, victim 2 returned to the front door of the unit and requested a thermal imaging camera. He climbed back to the second floor as another crew of firefighters was preparing to enter the fire unit. As the backup crew entered through the front door, an explosion occurred, forcing them down the porch stairs and trapping the victims in the bedroom on the second floor. The fire, which intensified after the explosion, had to be knocked down before the victims could be located and removed.

The town house was a tri-level, with two stories above ground and a daylight basement, which extended above ground in the rear of the structure. This structure was located on the northern end of a five-unit complex. The town house was built in the 1980s, of ordinary construction, and encompassed approximately 1,600 square feet. The residence had smoke detectors that were working at the time of the fire.

The fire originated in a storage area behind a knee wall in the bedroom on the upper floor. The bedroom was the only major room on the second floor and was framed by room-in-attic trusses. This construction allowed the storage areas behind the knee walls to connect to the attic space above. The void space provided by the room-in-attic trusses allowed the bedroom to be surrounded by fire and hot gases (fig. 4–14). The medical examiner listed the cause of death for both victims as smoke inhalation and thermal burns over 50% of their bodies.

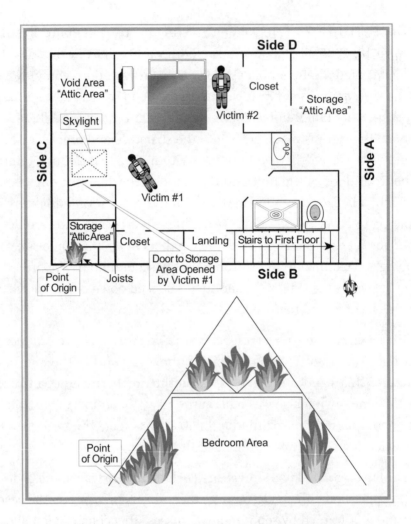

*Figs. 4–14. Diagrams of the second floor, showing where
the firefighters were overcome by flashover conditions.
Note the many void spaces that the fire occupied.*

STRUCTURAL COLLAPSE

From the time any structure is built, it begins to deteriorate. Depending
on such variables as the materials used to build the structure, the building
design and engineering, and the quality of work, structures all reach the

critical danger point of collapse at some point in their life span. Francis Brannigan was fond of telling us that buildings are nothing more than "gravity resistant systems" that eventually fail (fig. 4–15).

Fig. 4–15. Turn-of-the-century ordinary construction building. As Brannigan taught the fire service, all structures eventually fail.

Several building hazards are known to contribute to early collapse (fig. 4–16 and table 4–3). Examples include

- Unreinforced masonry
- Balloon-frame construction
- Unprotected steel support members
- Bearing walls and beams removed during remodeling
- Overloading of roof and floor systems
- Lightweight truss systems holding up roofs and floors

Fig. 4–16. Typical lightweight components found in modern construction. (Source: NIOSH. 2005. Preventing Injuries and Deaths of Fire Fighters Due to Truss System Failures)

Table 4–3. Chart depicting hazards that the fire service should be aware of in buildings. Working over or under lightweight construction is a risky venture; firefighters have been injured or killed due to lightweight construction. (Source: NIOSH. 2005. Preventing Injuries and Deaths of Fire Fighters Due to Truss System Failures)

Report ID	State	Truss Type	Number of Injuries and Fatalities	Event Leading to Death or Injury
98F005	Illinois	Heavy timber	2 F, 3 I	Backdraft
98F007	California	Heavy timber	1 F	Roof collapse
98F020	Vermont	Heavy timber	1 F	Roof/wall collapse
98F021	Mississippi	Lightweight wood	2 F	Roof collapse
99F002	Indiana	Lightweight wood	1 F	Roof collapse
F2000-13	Texas	Lightweight wood	2 F	Roof collapse
F2000-26	Alabama	Lightweight wood	1 F	Floor collapse
F2000-43	Delaware	Lightweight wood	3 I	Fire spread through truss voids
F2001-03	Arkansas	Lightweight wood	4 I	Roof collapse
F2001-09	Wisconsin	Heavy timber	1 F, 1 I	Roof/wall collapse
F2001-16	Ohio	Lightweight wood	1 F	Floor collapse
F2001-27	S. Carolina	Lightweight wood	1 F	Roof collapse
F2002-06	New York	Lightweight wood	2 F, 1 I	Floor collapse
F2002-50	Oregon	Heavy timber	3 F	Roof collapse
F2003-18	Tennessee	Lightweight metal	2 F	Roof collapse

* F = Fatality, I = Injury

CASE STUDY 3: THE MARY PANG FIRE — SEATTLE, WASHINGTON.[9]

Four firefighters died when a floor collapsed without warning during a commercial building fire in downtown Seattle on January 5, 1995. The fire was determined to have been arson. A suspect was apprehended and charged with four counts of homicide.

This fire resembles a number of other multiple-fatality incidents that have claimed the lives of more than 20 firefighters across the nation. The key similarities are as follows:

- Buildings that have access at different levels from different sides, resulting in confusion regarding the levels on which companies are operating
- Sudden and unanticipated floor collapses, which either dropped the firefighters into the fire area or exposed them to an eruption of fire from the level below

The structure involved in this incident was constructed primarily of heavy timber members. However, a modification to the structure had resulted in the main floor's being supported by a *pony wall,* an unprotected wood-frame wall, along one side (fig. 4–17). The sudden failure of this support caused the main floor to collapse without warning (fig. 4–18). Firefighters working on this floor believed that they had already gained control of the situation and were not aware that the main body of fire was directly below them or that it was exposing a vulnerable element of the structural support system.

The hidden flaw in the structure was the support for the ends of the new floor joists along the north wall. A ledge that was incorporated in the brick wall supported the original roof. Because the new floor was four to five feet higher than the roof, a wood-frame pony wall was fabricated from 2" × 4" wood members and was erected on top of the ledge to support the ends of the new floor joists. This assembly did not have the inherent fire resistance of the more massive wood members and could be expected to fail rapidly under fire conditions. The failure of this assembly would release the ends of the floor joists and result in a sudden floor collapse.

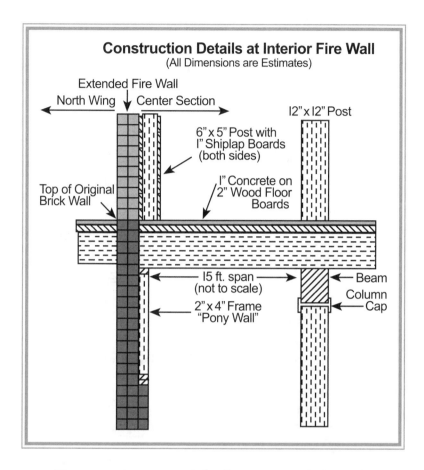

Fig. 4–17. Diagram of the floor section held up by a 2" × 4" pony wall that was installed during one of several remodelings during the building's many changes in occupancy (Source: USFA report 077, "4 Firefighters Die in Seattle Warehouse Fire")

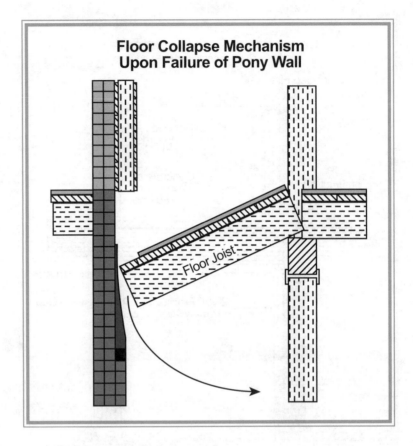

Fig. 4–18. Diagram of the failure of the floor section after fire impingement on the 2" × 4" pony wall. The floor section cantilevered into the basement area, killing four firefighters. (Source: USFA report 077, "4 Firefighters Die in Seattle Warehouse Fire")

CASE STUDY 4: CHURCH FIRE COLLAPSE DURING OVERHAUL.[10]
On March 13, 2004, a 55-year-old career battalion chief (victim 1) and a 51-year-old career master firefighter (victim 2) were fatally injured during a structural collapse at a church fire. Victim 1 was acting as the incident safety officer, and victim 2 was performing overhaul, extinguishing remaining hot spots inside the church vestibule, when the bell tower collapsed on them and numerous other firefighters. Twenty-three firefighters injured during the collapse were transported to area hospitals. A backdraft occurred earlier in the incident, injuring an additional six firefighters. The collapse victims were extricated from the church vestibule several hours after the collapse. The victims were pronounced dead at the scene.

The church was a National Historic Landmark that had been built in 1875. The building was still in use as a house of worship and a school at the time of this incident. The exterior construction was masonry with several courses of red brick covered with stone. The building foundation was approximately 120 ft by 70 ft and approximately 50 ft to the roofline. The pitched roof was covered with asphalt shingles and supported by heavy timber roof trusses. The stone facade exterior of the structure was added as part of a renovation in the 1930s. This renovation also included the addition of a 115 ft bell tower capped with four spires. The bell tower was not a stand-alone structure but was supported by steel I-beams with a brick-and-stone facade that was connected to the southwest corner of the original church.

The report of the county coroner's office listed the cause of death of victim 1 as asphyxiation due to compression of the body by building debris. Blunt-force trauma to the head, neck, pelvis, and extremities were contributory causes of death. The coroner's report listed the cause of death of victim 2 as asphyxiation due to compression of the body by building debris, with blunt-force trauma to the head and extremities (figs. 4–19 and 4–20).

Fig. 4–19. The structure just prior to the collapse
(Source: NIOSH Fire Fighter Fatality
Investigation 2004-17)

Fig. 4–20. The structure after the collapse
(Source: NIOSH Fire Fighter Fatality
Investigation 2004-17)

LIGHTWEIGHT CONSTRUCTION

Over the past 30 years, lightweight construction methods have become the standard for all types of building construction. Today, residential and commercial structures are built with lightweight roof assemblies, lightweight floor assemblies, and lightweight wall assemblies. Firefighters are aware that even while these lightweight systems may follow sound engineering principles of construction, these systems can and do fail rapidly under fire conditions, and anyone on top of or under these lightweight systems can be seriously injured or killed.

One of the primary reasons for lightweight construction collapse is the gusset plate connector, found holding together the webs and chords of trusses used for roofs and floors. Gusset plates are typically metal, with small teeth that penetrate only 3/8" into the wood members of the truss. It does not take much heat and fire to weaken these gusset plates, which cause the truss members to pull apart and fail quickly (fig. 4–21)

Fig. 4–21. The gusset plate is the weakest link in a truss system (Source: NIOSH. 2005. Preventing Injuries and Deaths of Fire Fighters Due to Truss System Failures)

Structures will continue to collapse under fire and other emergency conditions. The fire service needs to continually remind itself of the dangers of the boxes that we enter to fight fire. Firefighters should be well versed in and continually trained on building construction techniques. Officers need to be proactive and take their crews around their districts to look at and learn from all the new construction projects. Furthermore, at structure fires, officers and firefighters must take the time to evaluate the risk profile of the structure that is on fire and trying to kill them (fig. 4–22).

Fig. 4–22. Continual size-up and establishment of safe zones, a necessity when a defensive strategy has been declared. Structures will continue to collapse owing to fire and other emergency conditions. (Photograph courtesy John Lewis, Passaic [NJ] Fire Department, and Chief Rob Moran, Englewood [NJ] Fire Department)

When firefighters do get caught in a collapse and become trapped, they must have enough air to survive until they are rescued by other firefighters. Firefighters caught in a structural collapse will have a low survivability profile if they run out of air. This is why having a reserve air supply is so important. This emergency reserve allows trapped firefighters to stay alive that much longer, so that they can still be saved once found.

However, when firefighters breathe into this emergency reserve for fire operations, as is common practice throughout the U.S. fire service, they run the risk of not having any air left for such emergencies.

GETTING LOST

The vast majority of all firefighter fatalities have been precipitated by becoming lost and disoriented in heavy-smoke conditions. Firefighters work in very hazardous conditions and lose one of the key senses on which human beings rely—namely, their visual capability. Becoming disoriented is as easy as taking one wrong step.

Bart Bradberry, a Fort Worth, Texas, firefighter nearly lost his life when he became disoriented during a structure fire in a two-story residence. He missed the stairway exit by the width of a partition wall, then got turned around and found himself separated from his partner in a bedroom. He ran out of SCBA air and was forced, for self-survival, to bail out of the second-floor window owing to rapidly deteriorating conditions.

The U.S. Fire Administration (USFA) completed a study of firefighter disorientation in 2003, involving 23 firefighter fatalities occurring between 1979 and 2001.[11] The study indicated that disorientation is one of firefighting's most serious hazards and, according to NIOSH, usually precedes firefighter fatalities. The USFA study targeted 17 separate events, in which disorientation played a major role in 23 fatalities. The following are the significant findings of this study:

- In 100% of the cases, firefighters became separated from the hand line or became confused when they encountered loops of hose or tangled hand lines.

- In 51% of the cases, firefighters became confused when encountering tangled hand lines in zero visibility as they attempted to evacuate the structure.

- In 100% of the cases, company integrity was lost.

- In 65% of the cases, firefighters exceeded their air supply while attempting to evacuate.

- Disorientation occurred in distances of as little as 10 ft, 10–20 ft, 25 ft, and 30–40 ft from the point of entry. In two overly aggressive attacks, advances were made over distances of as much as approximately 80 and 200 ft, resulting in disorientation.

A sequence of events that caused firefighters to become disoriented occurred in each of the 17 incidents studied. The following is an outline of the sequence of events:

1. A fire occurred in an enclosed structure with smoke showing.

2. The arriving fire company immediately initiated an aggressive interior attack to search for the seat of the fire.

3. During the search, the seat of the fire could not be located, and conditions deteriorated with the production of heat and smoke and prolonged zero visibility.

4. As companies performed an emergency evacuation owing to deteriorating conditions, hand line separation occurred or tangled hand lines were encountered.

5. Disorientation occurred as firefighters exceeded their air supply, were caught in flashovers or backdrafts, or were trapped by a collapsing floor or roof.

6. When the firefighters were not located quickly enough, the outcome was fatalities or serious injuries.

The disorientation sequence is usually played out in a structure that does not have a sprinkler system or has an inoperable sprinkler system.

Prolonged zero visibility means heavy-smoke conditions lasting longer than 15 minutes. Since prolonged zero-visibility conditions exceed the approximate breathing time of a working firefighter's SCBA, these conditions should be considered extremely dangerous should disorientation occur.

READING SMOKE

Let's get to the point: The building fires you respond to today and the ones you'll respond to tomorrow are producing some of the most explosive, toxic, and volatile smoke that we, as a fire service, have ever experienced. Thanks to the work of the National Institute of Standards and Technology (NIST), the Bureau of Alcohol, Tobacco, and Firearms (ATF), and numerous product-safety test laboratories, we are beginning to document the new building-fire environment.

Many studies have already discerned that fire is the least of your problems. Instead, it's the smoke coming from low-mass synthetics, which are being heated without burning (*pyrolysis*), that demands your attention. According to some research, smoke production in modern fires is 500 times that of the wood-based fires of a few decades ago.

Noted fire chief, author, and firefighter safety advocate Billy Goldfeder said it best when he claimed, "The smoke today is not your daddy's smoke." The list of chemical products found in the smoke of a typical residential fire defies imagination: carbon monoxide (CO), hydrogen cyanide (HCN), benzene (C_6H_6), and acrolein (C_3H_4O) lead the "explosivity" list. In addition, smoke contains polyvinyl chloride, hydrogen chloride, hydrogen sulfide, polynuclear aromatic hydrocarbons (PAHs), formaldehyde, phosgene, nitrogen dioxide, ammonia, phenol, and many acids.

Further, the decomposition processes of burning and radiant heating are engaging rocket-science–like chemical morphing that can challenge most chemists. It gets worse. We used to measure the typical compartment (room) fire time-temperature curve by using a 10-minute cycle. Recent research tells us that the cycle is closer to 5 minutes, with nonflaming ceiling temperatures over 1,500°F.

Now we'll get personal. A review of firefighter news items, investigative reports, and incident data indicates the following:
- Firefighter cancer rates are escalating while the rates for the general public are declining.
- Firefighters that get just a few breaths of zero-visibility smoke either die or face significant medical issues that can become chronic.
- Firefighters are being treated for cyanide poisoning from "just a little" smoke exposure outside burning buildings.
- The rate of firefighter flashover deaths is increasing even while the number of fires is decreasing.New communication systems and fire-alarm centers have the ability to identify a firefighter with the key of a microphone on a portable radio. At the beginning of the shift, members sign on to the unit's computer-aided dispatch (CAD) roster, which correlates the assignment of a portable radio to each member in the company by name and employee number. Whenever a transmission is made from that portable radio, the dispatcher immediately knows to whom the radio is assigned.

The firefighter who focuses on the fire (*flaming*) to determine fire behavior is being set up for a sucker punch. To predict fire behavior in a building, we need to understand the realities that govern fire outcome. This starts with the notion that open flaming is actually good; allegedly, the products of combustion are minimized because the burning process is more complete. In fact, the complete combustion of common materials renders heat, light, carbon dioxide, and water vapor.

Within a building, the heat from flaming (exothermic energy) is absorbed in other materials that are not burning (contents and the walls/ceiling). These materials can absorb only so much heat before they start to break down and *off-gas* without flaming. At that point, smoke becomes flammable. Within a *box* (a room), off-gassed smoke displaces air, leading to what is termed an *underventilated fire*. Underventilated fires do not allow the open flaming to complete a reaction with pure air, leading to increasing volumes of CO, as well as the aforementioned smoke products. Smoke is now looking to complete what was started; two triggers may cause accumulated smoke to ignite: the right temperature and the right mixture.

Never has it been more important to *read smoke*. The practice of reading smoke has been around for many decades, although few have tried to define the process for fire officers to predict fire behavior and gauge firefighting progress.

Smoke attributes

Reading smoke is not difficult, although for most fire officers, it takes an effort to break the "heavy smoke or light smoke" mentality, which has come from rapid size-up radio reports. As noted author, instructor, and retired Los Angeles City Fire Chief John Mittendorf has said, "Smoke is the fire talking to you." Smoke leaving a structure has four key attributes: *volume*, *velocity (pressure)*, *density*, and *color*. A comparative analysis of these attributes can help the fire officer determine the size and location of the fire, the effectiveness of fire streams, and the potential for a hostile fire event (e.g., a flashover). Let's look at each of the attributes and how they contribute to fire behavior understanding.

Volume. Smoke volume by itself tells very little but can create an impression of the fire. For example, a small fast-food restaurant can be totally filled with smoke from a small fire. Conversely, it would take a significant event to fill the local big-box store with smoke. Once a container is full of smoke, pressure builds if adequate ventilation is not available. This helps us understand smoke velocity.

Velocity. The speed at which smoke leaves a building is referred to as *velocity*. In actuality, smoke velocity is an indicator of *pressure* that has built up in the building. From a tactical standpoint, the fire officer needs to know *what* has caused the smoke pressure. From a practical fire behavior point of view, only two things can cause smoke to pressurize in a building: *heat* or smoke volume. When smoke is leaving the building, its velocity is caused by heat if it rises and then slows gradually. Smoke caused by restricted volume immediately slows down and becomes balanced with outside airflow.

In addition to the speed of smoke, you need to look at its *flow* characteristic. If the smoke flow is *turbulent*, a flashover is likely to occur (see photograph; other descriptions include *agitated*, *boiling*, and *angry* smoke). Turbulent smoke flow is caused by the rapid molecular expansion of the gases in the smoke and the restriction of this expansion by the box (compartment). The expansion is caused by radiant heat feedback from the box itself; the box cannot absorb any more heat. This is the precursor to flashover. If the box is still absorbing heat, the heat of the smoke is subsequently absorbed, calming hyperactive molecules and reducing gas volume; this leads to a more stable and smooth flow, referred to as *laminar*. The word "laminar" is used to describe the smooth movement of a fluid, and smoke is basically fluid when moving through a building. The most important smoke observation is whether its flow is turbulent or laminar. Turbulent smoke is ready to ignite and indicates a flashover environment, which may be delayed by improper air mix.

Comparing the velocity of smoke at different openings of the building can help the fire officer determine the location of the fire: Faster smoke is closer to the fire seat. Remember, however, that the smoke velocity you see outside the building is ultimately determined by the size and restrictiveness of the exhaust opening. Smoke follows the path of least resistance and loses velocity as the distance from the fire increases. To find the location of fire by comparing velocities, you must compare only like-resistive openings (doors to doors, cracks to cracks, etc.). More specifically, you should compare a crack in a wall to other cracks in walls. Do not compare the speed of smoke leaving a crack in window glass to that of smoke leaving through cracks in a wall; the resistance to smoke flow is different.

Density. While velocity can explain much about a fire (how hot it is and its location), density tells you how bad things are going to be. The density of smoke refers to its thickness. Since smoke is *fuel*—containing airborne solids, aerosols, and gases that are capable of further burning—the thickness of the smoke tells you the *amount* of fuel that is laden in the smoke. In essence, the thicker the smoke is, the more spectacular the flashover or fire spread will be.

One other point needs to be made regarding smoke density: Thick, black smoke in a compartment reduces the chance of life sustainability, owing to smoke toxicology. A few breaths of thick, black smoke renders a victim unconscious and causes death in minutes. Further, the firefighter crawling through zero-visibility smoke is actually crawling through ignitable fuel.

Color. Most fire service curricula teach us that smoke color indicates the type of material that is burning. In reality, this is rarely true since all materials can off-gas smoke colors ranging from white to black. Smoke color is really a reflection of heat and fire location. Virtually all solid materials emit a white smoke when first heated. This white smoke is mostly moisture. As a material dries out and breaks down, the color of the smoke darkens. Wood materials change to tan or brown, whereas plastics and painted or stained surfaces emit a gray smoke. Gray smoke is a result of the mixture of moisture with carbons and hydrocarbons (black smoke). As materials are further heated, the smoke leaving the material eventually becomes all black. When flames touch surfaces that are not burning, they off-gas black smoke almost immediately. Therefore, the more black that the smoke you see is, the hotter it will be. Black smoke that is high velocity and very thin (low density) is flame pushed. Thin, black smoke means that open (and ventilated) flaming is nearby.

Smoke color can also tell you the distance to a fire. As smoke leaves an ignited fuel, it heats up other materials, and the moisture from those objects can cause black smoke to turn gray or even white over a distance. As smoke travels, carbon (and hydrocarbon) content from the smoke is deposited along surfaces and objects, eventually lightening the smoke color. So the question becomes whether white smoke is the result of early-stage heating or of late-stage heating smoke that has traveled some distance? The answer is found by looking at the velocity. Fast-moving white smoke indicates that the smoke has traveled some distance. White smoke that is slow or lazy is most likely indicative of early-stage heating.

By combining the aforementioned attributes, you can create a mental picture of the fire. Compare smoke velocity and color from various openings to locate the fire. Faster and/or darker smoke is closer to the fire seat, whereas slower and/or lighter smoke is farther away. Typically, distinct differences in velocity and colors will be seen from various openings. When the smoke appears uniform—that is, the same color and velocity from multiple openings—you should start thinking that the fire is in a concealed space (or is deep-seated).

Other factors that influence smoke
Weather, thermal balance, and container size can change the appearance of smoke. Let's look how each of these can influence smoke.

Weather. It is important to know how weather affects smoke. Once smoke leaves a building, the outside weather can influence its appearance. Temperature, humidity, and wind change the look of the smoke. Cold air temperatures cool smoke faster and cause it to stall and/or fall. Smoke rises much higher on a hot day because cooling is more difficult. Humidity in the air increases resistance to smoke movement by raising air density. Low humidity allows smoke to dissipate more easily. Humid climates keep the smoke plume tight. When outside air temperatures are below freezing, hot smoke leaving the building can turn white almost instantly as a result of moisture condensing. Wind can keep smoke from leaving an opening. Wind can rapidly thin and dissipate smoke, making it difficult to fully view its velocity and density. In a well-ventilated building, wind can speed up smoke velocity and give a false read on heat or location, although it should fan flaming. Firefighters engaged in an interior fire attack downwind of a wind-fed fire are in danger of being overrun by the fire!

Thermal balance. Most buildings do not allow fires to maintain thermal balance. Simply stated, *thermal balance* represents the notion that heated smoke rises and in doing so creates a draft of cool air into the flame (heat) source. As more air is drafted into the flame, the fire should grow. As the fire grows, it consumes the oxygen within the air, creating more of a draw of air. Ceilings, windows, doors, and inadequate airflow disrupt thermal balance. If a fire cannot draft and draw air, then it soon drafts and draws smoke, thereby choking itself; this leads to more incomplete combustion and creates denser smoke. As explained already, the thicker the smoke is, the more explosive it becomes. Viewed from outside a building, a fire out of thermal balance shows signs that air is being sucked through the smoke. A sudden inflow of air can cause the fire to take off—trapping firefighters.

Container size. All smoke observations must be analyzed in proportion to the building (box). For example, smoke that is low volume, slow velocity, very thin, and light colored may indicate a small fire, but only if the building is small. The same smoke attributes from several openings of a big-box store or a large warehouse can indicate a large, dangerous fire. Historically, firefighters have been killed at fires that were reported as "light smoke showing." Remember, the size of the building is an important indicator of the significance of the smoke leaving. Light, thin smoke showing from more than one opening of a very large building is a significant observation.

Putting it all together

To read smoke, the arriving fire officer should follow three simple steps:
1. Compare smoke volume, velocity, density, and colors from various points.
2. Determine if the weather, thermal balance, or container size is influencing the attributes seen.
3. Judge the rate of change (getting better or worse within seconds or minutes).

These three steps should give you an impression about the location, size, and potential of the fire. Sounds simple, right? Actually, with a little practice, you can start recognizing smoke patterns and their meanings, speeding up your ability to make the read. To practice, study the information presented here and use raw fireground footage (many sources of which are available online [search using the key words "fireground video" or go to Readingsmoke.com]).

—Dave Dodson (Battalion Chief, retired, Eastlake [CO] Fire Department,
and Lead Instructor, Response Solutions)

CASE STUDY 5: TWO FIREFIGHTERS DIE AFTER BECOMING DISORIENTED.[12] On November 6, 1998, two firefighters died while trying to exit a burning automobile salvage storage building. Arriving on the scene of a metal pole building with light smoke showing, the department chief assumed command and discussed the possible origin of the fire with the owner of the structure. The IC then decided to ventilate by ordering a firefighter to open one of two small roll-up garage doors. Once ventilation was completed, firefighters advanced two 1½" lines through the front door of the building, which was filled with light smoke. As the firefighters proceeded to the rear of the structure to determine the fire's origin, heavy black smoke collected below the ceiling, and small flames trickled over the ceiling's skylights. Approximately 80 ft inside the structure, firefighters found what they believed to be the seat of the fire and began to apply water. As firefighting activities proceeded, firefighters transferred the lines to other firefighters because the low-air alarms on their SCBA were sounding.

Approximately 11 minutes into the attack, the IC ordered both crews to exit to discuss further strategy. As the crews began to exit, an intense blast of heat and thick, black smoke covered the area, forcing firefighters to the floor. The chief and assistant chief were knocked off the hoseline, and their SCBA low-air alarms began to sound as they radioed for help and began to search for an exit. The two departed in different directions, and the assistant chief eventually ran out of air and collapsed. He was found immediately and assisted from the burning building. As firefighters pulled the unconscious assistant chief to safety, the lieutenant reentered the structure to search for the chief. During his search, the lieutenant ran out of air, became disoriented, and failed to exit. The lieutenant was discovered to have been equipped with a PASS device; however, it was not turned on. The chief was known to have entered without a PASS device. Additional rescue attempts were made but proved to be unsuccessful, and both the chief and lieutenant succumbed (fig. 4–23).

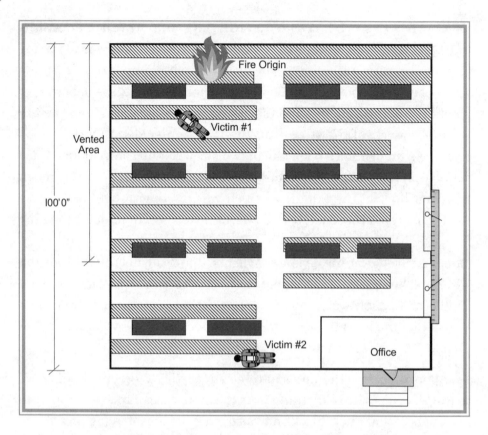

Fig. 4–23. Overhead diagram of where the two firefighters were found (Source: NIOSH Fire Fighter Fatality Investigation 98-32)

RUNNING OUT OF AIR

The limiting factor for firefighters working in an IDLH environment is the amount of SCBA air that is available for the task at hand. Firefighters are task oriented by nature and, at times, get tunnel vision, becoming focused on the work assignment to the exclusion of all else. Implementing air management, as outlined in chapter 10, is a pivotal part of firefighter safety when any of the aforementioned four disasters—exposure to toxic smoke, exposure to thermal insult, full or partial structural collapse, or becoming lost or disoriented—calls on your fireground operation. Firefighters' survivability in any of these catastrophic events will hinge on their air supply; therefore, managing that supply is critical. Know how much air you have before entering any IDLH environment, manage it along the way, and exit before your low-air alarm activates. These simple steps will increase both your situational awareness and your level of safety. In 16% of fatalities during the USFA study period (1990 through 2000), a total of 178 cases, firefighters were reportedly wearing SCBA at the time of their injury, and of these 178 cases 30% had reportedly depleted their SCBA air supply.[13]

Table 4–4 outlines the need for firefighters, company officers, chief officers, and fire departments to implement and enforce a comprehensive air management policy. Firefighters are needlessly dying year after year as a direct consequence of running out of air.

Table 4–4. The cold reality of running out of air in today's fire environment—death

Year	Number of Firefighter Deaths Due to Asphyxiation (Source: USFA Annual Reports)
2006	11
2005	8
2004	5
2003	6
2002	15
2001	18
2000	13
1999	16
1998	15
1997	15
1996	5
1995	20
1994	29
1993	4
1992	6
1991	10
1990	20
1989	11
1988	13
1987	13
1986	14

CASE STUDY 6: FIREFIGHTER DIES AFTER RUNNING OUT OF AIR.[14]
The following case study involves an experienced company officer in deteriorating fire conditions in a single-family residence. On January 20, 2005, a 39-year-old career captain (the victim) died after he ran out of air, became disoriented, and collapsed at a residential structure fire. The victim and a firefighter made entry into the structure with a hand line to search for and extinguish the fire. While searching in the basement, the victim removed his regulator for one to two minutes to see if he could distinguish the location and cause of the fire by smell. While searching on the main floor of the structure, the firefighter's low-air alarm sounded, and the victim directed the firefighter to exit and have another firefighter working outside take his place.

The victim and a second firefighter went to the second floor without the hand line to continue searching for the fire. Within a couple of minutes, the victim's low-air alarm activated. The victim and firefighter 2 became disoriented and could not find their way out of the structure. The victim made repeated calls over his radio for assistance, but he was not on the fireground channel. Firefighter 2 *buddy-breathed* with the victim until the victim became unresponsive. The victim became panicked and was yelling to the firefighter to get him out of the structure. The victim went down on his knees and took out his regulator, declaring that he couldn't breathe. Firefighter 2 once again began buddy-breathing with the victim, who was becoming more frantic. The victim collapsed to the floor and became unresponsive.

Firefighter 2 made his way to a window at the front of the structure and busted it out in an attempt to signal someone on the outside. He then went back to find the victim but became disoriented and low on air. Firefighter 2 heard the positive-pressure ventilation (PPV) fan and moved down the hallway toward the sound. He stumbled on the stairs and found his way back out of the structure. The fire intensified and had to be knocked down before the victim could be recovered.

The medical examiner listed the cause of death as smoke and soot inhalation. A carboxyhemoglobin level of 22.7% was noted on the report, consistent with death due to carbon monoxide poisoning (figs. 4–24 and 4–25).

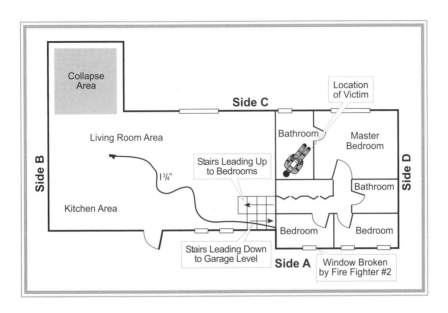

Fig. 4–24. Diagram depicting the main floor and the upper level, where the victim was found. (Source: NIOSH Fire Fighter Fatality Investigation 2005-05)

Fig. 4–25. Post-fire photograph taken from the front side of the structure. (Source: NIOSH Fire Fighter Fatality Investigation 2005-05)

FINAL THOUGHTS

All of the case studies examined in this chapter serve as graphic examples of the increased risk that we, as firefighters, accept by not managing our SCBA air supply. Put together with Murphy's law, this amounts to a wager that can crush our lives. Without an adequate reserve SCBA air supply, firefighters have very little opportunity for survival if they become exposed to the deadly smoke found in today's fire environment, if thermal dynamics rapidly changes in the work space, if structural integrity becomes compromised, or if disorientation rears its ugly head.

Having an air management policy is a great starting point for increasing firefighter safety on the fire ground. In chapter 10 we will give you our simple, easy to implement, air management solution, which is the "Rule of Air Management".

This type of policy will, without question, increase firefighters' survivability profile if they get into trouble (e.g., see fig. 4–26). It's critical that firefighters, officers, and department leaders take a proactive stance on the reduction of emergency-scene injuries and fatalities. One of the easiest and least costly changes that will save firefighters' lives is the implementation of an air management policy that is practiced as though lives depend on it—because they do!

Fig. 4–26. Turbulent smoke flow—a warning that flashover is imminent (Photograph, by Keith Muratori, from Firegroundimages.com).

MAKING IT HAPPEN

1. Have members research how firefighters are dying and the circumstances and types of structures in which they are dying.

2. Review your department's driving/apparatus operator policy and determine whether you're complying with local, state, and federal regulations.

3. Review your department's wellness/fitness program and determine whether you're being adequately screened for health and fitness.

4. Research your department's disabilities by type and determine whether there are any trends that can be improved upon, for both non-emergency and emergency operations.

5. What type of prioritized training program is your department providing for your firefighters? Evaluate this question and determine whether you're meeting the 75/25 practice, loosely defined as providing 75% of training on the prevention of firefighter mishaps (the getting-back-to-the-basics concept) and 25% on how to fix the problem when there is a firefighter mishap (e.g., RIT and get-out-alive training).

6. Have members research what carcinogenic products are contained in today's smoke.

7. Have members research how many firefighters have died in the past five years as a direct result of flashover or backdraft.

8. Have members research how many firefighters have died in the past five years as a direct result of structural collapse.

9. Have members research how many firefighters have died in the past five years as a direct result of being lost or disoriented.

STUDY QUESTIONS

1. About how many firefighters is the U.S. fire service losing in line-of-duty deaths each year?

2. True or false: Statistically, we know that the number of structure fires in the U.S. fire service is decreasing, while the rate of firefighter fatalities in structures is not.

3. True or false: In the late 1970s, traumatic deaths inside structures occurred at a rate of 1.8 deaths per 100,000 structure fires; by the late 1990s, this rate had risen to almost 3 deaths per 100,000 structure fires, and even more recently, this has increased to nearly 4 deaths per 100,000 structure fires.

4. The Phoenix Fire Department recently completed a study revealing that a typical room-and-contents fire in a modern structure produces what level of carbon monoxide (in parts per million)?

5. What is the only protection for firefighters in the toxic smoke environment of today's fires?

6. What occurred in the mid-1970s, initiating a significant change in building construction techniques?

7. True or false: Depending on such variables as the materials used to build the structure, the building design and engineering, and the quality of work, structures all reach the critical danger point of collapse at some point in their life span.

8. What primary reasons for lightweight construction collapse were discussed in this chapter?

9. When firefighters get caught in a collapse and become trapped, what must they have enough of to survive until they are rescued by other firefighters?

10. Fill in the blank: The vast majority of all firefighter fatalities have been precipitated by firefighters' becoming _____ _____.

11. What are the five ways firefighters are dying in structure fires?

NOTES

[1] NFPA. *U.S. Fire Service Fatalities in Structure Fires, 1977–2000.* Quincy, MA: National Fire Protection Association, Prepared by: Rita F. Fahy, Ph.D., July 2002

[2] Ibid.

[3] Dugan, Mike (Capt., FDNY). Personal communication.

[4] Phoenix Fire Department. In-house study. Kruchuk, Scott, Phoenix Fire Department study.

[5] Yob, Chris (Lt.). From his personal testimonial.

[6] Chiltern International Fire. http://www.chilternfire.co.uk.

[7] NIOSH Fire Fighter Fatality Investigation F2000-04.

[8] NIOSH Fire Fighter Fatality Investigation F2005-13.

[9] Routley, J. Gordon. Four firefighters die in Seattle warehouse fire. United States Fire Administration report 077.

[10] NIOSH Fire Fighter Fatality Investigation F2004-17.

[11] Mora, William. *U.S. Firefighter Disorientation Study, 1979–2001.* United States Fire Administration. Prepared by William R. Mora, Captain, San Antonio Fire Department San Antonio, Texas, July 2003.

[12] NIOSH Fire Fighter Fatality Investigation F1998-32.

[13] Firefighter Fatality Retrospective Study, 1990–2000. United States Fire Administration, Prepared by TriData Corporation, April 2002/FA–220.

[14] NIOSH Fire Fighter Fatality Investigation F2005-05.

CHAPTER 5

THE BREATH FROM HELL

INTRODUCTION

It is common knowledge among firefighters that the modern smoke environment is unforgiving and deadly. Many tactics classes taught across the nation include a section that deals with the expanding toxicity of smoke and the rapid development of zero-visibility conditions at working fires. The highly flammable nature of carbon monoxide (CO) and its ability to cause rapid asphyxiation are no secret to the professional firefighter. There is also a growing recognition that hydrogen cyanide may well be the culprit in many fireground deaths currently being attributed to CO. Taking this invaluable knowledge into the modern fire environment will allow firefighters to operate in a safe and effective manner, accounting for the many dangers of smoke that make it the *breath from hell* (fig. 5–1).

*Fig. 5–1. A fire officer surrounded by smoke,
attempting to radio the IC (Courtesy of Dennis Leger,
Springfield [MA] Fire Department)*

The question this leaves is why firefighters and fireground supervisors operate as if this poisonous smoke does not pose a threat. With all of the knowledge on the deadly components of even the average fire that has been gained in the past 10–15 years, why does this complacency persist? Many of these same firefighters and supervisors have visited hospitals to encourage veterans struggling against cancers or pulmonary diseases. Funerals across the country are sobering reminders that many brothers and sisters, most of whom were not given the information we possess today, aggressively fought fire and endured smoke only to experience deadly long-term effects. In light of this, why is there continued reluctance to recognize that toxic smoke at every structure fire demands a change in firefighter behavior?

Meaningful changes in behavior include actions that occur before, during, and after fires. Enhanced training, proper use of SCBA, safe overhaul, and effective rehabilitation are key areas that need to be addressed. However, the most-overlooked aspect of dealing with the modern smoke environment is the lack of a comprehensive air management program.

In the early 1990s, the first detailed work on the treacherous components of smoke surfaced. Dr. Deborah Wallace's book *In the Mouth of the Dragon: Toxic Fires in the Age of Plastics* detailed the true nature of the gases that emanate from plastics.[1] Chief Bobby Halton has described this as "the most important book in the fire service that no one has read." Chief Halton's pioneering work has helped firefighters to know the enemy they face. The torch has also been passed to other instructors, who have developed programs that assist us in our approach to and handling of smoke. Among these instructors are Dave Dodson (*The Art of Reading Smoke*) and Rob Schnepp (*Where There's Fire . . . There's Smoke!*). Research by these and other firefighters has provided us with a wealth of knowledge to evaluate how we're doing business.

WHERE THERE'S SMOKE, THERE'S WHAT?

Most definitions of smoke refer to particulate matter suspended in varying levels of heated air. Soot (carbon), fibers, dusts, and other materials add further detail, and varying colors are routinely present. Fire gases capable of asphyxiation, poisoning, and explosive flammability are increasingly being recognized as significant components of modern smoke. Oils and related materials contribute an aerosol component to smoke that can also be accompanied by extreme variations in heat.

There is a growing acknowledgment that what we're seeing—and not seeing—in the modern smoke environment is an exceedingly different animal than the one faced by our predecessors. It is a vicious beast, indeed (fig. 5–2).

Fig. 5–2. A large, black cloud of smoke issuing from a commercial fire (Photograph courtesy Seattle Fire Fighters Union, IAFF Local 27)

THE PAST

Even in the IDLH environments of the past, where smoke products were from primarily wood-related combustion (CO, nitrogen dioxide, and PAHs), breathing smoke presented an enormous risk to the short- and long-term health of firefighters. This has been made all too evident by the numerous cancers and pulmonary diseases that members have endured because of their service in firefighting.

The following list includes some of the major elements of the smoke of yesterday (fig. 5–3):[2]

- Carbon
- Nitrogen
- Polycyclic aromatic hydrocarbons (PAHs)
- Formaldehyde
- Acid gases:
    - Hydrochloric acid
    - Sulfuric acid
    - Nitric acid
- Phosgene
- Benzene
- Dioxin

Remarkably, this smoke is often considered to be breathable without respiratory protection. The preceding list, however, should demonstrate clearly that such preconceived notions are in error. Moreover, staggering statistics detail firefighter deaths due to the smoke of the past.

Carbon monoxide (CO) is a colorless and odorless gas that is responsible for the deaths of hundreds of people each year. Firefighters, sadly, are amongst its many victims. Its primary danger is the effect it has on our Hemoglobin and its ability to effectively carry oxygen. It combines with Hemoglobin 240 times faster than does oxygen and can cause respiratory distress, aggravated heart conditions, and death.

Nitrogen Dioxide is a common component of smoke that also causes respiratory distress and edema.

Polynuclear Aromatic Hydrocarbons (PAHs) are carcinogenic, organic compounds that result from the burning of materials that have carbon and hydrogen components that are burned. These materials are common and there are over 100 known PAHs that can be present in smoke.

Formaldehyde: You guessed it... The same stuff you dunked your Biology projects in can also find its way to your lungs. It can cause respiratory distress, headaches and other neurophysiological dysfunction.

Acid Gases (Hydrochloric Acid, Sulfuric Acid, Nitric acid) can cause any number of problems including respiratory distress, chemical burns, disorientation and death.

Phosgene is a poisonous gas that was used as a chemical weapon in World War One and is a by-product of burning Chlorinated Hydrocarbons. Its effects can range from blurred vision to edema and respiratory failure. It is heavier than air and, thus, will be in just the right place for advocates of "sucking the carpet".

Benzene is a commonly found component of cigarette smoke but is also found in fires. It is a known carcinogen and also can cause respiratory distress and asphyxiation.

Dioxin is a deadly by-product of some electrical equipment. It comes from the burning of Polychlorinated Biphenyls (PCBs) which are commonly associated with older transformers.

Adapted from the Seattle Fire Department's SCBA Training Guide

Fig. 5–3. Gases of the smoke of yesterday

THE PRESENT

In the modern firefight, smoke involves the toxic elements created by the burning of numerous synthetics, especially those from plastics. Also of note are the increased amounts of hazardous chemicals present even in the smallest of structures. With a smoke production level reaching 500 times that of fire burning a similar amount of wood, smoke has become a deadly cocktail that contributes to hundreds of deaths each year and is responsible for a wide variety of cancers that although difficult to tabulate, have had a very real toll on many lives.[3] The bill for this smoke exposure has not yet come due for the fire service, but it is inevitable. The tragedy will be compounded exponentially if we do not respond appropriately by using the knowledge right in front of us.

In his excellent book *The Combustibility of Plastics,* Frank Fire has detailed a few of the poisons, as well as their sources, that you may encounter in the fire you go to tonight:[4]

- *Acetals*—aerosol containers, combs, lighters, pens
- *Acrylics*—glues, food packages, skylights
- *Cellusose products*—video and audio tape, film, telephones, toys
- *Nylon*—various household containers, brushes, sewing thread, fishing line
- *Polyesters*—hair dryers, computers, kitchen appliances
- *Polypropylene*—bottles, diapers, furniture
- *Polyurethanes*—shoes, cushions
- *Polyvinyl chloride (PVC)*—carpet, clothes, purses, records, shower curtains
- *Thermosets*—televisions, coatings, toilets, buttons, flooring, insulation

And the list goes on. Among the nasty gases these and other plastics form are hydrogen cyanide, CO, carbon dioxide, formaldehyde, nitrogen oxides, ammonia, phenol, benzene, hydrogen chloride (hydrochloric acid), methane, and other crippling by-products—all of which make the breath from hell truly hellish.

The New York Telephone fire serves as a classic example of the effects of exposure to toxic smoke. On February 27, 1975, the Fire Department of New York (FDNY) was called to a fire that would change the lives of the 699 responding firefighters forever. The fire, in an 11-story telephone switching center, progressed to become a five-alarm inferno that involved tons of PVC-sheathed wiring and polychlorinated biphenyl (PCB)–laced transformers. (PCBs are highly toxic and carcinogenic chemicals.) The firefighters were exposed to extended doses of the toxic smoke as they courageously battled the blaze. There were widespread reports of respiratory problems developing immediately after exposure to the heavy smoke, as well as lingering difficulties that were not immediately noted (fig. 5–4).

Immediate Symptoms of Firefighters	
Injury	Percent Affected
sore throat, irritated eyes, dizziness, aching nostrils, confusion, weakness, and exhaustion	over 50%
chest pains, nausea, chest congestion, and headache	35–50%
irritated skin and faintness	20–30%
loss of control of arms and/or legs	10–20%

Wallace, Deborah. In the Mouth of the Dragon. Avery Publishing Group. Pg. 48

Fig. 5–4. Chart detailing the immediate symptoms of responders to New York Telephone Fire (Source: Wallace, In the Mouth of the Dragon)

Fire service instructor Dan Noonan, who was given a diagnosis of leukemia at 52 years of age, had this to say to reporter Bob Port of the *New York Daily News*:[5]

> We've explained again and again that all 699 men who fought the phone fire suffered some form of respiratory distress. Virtually every firefighter who responded to the phone fire's first two alarms has cancer.

As we ponder the ominous statistics of the New York Telephone fire, it is critical to recognize that the toxic environment of that blaze is now the *norm* for what can be expected in every fire that we extinguish. Firefighters are breathing these same poisons when not adequately protected by SCBA and trained in air management.

THE DEADLY DUO:
PVC AND HYDROGEN CYANIDE

Two of the least-recognized components of the modern smoke environment are PVC and hydrogen cyanide. The former is a product that is present in large quantities at most fires today, while the latter is a deadly by-product that is a silent killer on the fireground.

PVC is likely to be found in more abundance than any other product in today's homes. It is devastating to firefighters no matter where they are in the smoke, because its deadly fumes emanate from materials at every level of a structure. Deborah Wallace has noted that the emissions from PVC of benzene, chlorinated dioxins, and dibenzofurans, all of which are known carcinogens, may explain the high frequencies of several cancers, including leukemia, laryngeal and colon cancer, and rare soft-tissue cancers occuring at relatively early ages in firefighters.[6] Highly acidic hydrogen chloride and other gases are present in PVC fires and have an impact on respiratory and circulatory health.

Hydrogen cyanide is a colorless, odorless gas that emanates from both natural and synthetic sources. CO is commonly known to cause firefighter deaths, but hydrogen cyanide, once relatively unknown, is steadily assuming a more prominent place in the hierarchy of the causes of death at fires. It was employed as the gas of choice in Hitler's death camps (*Zyklon B*) and was used by terrorists of the Aum Shinrikyo religious cult in their attack on Tokyo's Shinjuku Station on May 5, 1995.

A recent article detailed the results of studies performed in Paris and in Dallas County, Texas.[7] In these two studies, blood samples were collected close to the time of smoke exposure, and blood cyanide levels were measured in survivors and fatalities, rather than only in fatalities. The following conclusions were reached by the studies:

- Cyanide concentrations were directly related to the probability of death.
- Cyanide poisoning may have predominated over CO poisoning as a cause of death in some fire victims.
- Cyanide and CO may have potentiated the toxic effects of one another.
- Elevated cyanide concentrations were pervasive among victims of smoke inhalation.

QUANTITATIVE DECOMPOSITION

Advocates of the practice known as *sucking the carpet* should note that carpet fibers are a large source of hydrogen cyanide gas, and those fumes develop long before it catches fire (fig. 5–5). As temperatures increase during the initial stages of the fire, plastics begin to give off large quantities of various gases. This is called *quantitative decomposition* and happens long before the materials actually burst into flame. Deborah Wallace describes it as follows:[8]

> *Unlike traditional materials, the temperature of quantitative decomposition for many plastics is less than half their ignition temperatures. Because of this large temperature difference, there is a long period of time when gas is emitted without the warning presence of flame. Generally, the gases emitted during the decomposition stage of a fire are more toxic than those emitted during actual burning. Thus, in many fires, the decomposition stage is the real killer. It is a killer because of its high toxicity, and the long period of time between attainment of quantitative decomposition temperature and ignition temperature.*

> *Later,* Wallace adds, *although most fire fatality victims have high levels of blood carbon monoxide, evidence shows that carbon monoxide is not the only factor in many fire deaths and that the victims had been incapacitated by the irritating products of decomposition long before their blood carbon monoxide levels became fatal.*[9]

Fig. 5–5. Firefighters ready for entry. Long before this fire broke through the top floor, hydrogen cyanide was most likely present in the smoke throughout the structure. (Photograph courtesy John Lewis, Passaic [NJ] Fire Department)

It is reasonable to postulate that *narcotic* effects of hydrogen cyanide are to blame for the bizarre and incongruent behavior associated with smoke inhalation. Further evidence comes from the cases in which victims, including firefighters, resisted aid and even fought with their rescuers until becoming totally overcome by the smoke. The Southwest Supermarket fire in Phoenix is a classic example. The courageous efforts of the rescue teams were thwarted when Bret Tarver fought with the firefighters trying to take him to safety. Since he was very big and strong, only after he had succumbed to the environment were firefighters able to pull him through the debris and out of the building.

CO: STRIKE THREE—YOU'RE OUT!

CO is probably the best known of the products of combustion and is viewed as a serious threat by firefighters. Many have encountered the deadly gas in the intentional deaths of suicide victims who have locked themselves in a garage with the car engine running. Unintentional deaths are caused by furnace or other equipment malfunctions that overcome sleeping families. Furthermore, it is a major threat at fires, since asphyxiation is among the leading killers of firefighters in structural emergencies.

SILENT KILLER

CO exhibits a deadly mixture of stealth and toxicity. Because of the following physical characteristics, it is not detectable by our basic senses:

- Colorless
- Odorless
- Tasteless

Moreover, CO possesses an incapacitating power that is as ruthless as it is efficient, owing to the following properties:

- Poisonous
- Oxygen displacing
- Gradually disorienting and causes confusion and fainting

Finally, because it is highly flammable and contributes to oxygen deprivation, creating backdraft situations, CO is considered as having a deadly, explosive nature.

For modern firefighters, CO is a fact of life at every fire to which they respond. As the fire begins, a mixture of complete and incomplete combustion will fill the air with a multitude of gases. CO is a direct by-product of incomplete combustion and will be present long before the flames associated with a typical structure fire become visible. Since CO is lighter than air, it has a tendency to move into areas not yet involved in the fire. The consequences of this can be devastating: since these areas may already have been deemed to be safe, firefighters may not feel the need to wear respiratory protection (fig. 5–6).

UNWANTED GUEST

CO does its damage by displacing oxygen in blood cells. The iron components in hemoglobin readily accept CO, leaving no room to carry oxygen. It is estimated that hemoglobin has 200–240 times the affinity for CO as for oxygen. Thus, our oxygen-carrying ability is compromised by CO, thereby reducing the ability of oxygen to get to our tissues. This is the physiological mechanism behind the following effects associated with CO poisoning:

- Difficulty breathing/shortness of breath
- Dizziness
- Headaches

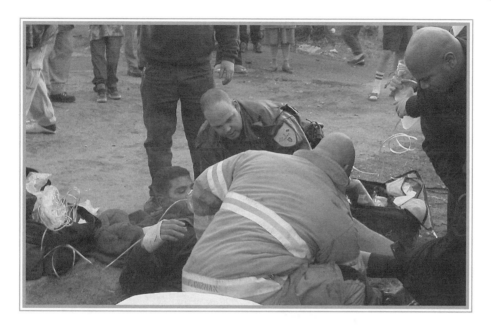

Fig. 5–6. A firefighter receiving medical attention for smoke inhalation (Photograph by Dennis Leger, Springfield [MA] Fire Department)

- Confusion/disorientation
- Drowsiness/fatigue
- Nausea
- Syncope/fainting
- Seizures
- Shock
- Unconsciousness
- Death

The scary thing about CO is that it can be there and not be obvious at all. It can be slowly dropping you into a state of confusion, disorientation, or unconsciousness, and you may not even know it. Think of a frog in the kettle; it doesn't feel the incremental temperature changes that precede boiling water—and ultimate death.

A SINKING SHIP AND A SINKING FEELING

An incident in Seattle highlights the true nature of CO and the threat it poses to firefighters. A number of units were dispatched to a sinking vessel under the Ballard Bridge. On arrival, they discovered a 100 ft yacht that was taking on water and beginning to sink, stern first. Firefighter Brian Mattson was among the first to enter the ship to evaluate. He found a foot of water in the lower crew deck and reported that it was rising.

Little headway was made during the initial efforts at dewatering, and the vessel continued to slowly sink until the stern made contact with the bottom of the bay. Because of concerns about the further stability of the ship, it was determined that operations should cease if immediate headway was not possible. Additional pumps and dewatering eductors were added to the operation. The crews were instructed to do their best to keep floating debris from clogging the eductors. The bedding and other items from the crew quarters posed a big problem, since they floated toward the dewatering devices.

The operation took a turn for the positive. Firefighters made headway on the rising water, as long as the debris was kept from clogging up the eductors. Firefighter Mattson descended a ladder into the flooded compartment to remove a blanket with a pike pole that had blocked one of the eductors. This was when he realized that there was problem and that it was totally unrelated to the water.

Until this time, the attention of the responding units had been focused exclusively on the stability of the ship and marking the rising water. When Firefighter Mattson hooked a blanket with his pike pole, its weight seemed out of proportion to what would normally be expected; in Mattson's own words, "It felt like it was made out of lead." Suspecting that he was probably just getting fatigued from the lengthy operation, Mattson wrestled the blanket out of the water and immediately became light-headed. His subsequent attempts to slow down and catch his breath brought a horrible realization: He could not get a good breath, and his vision was starting to go fuzzy. As those symptoms worsened, he knew that he was close to passing out.

Using every bit of strength left to him, Firefighter Mattson sought help and moved to where his partner was working. When he reached his partner, he found him crouched against a railing with his eyes rolled back in his head. He recounted afterward that all resilience then drained from his body, and a single thought went through his mind: "I am going to die." He recalls falling to the floor and mumbling help as everything faded to black. That call went unheard.

Some time later, a lieutenant entered the vessel to check on the crews and found the two members going in and out of consciousness. He immediately radioed for help, and the two firefighters were moved onto the bow of the ship. Firefighter Mattson remembers feeling like he was being left behind as he was set down on the bow. Even though the crew was right by his side, he started crawling frantically toward a plank that would lead him to the pier. Fortunately, the alert firefighters stopped him, and he was safely taken to medical help.

The ship was abandoned of all personnel, and subsequent air monitoring found 500–1,000 parts per million of CO and slightly decreased oxygen levels. Three firefighters sent to Harborview Medical Center were found to have 20% CO in their blood. Interestingly, only two of three were symptomatic, even though the CO levels were the same. Firefighter Mattson and his partner were placed in a hyperbaric chamber for six hours, to clear the CO from their bodies. All three of these firefighters are on the job today and doing well.

It was determined that an inversion layer, coupled with heavy fog, might have caused the CO to sink, rather than rise, in areas of the ship. Thus, pumping could have contributed by creating a vacuum that pulled in CO from gas-powered engines being operated in the vicinity. Also, air monitoring inside the vessel was an obvious solution that was overlooked owing to the exceptional nature of the operation.

Importantly, this incident could have easily resulted in fatalities, and it was, by most accounts, not a typical IDLH incident. The crew quarters, where the CO poisoning occurred, routinely housed members of the ship; the tanks below were empty; and there was nothing to indicate the problem that would develop. From incidents such as this, we can develop strategies to expect the unexpected. This is not the same as looking at a house fire and being surprised that there are elevated levels of CO inside. Every competent firefighter should be fully aware that there will be elevated levels of CO at every fire and, moreover, that every fire they fight has the potential to kill them—in many cases, within minutes or even with a few breaths.

TO FILTER-BREATHE
OR NOT TO FILTER-BREATHE

A common fallacy that opponents of the ROAM and of air management programs in general promote is that you can always fall back on *filter-breathing* when you get in trouble. This notion is a throwback from the days before SCBA were the norm, when sucking the carpet was a matter of survival for aggressive firefighters (fig. 5–7).

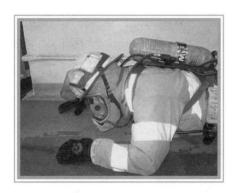

Fig. 5–7. A firefighter sucking the carpet
while attempting to filter-breathe

With the introduction of SCBA came the well-worn tactic of stuffing the low-pressure hose into the bunking coat when one's air ran out. This met with mixed results and often was forgotten in the anxious moments between the realization that one's air was gone and the onset of asphyxiation. After the *Ocean Pride* ship fire, Firefighter Ron MacDougall said that he didn't attempt to filter-breathe even though he had done so in the past.

There is no way to tell how you are going to react when things go wrong. Training will play a significant role, and we should be prepared to act in a manner that will be proactive, safe, and lead to positive results. Nevertheless, the stress of the situation—coupled with exposure to disorienting factors, such as obscured visibility—and the physiological consequences of smoke have a dramatic impact on our actions.

In most fire departments, low-pressure hoses are being replaced with mask-mounted regulators. While these represent a positive change, they provide an interesting dilemma for advocates of filter-breathing. Some versions of the SCBA ensemble include a filter that is moved into place if the regulator fails or if air is depleted. This allows the firefighter to filter-breathe and is dependent on the type of filter being utilized. Many departments train firefighters to pull their hood over their facepiece to provide a crude filter when their air is gone (fig. 5–8). While this may filter large particles and soot, it does nothing to stop the deadly gases mentioned earlier. Filter-breathing is nothing but a last resort to be used in a clear Mayday situation, when firefighters have failed to manage their air. Normally, such desperate measures do not yield positive results.

Fig. 5–8. A firefighter placing his hood over the air opening to demonstrate filter-breathing

PRESUMPTIVE LEGISLATION AND FIREFIGHTER CANCER: MARK NOBLE'S LEGACY

One of the biggest impacts of keeping the emergency reserve of SCBA air for use in firefighter emergencies only is the dramatic decrease in routine exposure to smoke. If firefighters do not run out of air in IDLH environments, they will greatly diminish their exposure to the carcinogenic hazards of firefighting. As will be shown in this section, that will be of tremendous importance for all firefighters in the future.

Mark Noble's incredible story was captured in a video presentation entitled *You Need It Like a Hole in the Head.*[10] The authors are committed to bringing this important message to firefighters across the country and have made it a part of our interactive seminar.

Mark served with the Olympia (WA) Fire Department until a brain tumor was diagnosed in 2002. Like most firefighters, he loved his job and was considered very good at what he did. Also like most firefighters, he was living his life without the expectation that a brain scan might forever alter the course of his remaining days. Mark was engaged to be married, had two teenage sons that he was eager to see do well in school, and did not expect to be told he had three months to live.

*Fig. 5–9. Mark Noble's legacy—courage
and desire that firefighters should be safer.
(Photograph courtesy of Rebecca Noble)*

Mark's goals changed dramatically after the visit to the doctor that ended in detection of the brain tumor. In the short term, he wanted to see his oldest son graduate from high school in two months. His other driving passion became sounding the alarm to his fellow firefighters that some of their actions on the fireground and in the stations were killing them (fig. 5–9). Chief among these was routine exposure to smoke and the products of combustion.

His research took him on an incredible journey that will ultimately save the lives of many of the brothers and sisters he loved. Many of the deadly gases that were detailed in the first part of this chapter were researched, as well as hazards in the station, such as asbestos and fire apparatus emissions. The following statistics, comparing the likelihood of cancer among firefighters to that of the general population, are the result of Mark's in-depth research:[11]

- *Brain cancer*—3.5 times as likely in firefighters with 10–19 years of service
- *Leukemia/lymphoma*—3 times as likely
- *Non-Hodgkin's lymphoma*—2 times as likely
- *Multiple melanoma*—2.25 times as likely; after 30 years, 10 times as likely
- *Bladder cancer*—3 times as likely
- *Kidney cancer*—4 times as likely
- *Prostate cancer*—2 times as likely
- *Testicular cancer*—2.5 times as likely
- *Colorectal (large-intestine) cancer*—2 times as likely
- *Liver cancer*—2 times as likely
- *Skin cancer*—3 times as likely

While studies across the country offer varying results as to how great our risk is compared to that in the general public, one thing is crystal clear: As a direct result of putting our lives on the line every day, we are exposed to elements that greatly increase our chances of contracting cancer. The primary culprit is inhaled smoke. To be sure, there are other cancer-causing agents. Skin contact with carcinogenic products is an area that certainly needs further research. The primary action we can take, however, is to stop inhaling compounds that we know are killing us.

While Mark Noble lost his battle with brain cancer on January 15, 2005, his courageous spirit prevailed over the initial diagnosis. In the three years following his initial doctor's visit, he saw not only his oldest son's graduation but also his youngest son's. Furthermore, he challenged us to take a good look at how we are putting ourselves and the lives of our comrades at risk. We are indebted to Mark for that challenge.

REMEMBERING MARK NOBLE

I am so honored that Mark and I get the opportunity to be part of this important chapter and very excited to see others fighting for the health and safety of those who have chosen to be firefighters. Mark's greatest hope during his illness was that his video would be viewed by those who could make the changes necessary to spare other line firefighters and their families from going through what he and our family experienced.

Mark and I both saw his illness as preventable had there been awareness of the hidden causes that we now are becoming more familiar with. We recognize that those who have been in the field for many years have been exposed to many of the elements discussed in this book and that we will see more firefighters with cancers in the near future. The good news is that because of early intervention together with solid safety policies that are being developed, we will see less toxic exposure and illnesses.

Mark also felt it was important that all firefighters receive the best health care coverage when injured on the job. To this end, we must work toward a *national presumptive law* that will recognize and support firefighters who have been affected by their exposure and give financial support for the family members left behind.

It is with these thoughts that Mark's sons and I hope for everyone who chooses this wonderful field to be safe, healthy, and feel supported by their communities. It is up to us to make noise to get the changes in place that will help to ensure a long and healthy life.

Be healthy and safe.

—*Rebecca, Luke, and Shane Noble*

WHO'S GOT YOUR BACK?

Because of the obvious susceptibility of firefighters to cancer, there has been a big push by firefighters, labor organizations, and concerned citizens to enact protections for those affected. This long and difficult struggle achieved vaguely positive results in the form of *presumptive legislation* that covered firefighters in some states who contracted cancer. These laws presume that certain cancers were caused as a direct result of firefighting, thus allowing the firefighter to receive benefits such as medical coverage. Most sober observers would view this as a fair and honest approach to the sacrifices being made on a daily basis; however, that has not proven to be the case.

As an example, we will focus on the struggle to get firefighters presumptive coverage for cancer in Washington State. The initial attempts at passing this legislation focused on coverage for the following types of cancer:

- Breast cancer
- Central nervous system cancer
- Oral and digestive system cancer
- Hematological system cancer
- Lymphatic system cancer
- Reproductive system cancer
- Skeletal system cancer
- Skin cancer
- Urinary system cancer

In what can only be viewed as direct slap to the face of firefighters, the Washington State Senate Ways and Means Committee whittled the original list down to the following:

- Bladder cancer
- Brain cancer
- Kidney cancer
- Leukemia
- Malignant melanoma
- Non-Hodgkin's lymphoma
- Ureter cancer

CANCER AMONG FIREFIGHTERS

Despite rising evidence that demonstrates exaggerated levels of cancer among firefighters, these seven were all that received coverage. The Seattle Fire Department recently walked with the courageous family of firefighter James Scott ("Scotty") Barnard through his fight with colon cancer, from which he ultimately died. Scotty and his family received no coverage under the current presumptive legislation throughout their ordeal.

To add insult to injury, the senate committee put into place other restrictions and rules that every Washington firefighter should ensure they know about. The coverage extends to firefighters on the basis of the number of years served. It provides coverage for 3 months for every year served, up to 60 months. Thus, firefighters with 20 years' service will have maxed out their available coverage, despite the continued threat of exposure over the rest of their careers. This means that 20-year firefighters have 5 years after retirement to receive a diagnosis of cancer and be covered under the legislation; otherwise, they are on their own. Ten-year firefighters have roughly 2½ years to receive a diagnosis, or their service will be ignored. No coverage; no recourse; no help!

Additional variables have been incorporated that further question whether the cancer is job-related, including smoking history, fitness, and so on. Moreover, just as investigators want to know whether those hurt in apparatus accidents were wearing their seatbelts, the same questions will be asked about precautions taken to avoid smoke exposure. If PPE is not routinely worn and respiratory protection is not mandated for all IDLH exposures, then the firefighters should pay the price in the view of lawmakers. It will not matter that you were part of an aggressive crew that sometimes didn't have time for SCBA; it will not matter that you saved many lives and untold millions in property damage; it will not matter that your family is left without their loved one and without financial support despite enormous medical costs. According to the presumptive legislation of Washington State, your disease will be considered as *non–duty related.* For most firefighters in Washington State, that means not only retiring with no medical coverage but also getting to foot the bill alone for cancer treatment.

As a result of intensive lobbying by the Washington State Council of Firefighters and pressure from the citizens, further changes were made in 2007. The following illnesses were added to the existing presumptive legislation:

- Prostate cancer diagnosed prior to 50 years of age,
- Colorectal cancer
- Multiple myeloma
- Testicular cancer

The gaps that remain are extensive. A recent study conducted in Cincinnati highlights the extreme risks that firefighters face in their efforts to protect citizens. Those risks are not truly being acknowledged by legislators as is evident from the current presumptive legislation.

Not every state has presumptive legislation. There is currently no protection for firefighters employed by the federal government. While Washington State's version is miniscule and insufficient, it at least gives firefighters there a chance of coverage, and it is improving. Many states have ignored their firefighters completely. There are many sad tales of firefighters going to great lengths—and tremendous cost—to establish that a disease is obviously duty related and being rebuffed.

THE FIRE SERVICE NEEDS TRUE LEADERS

The struggle to get firefighters to use appropriate respiratory precautions is an old one. The first step is to get responders to wear their SCBA in the first place. Sadly, in some departments, it is still not a requirement, and the toll from those decisions will be great. The smoke of modern fires is deadly and will exact its fury in multitudes of early deaths for firefighters who disregard their SCBA (fig. 5–10).

Fig. 5–10. Firefighters outside a modern structure fire.
Firefighters need to protect themselves from all aspects
of the IDLH environment. (Photograph by Dennis Leger,
Springfield [MA] Fire Department)

The next hurdle to overcome is staying in the smoke too long and running out of air. Great lengths have been taken in the previous chapters to establish incontrovertible evidence that many fireground deaths are caused by disorientation and/or asphyxiation due to smoke inhalation. Running out of air is killing us on the fireground. It is also causing us to contract diseases. Unless you stick your head in the sand and ignore the obvious, routine exposure to smoke increases the likelihood that you or your crew will get cancer.

One objections repeated time and again is that although firefighters believe these messages to be true, the concepts could never be put into practice in their department. The reason they give is that their firefighters are too aggressive and would never do it. In answer, I would ask, very respectfully but firmly, Who is in charge?

What the fire service needs are leaders who are willing to be as aggressive about protecting the lives of their firefighters as they are about protecting the lives of citizens or protecting their own egos. A straightforward reply to firefighters who refuse to keep their emergency reserve of air intact on exiting is easy: You will not be on my fireground. That is the position of chiefs and company officers who care about their crews.

Firefighters are, by nature, aggressive and gung-ho. This is a part of our culture that is admirable and essential. Nevertheless, that same determination needs to be monitored and controlled by fireground leaders, who should be similarly aggressive about mitigating the emergency while taking care of those fighting it.

THINKING HARD ABOUT HOW WE DO WHAT WE DO

For many years, I did my best to avoid wearing an air pack. It wasn't that I liked the taste of smoke, more that I have crappy eyesight and I faced a choice: seeing or breathing. Naturally, I chose seeing. Just after that, the term "duh" was invented.

Also, back then, in the 1970s and 1980s, there seemed to be something wonderful about coming back covered in soot, blowing out black snots, and stuff like that. We *felt* like firefighters and had the marks to prove it—smoke, soot, black snot, and a few burns.

I was never a metropolitan firefighter. Instead, I worked in areas where I would go to about two to three dozen working building fires annually. Not bad, but hardly one every day. Actually, few—very few—firefighters I know from anywhere go to a working fire every day, and I know a lot of big-city firefighters and bosses. I guess my past 34 years have been about average, with a couple of dozen working fires every year. Just as time has taken its toll on my previously black mustache, now gray, it takes its toll on lots of other stuff as well.

Time has also taken its toll on several dear and close friends of mine. How many of you know or knew a firefighter with cancer? In just the past few years, *good* friends of mine—long-time firefighters Louie S., James S., Mike M., Lee S., Dennis H., and several others—have passed away. All were firefighters; all were characters; all were firefighters with cancer.

Now let's fast-forward to the recent study by University of Cincinnati (UC) environmental health researchers. Their work—which has a direct impact on you—has determined that firefighters are much more likely to develop four different types of cancer than workers in other fields.

Their findings discuss the fact that the protective equipment we have used in the past didn't do a good job in protecting us against the cancer-causing crap (read *smoke*) that we encounter every time there is any kind of smoke showing. The researchers found, for example, that we are twice as likely to develop testicular cancer and have significantly higher rates of non-Hodgkin's lymphoma and prostate cancer than nonfirefighters. The researchers also confirmed previous findings that firefighters are at greater risk for multiple myeloma.

The doctors leading the research were Grace LeMasters, PhD, Ash Genaidy, PhD, and James Lockey, MD. This is the largest comprehensive study to date that has investigated cancer risk associated with working as firefighters. "We believe there's a direct correlation between the chemical exposures firefighters experience on the job and their increased risk for cancer," says LeMasters, professor of epidemiology and biostatistics at UC.

Firefighters are exposed to many compounds designated as carcinogens by the International Agency for Research on Cancer (IARC), including benzene, diesel engine exhaust, chloroform, soot, styrene, and formaldehyde, LeMasters explains. These substances can be inhaled or absorbed through the skin and occur both at the scene of a fire and in the firehouse, where idling diesel fire trucks produce diesel exhaust.

The UC-led team analyzed information on 110,000 firefighters from 32 previously published scientific studies to determine the comprehensive health effects and correlating cancer risks of their profession. They found that half the studied cancers—including testicular, prostate, skin, brain, stomach, and colorectal cancers, non-Hodgkin's lymphoma, multiple myeloma, and malignant melanoma—were associated with firefighting to varying levels of increased risk.

We need enhanced and additional protective equipment to help us avoid inhalation and skin exposures to these occupational carcinogens. The good news is that there new features will be introduced to our PPE very soon. One example is the International Association of Fire Fighters' "Project Heroes" PPE, made by Morning Pride, which actually forces the products of combustion out of the PPE through a simple but very effective positive-pressure design.

Inhalation is a serious problem. Breathing that crap represents a direct path to trouble—in the short or the long term. In addition, there is the soot on our skin. Remember when you took your dirty gloves and wiped your face with them so that others could see that you were at a fire? Yeah, that stuff. What does that stuff do? It directly enhances your body's intake of that crap—through your skin. Wonderful.

The solution? This isn't that difficult. Thoroughly wash your entire body, to remove soot and other residues accumulated on fire calls to avoid exposure through the skin. How often do we just let it sit there—sometimes cleaning up after a fire only after a few hours' sleep. What do you think that crap is doing while you sleep? Entering you through your pores. Welcome home.

Surprised? Their findings suggest that the protective equipment—SCBA, masks, turnout clothing, and so on—that firefighters have used in the past does not do such a great job in protecting them against cancer-causing agents. I mean, did you ever think a little soot on your face or hands is a problem? Is the issue as simple as our using what we have properly? These days we do have full protective clothing and equipment that does protect us to some extent. Maybe we just ought to wear it? And maybe we need fire officers who will make us wear it without worrying about who likes them or who they might offend?

Quite honestly, that study blew us away—sort of. It's not as if we didn't *know* about the problems; we just didn't really *worry* about them. Well maybe now, though, we are worried. The UC-led team analyzed information on 110,000 firefighters from 32 previously published scientific studies to determine the comprehensive health effects and the correlated cancer risks of our profession. Heartwarming news.

So why now? We always knew to wear our SCBA, but we didn't always listen. In some towns, we rarely listened. We knew that wearing SCBA made sense, but we didn't really worry about it when there was just some smoke. We also never thought that a little soot on our face or hands was a problem. We never really thought about those outside, such as the chiefs, the pump operators, the safety folks, the sector bosses, the emergency medical services (EMS) personnel, and even the fire photographers and others who were actually *in* the smoke—even though it would clear "in just a little while."

So now what? This book is the answer. This book is going to show you how to minimize the problem, as well as the risk. Will it eliminate the risk? No, never. This is firefighting, and firefighting is risky—especially when we have to get someone out. Saving property is one thing; saving a life is another. If you had to die in the line of duty, would you rather have your family remember you as trying to save a warehouse filled with automobile parts or children trapped inside a house?

Not everyone goes home

The slogan "Everyone goes home" represents an attitude—an attitude within a fire department that we'll do all we can to try and bring all our firefighters home. It possesses several important meanings:

- If we don't drive like idiots, we'll probably make it home.

- If we follow standards such as NFPA 1403, firefighter trainees will probably make it home.

- If we put our seat belts on, we'll probably make it home, even if we collide on the way to a fire.

- If we are a hundred pounds overweight but eat more salads, we'll probably make it home.

- If we religiously wear our PPE and SCBA and learn to manage our air, we'll probably make it home.

- If we stay out of the way when it is obvious the building will collapse, we'll probably make it home.

- If we have the right amount of trained staff and good bosses at a fire, we'll probably make it home.

- If we drill and train on actions we need to perform regularly, such as quickly getting water on the fire, we'll probably make it home....

- Finally, if we take the messages of this book seriously, we will greatly minimize our chances of *not* going home.

To be clear, sometimes we do have to take risks. Not stupid risks—those we do not have to take. However, we are expected to take smart risks when the odds are on our side. When we know there are people inside, unless there is clearly no chance for us to save a life, we go in and get those people. That's why 343 FDNY members died on 9/11. There were people inside, and they went in to get them. No one else and no other government service goes in to rescue people; that's what we do, and sometimes we don't go home.

We can greatly reduce the amount of firefighter line-of-duty deaths (LODDs) each year when we look at things like health, driving, training, leadership, and wearing PPE. Sometimes all of that is covered, yet we still die. That's generally called a heroic LODD. Many LODDs are not heroic deaths.

Unfortunately, a majority of the all very tragic losses of firefighters in the line of duty each year in North America were not heroic LODD's. Often, their LODDs were avoidable and predictable as well. They all lived heroic lives because they were firefighters, but so many didn't die heroically or by performing acts of heroism.

If you are anything like me, you love being a firefighter, and you are and have been surrounded by wonderful opportunities. Taking the words in this book and actually applying them is, once again, another great opportunity.

One organization we strongly support and urge firefighters to contact is Firefightercancersupport.org. This is a 100% firefighter-operated organization that provides prevention education and support to firefighters and their families. Also, Firefighterclosecalls.com promotes firefighter safety and survival information.

—Billy Goldfeder, E.F.O. (Deputy Chief, Loveland-Symmes [OH] Fire Department; Chairman of the IAFC Safety, Health, and Survival Section; member of the Board of Directors of the NFFF, the Board of Directors of the September 11th Families Association and the National Firefighter Near Miss Reporting Task Force)
http://www.firefighterclosecalls.com/

EXPERIENCE CAN BE THE BEST AND WORST TEACHER

Some of the descriptions given by firefighters who have been lucky to escape the breath from hell are notable. Lieutenant Chris Yob experienced it firsthand, just before he jumped from a three-story building, barely escaping with his life. As he was desperately fighting the effects of CO (and probable hydrogen cyanide) poisoning due to filter-breathing, he attempted to place his low-pressure hose out a window of the fire building for a breath of fresh air. The eddying smoke came back around and he got a full breath that he described as tasting like "liquid, molten plastic." He said that it was the worst thing he ever breathed, and he knew he couldn't stand another breath. In his dazed state, he attempted to jump to what he thought was an adjoining roof, only to fall three stories (fig. 5–11).

Fig. 5–11. The Sunset Hotel, where Lieutenant Chris Yob ran out of air and nearly lost his life (Photograph courtesy Seattle Fire Fighters Union, IAFF Local 27)

Another fire in Seattle caused the fatality of a veteran firefighter, after the fire was out. Bob Earhardt was found on the second floor of the fire building, where he had been operating alone and was overcome by what was believed to be CO poisoning. Captain Dana Caldart was the first to find Firefighter Earhardt, in a room that was remote from the fire and down a dark narrow hallway. The entrance to the room was partially blocked, and the door closed behind Caldart as he entered. Earhardt was found kneeling against a wall with his head cupped beneath his hands. His gloves were lying to each side, and he had removed his helmet and facepiece. His bunker coat appeared to have been pulled off and left to fall behind him, as it was across the backs of his legs and boots, with the liner facing up.

His SCBA was in the middle of the room and displayed a zero reading. Despite the fact that no fire was present, the probable accumulation of gases like hydrogen cyanide, hydrogen chloride, and CO caused a veteran firefighter to become disoriented and lose control of his actions. Later testing revealed high levels of CO in the air, and it is unknown whether any testing was done for other gases.

Lessons learned lead to ROAM

These lessons—and thousands just like them—provide stark reminders that we are not playing heroes in a Hollywood movie. Every emergency that we go to is simply the next one that is trying to kill us. We are trained, ready, and willing to do our jobs and serve the citizens of our jurisdictions. However, with that willingness must come the recognition that it is our responsibility to deal appropriately with the hazards that routinely confront us. The deadly smoke of today's fireground is certainly one of those hazards.

The only reasonable solution is to exit the IDLH prior to exposure to the breath from hell. The simplest and most effective way to aid our exit is to follow the ROAM: Know how much air you have in your SCBA, and manage the amount of air you have so that you leave the hazardous environment *before* your SCBA low-air alarm activates.

By keeping your reserve air strictly for emergencies, you greatly increase your odds of never having to filter-breathe or sample the breath from hell. The immediate consequences of exposure to the breath from hell are becoming disoriented, compromising your team, and drawing on other resources to bail you out. The short-term consequences are injury, scarring of the lungs, and reduced respiratory capacity. The long-term consequences are evidenced by the diverse and insidious cancers firefighters develop, along with an array of other health issues that destroy lives of firefighters and haunt their families. Finally, there is always the possibility that you will immediately join the long list of fireground deaths directly due to smoke inhalation and asphyxiation.

The conclusions are sobering:

- Allowing yourself or anyone else under your supervision to inhale the smoke of the modern fireground is a dereliction of duty.

- Ignoring the need for air management training increases the chances that your members will be involved in close calls, near misses, and actual tragedies.

- Staying in the hazard area until your low-air alarm activates virtually guarantees that your crew will be exposed to the breath from hell.

- Using filter-breathing and/or sucking the carpet as anything other than a last resort is foolish and deadly.

In the tradition of all warfare, it is imperative to know your enemy. There are many enemies attempting to harm the modern-day warriors who bunk up every day and go to battle fire. Smoke is among the most insidious. It is the responsibility of every firefighter to take this threat seriously; by preparing themselves for the threat, firefighters can ensure that the breath from hell is not allowed to affect their lives.

THE NEW YORK TELEPHONE FIRE

The New York Telephone Company Main Switching Center building is a virtual fortress. The structure, designed to be earthquake and riot proof, has windows constructed of heavy wire glass in reinforced steel frames that were mounted with ¼" Lexan—a bullet resistant plastic. At the time of the fire, all windows at street level and on the second floor were covered with heavy metal cages to protect them from vandals.

For nearly a half-century, the art deco, 11-story structure served as the main switching center for the Lower East Side. It served a 300-square-block area and was equipped to handle 10,000 calls an hour. Its customers included major business, six hospitals, nine housing projects, three universities, 11 secondary schools, several police precincts, almost all units in the FDNY's First Division, and 92,000 residential phones.

On February 26, 1975, at approximately 4 p.m., a fire alarm rang in the New York Telephone Company lobby at 204 Second Avenue. A building foreman discovered that the glass on an alarm box in the first-floor stair landing had been shattered. Since the accidental breaking of the glass by workers carrying equipment through the first floor had occurred before, the foreman manually reset the alarm and reset the fan system, which had automatically shut down. With no replacement glass readily available, the foreman inserted a rectangular piece of plywood.

Meanwhile, a manhole crew working outside was trying to locate troubles in a trunk line leading into the Second Avenue side of the building. A card game was in progress in the 10th floor locker room. A craftsperson took a nap on the couch.

Around 10 p.m., the lights over three public telephone booths in the lobby went out. The same building foreman reset the circuit breakers that controlled the lights. The lights went back on.

At midnight, an employee on his way home observed that the lobby clock did not agree with his wristwatch. A building service woman, also on her way home, noted that the clocks had, in fact, stopped.

At 12:05 a.m., there were 23 employees in the building: 10 on the fifth floor, 2 on the fourth floor, 5 on the third floor, and 4 on the second floor. On the first floor, a worker was finishing a lengthy telephone call. The guard stood at his post in the lobby.

At 12:11 a.m., a telephone company employee entered the building, walked through the lobby, stepped into another room and saw heavy smoke seeping out from under a door. He looked down a 2 ft-by-2 ft hole in the floor—a cable conduit—and saw a mass of burning cables glowing bright orange. The employee tried to telephone a fire alarm.

The telephones were dead.

At the same time, the fifth floor began to fill with heavy blackish gray smoke. "Fire!" a worker shouted, as fire alarms started to sound all around him. He grabbed a telephone and dialed 911; however, the telephone was not working.

Workers on the third floor also tried to call for help. Their telephone lines were dead as well.

As panicky employees began congregating in the lobby, the lobby guard tried his telephone—also dead.

"For the love of God," the guard screamed to a worker, "get the fire department. We have a serious fire." The worker ran to the corner and pulled the fire-alarm box.

At 12:25 a.m., a pull-box alarm was received at the fire dispatching headquarters in Central Park, transmitting Manhattan box 465.

We had experienced a typical February night at Ladder 3, located at Third Avenue and 13th Street, on Manhattan's Lower East Side. We were worn ragged from the crushing workload and lack of sleep. There had been a smoldering mattress fire in an East Side apartment, a car crash with fatalities, a dumpster fire, a heart attack, calls from residents claiming they had no heat or hot water, a lost child, a stuck elevator, and the usual dose of false alarms. In other words, it was business as usual for a truck company in the FDNY.

At 13 minutes past midnight, I left the warmth of the firehouse kitchen and relieved another firefighter at the cold drafty house watch desk in the front of the firehouse. I hated working through the night. As a single person living on Manhattan's Upper East Side, I knew that the dating scene would be in full swing right about now—and I longed to be out there in the middle of the action. I thought about calling my girlfriend of two years, who was waiting tables at Suspenders, a firefighters' hangout at Second Avenue and 38th Street, to find out what my crowd was up to—which fireman was hitting on which stewardess, who had had too much to drink, if the guy who owed me $100 from a football bet had shown his face yet. Instead, I broke open a paperback that I'd been trying to finish for months, sipped a cup of black coffee, and settled in for a long and hopefully uneventful night.

At 12:26 a.m., a radio dispatcher's broken transmission blared through quarters: "Ladder 3, respond."

I laid down my paperback and replied, "Ladder 3."

Although the dispatcher's subsequent words were unusually difficult to decipher, this was what I understood: "Report of a working structural fire, box 465, Second Avenue and 13th Street. The New York Telephone Building. Acknowledge, Ladder 3."

Having had only two years' experience as a New York City firefighter, during which I'd fought all the typical fires—apartment, dumpster, mattress, car. A working *structural* fire meant that an entire building might be ablaze. A chill raced up my back.

"Ladder 3, 10-4." I reached to my left, struck the fire-alarm buzzer—its loud, harsh sound always startling—and shouted, "Everyone goes!"

Men stopped whatever it was they were doing—cleaning up after a stale meal that had been interrupted by three separate fire alarms, checking tools, repairing equipment—and ran downstairs, slid down poles, slipped into fire gear, and jumped onto the trucks. As I ripped the paper from the printer that contained the hard copy of the alarm information, I kept yelling, "Everyone goes! Chief goes too."

I donned my own fire gear, handed Lieutenant Pullano—a former stockbroker—the printout. In moments we were racing down 13th Street, our bare faces stung by a frigid wind as our air horn and sirens shattered the wintry night.

"Smell that?" Patty Sullivan stood beside me as our rig approached the scene of the alarm. "I've never smelled anything like that before."

I took a breath. The odor was sharply acrid. I looked and saw Telephone Company employees, faces smudged by soot, running from the building and retching.

Engine 5 had been returning to their firehouse following a false alarm and were the first on the scene. They had begun pulling hose lengths to the entrance to the structure.

"Engine 5 to Manhattan, K., 10-75," I heard them broadcast over our truck radio. "Working fire, box 465, New York Telephone Building."

Engine 33, Ladder 3, and Tower Ladder 9, all under the direction of Battalion 6, were also at the location.

As we turned the corner on 2nd Avenue, Lieutenant Pullano got our attention by pounding his fist on the glass partition separating the cab from the crew's assigned positions. "Bring Scott masks." These clumsy, yellow 30-pound breathing apparatus are carried on firefighters' backs and look exactly like scuba gear.

The rig came to a halt in front of the building, and I saw black smoke billowing from the New York Telephone Company lobby and more employees stumbling from the building, coughing and choking. We grabbed our gear, axes, halligans, and other forcible-entry tools and rushed inside.

We assembled in the lobby and were told that the fire had grown to a second alarm and the battalion chief had put in a call for an additional engine and ladder company. The telephone service in a 300-square-block area on Manhattan's Lower East Side was out of service, and the department was communicating on its emergency backup system—that must have been the reason the dispatcher's firehouse voice transmission had been broken up.

A moment later we learned that the New York Telephone Company's fire-alarm display board indicated a fire three floors below, in the vault of the subcellar. The chief determined that although flames were not yet visible, an assault on the underside of these vaults had to be made. As the first due ladder company, our responsibility was to make that assault and locate the seat of the fire—an essential step in the foundation of all firefighting strategies. Some firefighters were assigned vertical smoke ventilation duties, others horizontal.

"Let's get to it," Pullano said.

Patty Sullivan turned to me. "Remember to stay low, Noonan." He helped me check my gear. "This could be a bad one."

And then Patty did something I'd never seen him do before. He blessed himself, making the sign of the cross—three times. Watching the fearless, gray-haired firefighter, I flushed with apprehension. What the hell were we walking into?

I was a 17-year-old shucking clams at Fitzgerald's Ocean Front Clam House in New York's Rockaway Beach the first time I'd met Patty Sullivan. I'd been raised in that small, suburban community—on a peninsula separating Jamaica Bay from the Atlantic Ocean. I was, by my parents' design, isolated from the rest of the five boroughs. My father—a detective in the New York Police Department—and my mother wanted to protect their children from the perils of the inner city.

Summers in Rockaway Beach were a mix of blue-collar and white-collar families, all escaping the stifling crowds and heat of the city. Every Memorial Day, the area would metamorphose from a sleepy village into a bustling seafront community.

During one of those early-summer months, I began working at Fitzgerald's and got to know a group of regulars, men whose boisterous banter and fun-loving humor always livened up the place. I soon learned that they were New York City firefighters. Since many of them were skilled in more profitable professions—accounting, physical therapy, construction—I learned that firefighting was more like a religious calling than a regular vocation. They didn't do it for the money—which explained why they were held in such awe by the bartenders, waitresses, customers, and as the summer progressed, me.

"What's it like to be a fireman?" I asked Patty Sullivan one August afternoon.

Patty, father of three and volunteer firefighter on his days off from the FDNY, dressed boat-shoe casual and shaved only when absolutely necessary. "Thought you were gonna be a cop." He puffed on his cigar and sipped his beer. "Like your dad."

"He wants me to go to college."

"Good idea," Patty said. "Go to college."

"You didn't answer my question."

Patty looked at me thoughtfully, rubbed the stubble on his cleft chin. "Best job in the world," he said, "but a dangerous one. No margin for error."

No margin for error. Now those words pinballed around my brain as Patty finished checking my gear. He patted me on the back, flashed me what he tried to pass off as a self-assured smile. I could see fear in my friend's eyes, though.

"Ready?" Pullano said.

We donned our Scott masks and descended into the basement of the building. The halls were long and tiled, with several steel doors. The smoke had gathered at the top of the corridors, creating a dim mist that caused the fluorescent lights to give off an eerie glow. The constant ringing of the fire Klaxons—identical to a submarine dive alarm—added to the confusion.

We walked through a door marked "Extremely high voltage" and negotiated a narrow ladder that descended at a sharp angle to the smoke-filled basement below. Officers experienced extreme difficulties in communicating. We could barely see or hear simple commands like "Don't let go of the hoseline" and "Stay together."

We searched for fire in the pitch-black smoke—and found none. Then we hit pockets of glowing cable and doused them with streams of water. But the fire was burning the PVC insulation of the wires in the vertical piping of the 11 story building. Our efforts were futile.

"Holy Mary, mother of God," someone said behind me. And then I could clearly distinguish a voice, the engine officer's, transmitting over his radio to the IC that our position was untenable and perilous. I realized he must have removed his facepiece and breathed poisonous air to make the communication.

"Two telephone employees are unaccounted for," I heard announced over the radio. I overheard them assign a rescue company the task of searching the building for the missing employees.

"Come on, lads," Patty said as he led us along the inside of the 13th Street wall, searching in vain for the fire.

Fifty or so steps later, we found ourselves enveloped by 10 million cubic feet of zero-visibility toxic smoke. After several minutes of blindly cracking into walls, iron railings, and each other, we were forced to retreat.

As we pulled back along the same route, someone asked, "Hey, where's the lieutenant?"

"Pullano?" Patty said into his radio.

No answer.

Since I was beginning to feel a bit disoriented, I imagined that Pullano did too and had wandered off. We spread out, searching as best we could. Pullano was no longer anywhere near us.

We retraced our steps along the interior wall and through the subcellar, down a long, narrow corridor through nearly impenetrable, toxic black smoke. There were no windows, no doors, and no other source of breathable air or outside ventilation.

"Bang on something, Pullano," Patty said into the walkie-talkie, a last ditch effort to locate the missing fire lieutenant.

We heard three loud bangs. Pullano was somewhere ahead of us shrouded in deadly blackness.

We inched forward, felt along the walls, and discovered a metal door. I used my fist to bang on it. Someone banged back.

"Get me the hell outta here," Pullano's muffled voice came from the other side.

I tried the doorknob, twisted it. Pushed. Shoved. It didn't budge. Patty elbowed me aside, jammed the working end of a forcible-entry tool into the door jamb above the lock, pressed forward, then yanked back with all his considerable strength.

No movement.

Sullivan quickly reset the forcible-entry tool. This time, he drove it deep into the doorjamb, below the lock. He and Joe Rosa, a family man and restaurant co-owner who wore his love for Italian food around his waist, heaved their full weight against it; finally, the lock shattered, and the door ripped open.

"Thank God," Lieutenant Pullano said, as he stumbled out.

Then—*flashover*.

A blue-green lateral flame exploded from the room in which Pullano had been trapped and raced though the subcellar. I dove to the ground, rolled to my knees, crawled as fast and as far as possible from the inferno. When I looked back through the blowtorch flames and bellowing soot, Sullivan, Rosa, and Pullano were gone.

"Lieutenant, Patty, Joe?" My words were muffled by my face mask and obscured by the roar of diesel turbines. I crawled back to the flashover site, using my hands as eyes, felt around for my comrades, and found no one.

I was alone.

My Scott mask low-oxygen alarm sounded. I clicked on my six-volt searchlight to check the oxygen gauge, but the powerful light didn't seem to work. I jiggled the switch, clicked it on, off, banged the damned thing against my leg. Pointed it directly into my eyes through my face mask—and saw light. It was working but could not penetrate the lethal smoke.

My Scott mask alarm continued to sound, indicating my air was fast dwindling. I had two minutes to make it to the stairwell and climb four flights. Two minutes to reach fresh air. Two minutes of life. I breathed a prayer.

More Scott masks, low-oxygen alarms sounded somewhere in front of me—or were they behind me? I realized that I was lost, turned around, with no idea which way I'd come, which passageway and stairwell I'd taken. I felt the first twinges of panic.

Following a fire hose, as my lifeline to the world above, I backtracked and crashed into a wall, a dead end. The hose ran straight up a wall! I shuffled to the left and realized that, owing to the bubbling tar and melting polyethylene, my boots were adhering to the surface of the floor. God, it was hot.

Following the wall, I finally came upon another firefighter. Somehow he had a sense of where he was and pulled me across a six-foot abyss, into a narrow staircase clogged with other firefighters coming in the opposite direction, sent to relieve us. I was forced to give way to the reinforcements and heard more Scott mask alarms sounding behind us. And then I heard a single ping—I had only seconds of air left. I pushed forward, clawing my way, hit the deck, and crawled as fast as I could up the slippery waterfall of a stairwell.

As the smoke became less dense, I could see other firefighters—some disoriented, others felled by the noxious fumes—all coughing, choking, and vomiting. I took a deep breath, held it, and ripped off my mask.

I dashed up the last few steps, through the lobby—my lungs burning—and out, onto the street. I exhaled, fell on all fours, heart pounding, sweat stinging my eyes, blinding me. I drew a series of deep breaths and realized the air outside was, in some measure, also contaminated.

"Sweet Jesus," Rosa said as he collapsed beside me and slipped off his empty air tank. "I've never seen such blackness."

"Where're the lieutenant and Sullivan?"

Rosa gestured to the telephone company building.

Lieutenant Pullano came out of the lobby next, sat down beside me. "I owe my life to Patty Sullivan." He gasped for breath. "Patty shared his oxygen with me."

I wiped the sweat from my eyes as best I could, scanned the mass of fallen firefighters. Patty Sullivan was not among us. "Where is he?"

"He was behind me." Pullano looked to the building entrance, waiting for Sullivan to come out.

He did not.

Pullano pulled out his radio. "Ladder 3 Alpha, What is your location, K.?" (Ladder 3 Alpha was the radio code assigned to Sullivan.) "Ladder 3 Alpha, come in." We waited. No response. "Patty Sullivan, talk to me."

"Patty's in trouble," I said.

"Patty Sullivan owes me five dollars," Rosa added.

Rosa, Pullano, and I struggled to our feet, raced back to the New York Telephone Company building and were nearly trampled by a surge of frenzied firefighters escaping the lobby.

"What's going on?" Pullano asked a fire captain as we changed our oxygen tanks.

"South wall's cracked," the captain said.

"But that's reinforced concrete," Pullano said.

"Temperature's 1,400 degrees Fahrenheit on the first floor, 1,300 on the second. Chief's predicting a possible catastrophic building collapse."

"One of our men is missing," Pullano informed the captain.

"He's low on oxygen," I said.

"Hell, he's *out* of oxygen," Pullano said.

"Right." The captain took out his radio. "I'll see if the chief will authorize a rescue."

"There's no time," Pullano shouted back to the captain as we pushed past, sliced through the escaping firefighters, and entered the lobby.

The first thing I noticed was that a plume of thick, black smoke was billowing from the main stairwell, which had become a literal chimney to the subcellar. The stairwell was impassable.

"He won't even know street level," Rosa said.

"Then he's finished," Pullano declared.

I felt sick to my stomach. It would take a miracle for Patty to survive.

"Even if we could make it down," Rosa said, "we could pass him on that stairwell and never know it."

"We could try spreading out," Pullano suggested.

"Lock arms?" I said.

"Yeah," Pullano agreed, "that might work." Pullano pulled out his radio. "Ladder 3 Alpha. Patty Sullivan, what is your location, K.?"

Noise—a grating sound like twisting steel came from somewhere overhead. I looked at Pullano and Rosa. They stood dead still, eyes wide as Frisbees watching the ceiling above us.

"C'mon, Patty," Pullano said into the radio. "Answer me for Christ's sake. What is your location?"

Silence.

Then static, a faint transmission—"3 Alpha to 3."

"That's him!" Rosa yelled.

"Patty, where the hell are you?" Pullano asked.

"I'm somewhere on the main staircase heading up to the lobby. Do *not* attempt rescue." Patty Sullivan coughed uncontrollably—his Scott mask had to be off. "Repeat," he gasped, his voice growing weaker with each word. "Do *not* send anyone. I'm on my way up."

A sudden rumbling sound followed by a minor ground tremor came from the direction of the stairwell. A surge of soot bellowed out—a smoke explosion had detonated somewhere below.

Patty Sullivan's radio went dead.

Pullano, Rosa, and I charged into the stairwell.

The heaving plume of searing black smoke struck us like a gusting wind, forcing us to step back. We grabbed onto the stairwell handrails. Reset our footing. Leaned into the current of hot air. Locked arms. Fanned out. And took the first, tentative, blind step down into hell.

Thirteen steps below, Pullano tripped over a prone firefighter—Patty Sullivan. I pulled my glove off, felt for a pulse. Sullivan was alive. His face mask lay beside him. Pullano took a deep breath, held it, slipped off his own Scott mask and fixed it to Sullivan's face. Rosa and I took turns sharing our oxygen with Pullano.

Then the rising current of muddy air became markedly stronger.

"Hurry," Pullano said.

We lifted Sullivan, carried him, slipping and sliding, up the wet stairs, through the lobby, and out to the street as a force-five rush of inky soot surged around us.

—*Danny Noonan (Firefighter, retired, FDNY); http://staylow.com*

Adapted from "Red Star of Death"

MAKING IT HAPPEN

1. Read Deborah Wallace's book *In the Mouth of the Dragon: Toxic Fires in the Age of Plastics*.

2. Visit the Web site of the Cyanide Poison and Treatment Coalition (http://www.cyanidepoisoning.org) and review the case studies and information available.

3. Check your local presumptive legislation and drill your company on the ramifications of their coverage. The IAFF's Web site (http://www.iaff. org/HS/PSOB/infselect.asp) can help you to determine coverages.

STUDY QUESTIONS

1. What is the most overlooked means of dealing with the modern smoke environment?

2. What type of products were the primary sources of smoke production in the fires of the past?

3. Name at least three of the products of wood-based combustion.

4. What is the typical color and odor that emanates from hydrogen cyanide gas?

5. Historically, how has cyanide gas been used as a tool of death or terrorism?

6. According to Deborah Wallace, do gases have a higher toxicity during decomposition or actual burning?

7. Provide two reasons why CO can cause a firefighter to become incapacitated.

8. Why does hemoglobin become compromised by CO exposure?

9. What components of smoke will be decreased by filter-breathing?

10. Will filter-breathing eliminate gases such as CO and hydrogen cyanide?

11. What is the best way to describe a firefighter's allowing him- or herself or any crew member to inhale products of combustion at fires?

NOTES

[1]   Wallace, Deborah. *In the Mouth of the Dragon: Toxic Fires in the Age of Plastics.* Avery Publishing Group, Garden City Park, New York.

[2]   The Seattle Fire Department's SCBA training guide.

[3]   Dugan, Mike. The Complete Truck Company Lecture.

[4]   Fire, Frank L. *Combustibility of Plastics.* Van Rostrand Reinhold, New York, New York.

[5]   Port, Bob. 2004. Red star of death. *New York Daily News.* March 14.

[6]   Wallace, *In the Mouth of the Dragon,* p.13.

[7]   Alcorta, Richard. "Smoke Inhalation and Cyanide Poisoning". JEMS Communications. Summer 2004. Pp. 6–15.

[8]   Wallace, *In the Mouth of the Dragon,* p. 9.

[9]   Ibid., p. 35.

[10]   *You Need It Like A Hole in the Head.* 2005. Ergometics Applied Research Personnel Reference (DVD).

[11]   Ibid.

[12]   Certain material in this section has been adapted, with permission, from an article by Tim Valencia. We are all indebted to him for his continued work on behalf of firefighter safety. Valencia, Tim. 2005. Presumptive medical conditions for female fire fighters: Initial consideration. May 25.

PART 2

THE MANDATE

CHAPTER
6

NFPA 1404 AND AIR MANAGEMENT

INTRODUCTION

The NFPA is the guiding light that has provided standardized methodologies for how the U.S. fire service does business. These standards address how we dress for emergencies, build fire stations, provide adequate staffing, and safely provide training—to give only a few examples. There is far too much historical evidence, in the form of NIOSH, NFPA, and USFA reports, demonstrating that firefighters either have been injured or died as a result of noncompliance with the standards set forth by NFPA. Cited in this chapter are examples of fire department errors and the price paid.

This chapter provides insight into how the NFPA has made changes to improve firefighter safety through significant revision of the respiratory standard utilized for training on SCBA. The most important issue raised in NFPA 1404 (2006 edition) is a requirement that departments establish an air management program.

WHAT THE NFPA STANDARDS MEAN FOR YOUR DEPARTMENT

Fire department standard operating procedures (SOPs) are developed on the basis of a number of factors, including federal, state, and local laws; past practices; and national standards. The most recognized of these standards are those of the NFPA. The NFPA standards run the gamut from how to build fire stations to the protective garments that firefighters must wear for structural firefighting and more.

All regulations are, by nature, reactive documents. In general, something negative has to happen for these documents to be revised or updated; this holds true for the NFPA standards. For the most part, the NFPA updates the standards on a cyclic schedule and addresses firefighter tragedies and other catastrophic events in its changes. A cause for the tragedy or event is recognized, and the corresponding standard is changed in the hope that history doesn't repeat itself (fig. 6–1).

The NFPA's mission is to "reduce the worldwide burden of fire and other hazards on the quality of life by providing and advocating consensus codes and standards, research, training, and education."[1] The NFPA was established in 1896 and has since served as the world's leading advocate of fire prevention and an authority on public safety. In fact, the NFPA's 300 codes and standards influence every building, process, service, design, and installation in the United States, as well as in other countries.[2]

Although these standards are not recognized as laws, for the most part, fire departments across the nation follow the NFPA standards as closely as possible. In the wake of any serious firefighter injury or death with the possibility of litigation, these standards carry legal weight when presented as evidence. For those fire departments that choose not to follow the NFPA's recommendations, it's only a matter of time before the legal ramifications arrive at their doorstep. Never has a judicial system made a legal ruling against any NFPA standard.

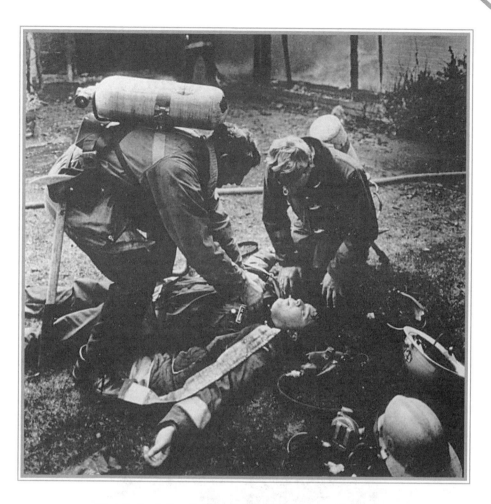

*Fig. 6–1. A firefighter injured at a working fire,
receiving medical aid from his crew (Photograph
courtesy Seattle Fire Fighters Union, IAFF Local 27)*

FIRE DEPARTMENT TRAINING AND NFPA STANDARDS

The importance of training is recognized by all fire service members, from the newest firefighter to the fire chief. Without training, we cannot safely and effectively solve Mrs. Smith's problems, in the terms of Chief Alan Brunacini. Learning is a cycle providing positive reinforcement of the basics. Firefighters learn as human nature dictates—first by hearing, then by seeing, and finally by doing. This step-by-step process is the time-proven methodology for learning new skills pertaining to the fire service profession, as well as reinforcing retained skills (fig. 6–2).

Fig. 6–2. Firefighters in a classroom setting, learning new air management techniques through the tried-and-true step-by-step method of first hearing, then seeing, and finally doing (Photograph by Lt. Tim Dungan, Seattle Fire Department)

The NFPA recognized the need for standardization of several elements of training since firefighters were getting injured or dying in training events. Although well intentioned, many fire departments were not following recognized safe practices that other fire departments had learned the hard way. This is where the NFPA came in: the organization compiled records of firefighter injuries and deaths related to training and provided standards to address unsafe practices. For example, the NFPA established the following training standards:

- NFPA 1403—Standard on Live Fire Training Evolutions
- NFPA 1404—Standard for Fire Service Respiratory Protection Training
- NFPA 1410—Standard on Training for Initial Emergency Scene Operations

WHAT HAPPENS WHEN STANDARDS ARE NOT FOLLOWED?

NFPA 1403 was initiated in 1986, after a tragic live-fire training accident in 1982 that resulted in the deaths of two firefighters. The Committee on Fire Service Training was urged to address the issue of live-fire training evolutions in structures. NFPA 1403 has undergone several revisions over the years, and many fire departments have faced horrific losses and litigation due to noncompliance with this standard. The most effective tools the fire service can utilize in regards to firefighter safety are learning from history and complying with standards based on firefighter tragedies (fig. 6–3).

*Fig. 6–3. Firefighters complying with NFPA 1403, receiving a
live fire preburn briefing from the IC and the safety officer
(Photograph by Mike Wernet, South Kitsap [WA] Fire and Rescue)*

Yet another grievous example of not following NFPA 1403 comes from the death of a recruit who died in live-fire training in the City of Baltimore in early 2007. The preliminary internal report from the fire department provided overwhelming evidence that NFPA 1403 (2002 edition) was not followed. The internal report goes on to list over 30 items of possible noncompliance within the standard. For example,

- Multiple fires set in the acquired training structure
- Lack of sufficient safety personnel strategically positioned throughout the building
- Failure to conduct a walk-through of the building with all participants prior to the exercise

- Lack of a preburn plan and a preburn briefing
- Lack of radio communication between units, the incident command system, and the safety officer

As of this writing, one dismissal at the division chief level and two time-off-without-pay penalties at the company officer level have been issued in connection with this training accident. Litigation is still pending.

What would be obvious in normal circumstances was lost during the course of events that unfolded during this training situation. Failure to follow the recognized national training standard was certainly a factor contributing to the firefighter fatality. This department will suffer emotional and financial repercussions as the facts surface and the litigation progresses. Regardless of how the department changes, how the legal cases are resolved, or how the money is spent, the sad fact remains that her eleven year-old son and eight year-old daughter will grow up without their mother as a result of predictable and preventable mistakes.

AIR MANAGEMENT TRAINING

I am a training officer with Portland (OR) Fire and Rescue and have been with the fire service for 27 years. Most of that time has been as a line firefighter or a line officer. I am honored to have served as chair of the NFPA 1981 and as a member of NFPA 1404.

After years spent trying to develop the perfect RIT system and SOPs to respond to that Mayday we all hope will never come, it became clear that many of those near misses and Maydays were not the result of being trapped by an aggressively moving fire, a catastrophic collapse, or facilitation of an imminent rescue. Although these incidents are part of what we face on a daily basis as firefighters, it is not why so many of us have come so close to our own deaths. *Unfortunately, it is simply the result of running out of air.*

How can this be? With all the risks we face, with all the good we do, how can we accept that our brothers and sisters are dying just because they run out of air? What a terrible shame. In scuba diving, there is a culture of discipline that would never allow the use of reserve air for anything other than a true emergency. We in the fire service also have emergency air, but regrettably our culture has endorsed using this precious reserve as "get-out air."

Generally, no air is left for any unforeseen mishap. The low-air alarm has become the indicator that it is time to start thinking about getting out. We first check to see if it is our own alarm that is ringing. If it is, we pull a little more Sheetrock, do just a little more work, and then we head out. That is when it happens: We get a little twisted up—not much, just a little—and have no air left.

Is it possible to change a whole culture? I thought that I was alone in my convictions until I received a call from my colleagues in the Seattle Fire Department. What encouragement and hope they brought! I was asked to attend one of their Point of No Return air management classes. I was deeply impressed. My ideas were still rattling around in my head, whereas they had already developed a training program that really works.

The next step, completed with help from the Seattle Fire Department, was to write and submit a proposal to NFPA 1404 requiring that fire departments develop an individual air management training program. This training program also identified the need for firefighters to exit the IDLH environment before activation of their low-air alarms.

The proposal was received by NFPA 1404 and brought to the floor for discussion. I think you can picture how it was first received. I was not very popular, but after many very long and heated discussions, we finally just faced the facts, knew the proposed standard was right, and put aside our emotions. We ended up voting unanimously in favor of the new proposal.

We must not rely on emergency air as get-out air. If we do, we will continue to see thousands of close calls, injuries, and deaths because of the dangerously reduced safety margin of time when the unexpected occurs. We must train for worst-case scenarios, especially when the surrounding environment is so unforgiving. Although new technology can help as an air management tool, we must remember that every piece of technology that we carry into the IDLH environment can fail at one time or another. This includes low-air alarms and heads-up displays.

With my own career reaching an end, I am thankful that the next generation of firefighters will inhabit a new culture. Unfortunately, they will still face the possibility of giving their lives for those they serve. However, with this new focus on air management, the ultimate price will hopefully be paid only when absolutely necessary.

To my dear brothers Mike, Steve, Phil, and Casey, thank you for displaying the courage to do what you knew was right. God bless you.

—Dan Rossos (Portland [OR] Fire and Rescue)

NFPA 1404: ARE YOU COMPLYING?

The training standard for SCBA utilization is NFPA 1404.[3] As with all NFPA standards; this pivotal training standard has undergone recent revision. NFPA 1404 was initiated in 1989 and updated in 1996, 2002, and 2006. Implementation of the 2006 edition has instigated many significant changes in SCBA usage. This section outlines these changes in how the fire service should be training on and utilizing SCBA.

Chapter 5 of NFPA 1404 states (in 5.1.4),

> *The Authority Having Jurisdiction (AHJ) shall establish and enforce written standard operating procedures for training in the use of respiratory protection equipment that shall include the following:*
>
> *(1)   When respiratory protection equipment is to be used*
>
> *(2)   Individual air management program*
>
> *(3)   Emergency evacuation procedures*
>
> *(4)   Procedures for ensuring proper facepiece fit*
>
> *(5)   Cleaning of respiratory protection equipment components*
>
> *(6)   A policy for changing respiratory filters and cartridges*
>
> *(7)   A policy defining the end of service life for all types of filter or cartridge-type respirators*
>
> *(8)   A policy defining the proper storage and inventory control of all respiratory protection equipment*

Numbers (2) and (6)–(8) are all additions to the 2006 edition. Notably, number (2) requires an air management program. Appendix 5.1.4(2) states,

> *This program will develop the ability of an individual to manage his or her air consumption as a part of a team during a work period. This can require members to rotate positions of heavy work to light work so air consumption is equalized among team members. The individual air management should include the following directives:*
>
> *(1)   Exit from an IDLH atmosphere should be before consumption of reserve air supply begins*
>
> *(2)   Low-air warning is notification that the individual is consuming the reserve air supply*
>
> *(3)   Activation of the reserve air alarm is an immediate action item for the individual and the team*

In summary, firefighters need to be out of the IDLH atmosphere with their emergency reserve intact, and if firefighters work into their emergency reserve, it is an immediate-action item. The emergency reserve is the remaining 25% of any SCBA cylinder, regardless of size or manufacturer.

What other immediate-action items are there on the fireground in regards to firefighter safety? The answer is that in addition to the emergency reserve, there are only two other immediate-action items: a Mayday situation and an activated PASS device. Together these three items mark a significant change in behavior in SCBA use by the fire service. As mentioned in chapter 4, asphyxiation is the cause in 63% of firefighter fatalities on the fireground (table 6–1). Leaving the IDLH atmosphere with the emergency reserve intact represents an enormous step toward reducing this staggering figure.

Table 6–1. The hard, cold facts regarding the number of firefighters that have died due to asphyxiation in structure fires (Source: USFA, annual reports)

Year	Number of Firefighter Deaths Due to Asphyxiation
2005	6
2004	5
2003	6
2002	15
2001	18
2000	13
1999	16
1998	15
1997	15
1996	5
1995	20
1994	29
1993	4
1992	6
1991	10
1990	20
1989	11
1988	13
1987	13
1986	14

Chapter 5 of NFPA 1404 states (in 5.1.6 and 5.1.7),

> *The Authority Having Jurisdiction (AHJ) shall establish written training policies for respiratory protection. Training policies shall include but not be limited to the following:*
>
> *(1) Identification of the various types of respiratory protection equipment*
>
> *(2) Responsibilities of members to obtain and maintain proper facepiece fit*
>
> *(3) Responsibilities of members for proper cleaning and maintenance*

(4) *Identification of the factors that affect the duration of the air supply*

(5) *Determination of the air consumption rate of each member*

(6) *Responsibilities of members for using respiratory protection equipment in hazardous atmosphere*

(7) *Limitations of respiratory protection devices*

(8) *Battery life limitations and recharging requirements of PAPRs*

The parts that have changed from the 2002 edition are numbers (5) and (8). The 2006 edition requires that firefighters determine the air consumption rate of each member. Appendix 5.1.7(5) states,

> *The air consumption rate will be different for each individual. Some factors that determine the individual's air consumption rate are as follows:*
>
> (1) *Physical fitness and condition*
>
> (2) *Size and weight of the individual*
>
> (3) *Work being performed*
>
> (4) *Environment where work is being performed*
>
> (5) *Other stressors (e.g., people trapped, difficult access, outside temperatures)*
>
> (6) *Type of clothing used*
>
> (7) *Training*

Training is essential. The air consumption rate is recognized through hands-on training during task-related firefighting activities in an IDLH atmosphere. The only method firefighters have to determine air usage rates is to utilize SCBA in a wide variety of training activities while regularly monitoring air usage (fig. 6–4).

Fig. 6–4. Firefighters receiving SCBA air management training during hands-on training in an acquired structure. Through regular SCBA training, firefighters determine their air usage rate, as well as other team members' use rates.

Additionally, the 2006 edition requires that firefighters exit the IDLH atmosphere *prior* to utilizing the SCBA emergency reserve. This will effectively eliminate the low-air alarm, as well as the resultant false-alarm mentality, from the fireground. NFPA 1404 also establishes that the low-air alarm indicates the use of the emergency reserve, rather than a time-to-exit alarm. Use of the emergency reserve is an immediate-action item for the individual and the team.

In the event that a low-air alarm is activated, the first line of defense should be members of your team or other teams operating in close proximity. There have been numerous close calls and deaths because low-air warnings were ignored owing to the false-alarm mentality. What was once common practice on the fireground is no longer acceptable; low-air alarms should not be ignored.

HAZARDS OF NONCOMPLIANCE WITH NFPA 1404

As detailed in chapter 18, the fire service knows that it takes time for the efforts of Rapid Intervention Teams (RITs) to be successful. The reality of rapid intervention is that if NFPA 1404 is not followed and firefighters work into their emergency reserve, they significantly reduce their survivability profile. An intact emergency reserve provides critical time for RIT notification, activation, entry, location, protection, and extraction.

In addition, the same holds true for other catastrophic events—such as structural collapse, getting lost or separated, rapid change in thermal conditions, or toxic smoke (see chap. 4). In these rapidly changing situations, firefighters with an intact emergency reserve have significantly increased potential to survive.

NFPA 1404 AND BUDDY-BREATHING

Another interesting change to the 2006 edition of NFPA 1404 is the following quotation, from Appendix 4.2.7, that addresses the practice known as *buddy-breathing*:

> *Several manufacturers of SCBA currently market "buddy" or rescue breathing devices as a component of their SCBA. NIOSH has no guidelines and does not test or certify respirators when being used to give or receive breathing support. However, OSHA 29 CFR 1910.156(f)(l)(iii) states, "Approved self-contained breathing apparatus may be equipped with either a buddy-breathing device or quick disconnect valve, even if these devices are not certified by NIOSH. If these accessories are used, they shall not cause damage to the apparatus, or restrict air flow of the apparatus, or obstruct the normal operations of the apparatus." Until and unless NIOSH approves an auxiliary device for buddy breathing, the practice of buddy breathing is not endorsed in any way by the Technical Committee on Fire Service Training. The practice of passing the SCBA facepiece back and forth between two users is considered unsafe because highly toxic air contaminants can enter the facepiece during the exchange of the facepiece.*

As noted in chapter 5, the toxic smoke in today's fires is potentially deadly. Buddy-breathing is an extremely risky task because of possible exposure to carbon monoxide, hydrogen cyanide, benzene, and other disabling or even deadly products of combustion. The easiest way to not become a victim of these deadly smoke products is to manage the air in your SCBA and follow the ROAM (see chap. 10) and the parameters of NFPA 1404, which requires that firefighters exit the IDLH atmosphere with the emergency reserve intact.

MAKING IT HAPPEN

1. Train all firefighters on the newest requirements of NFPA 1404, as outlined in this chapter.

2. Have firefighters perform a variety of tasks while using their SCBA, so that all individuals will have a clearer recognition of their air consumption rate.

3. Have firefighters practice the ROAM through training exercises in which all members must exit the simulated IDLH area with the emergency reserve intact.

4. Train all firefighters on the advantages and disadvantages of buddy-breathing.

STUDY QUESTIONS

1. How are fire department SOPs developed?

2. Why did the NFPA develop training standards?

3. Name the parts of NFPA 1404 that require fire departments to have an air management program.

4. According to appendix 5.1.4(2) of NFPA 1404, how can members maximize the team's effectiveness when working in an IDLH environment?

5. What is the emergency reserve of any SCBA cylinder?

6. What are the three immediate-action items pertaining to firefighter safety on the fireground?

NOTES

[1] National Fire Protection Association. http://www.nfpa.org.

[2] Ibid.

[3] NFPA 1404—Standard for Fire Service Respiratory Protection Training. 2006 edition.

CHAPTER 7

THE PHOENIX FIRE TRAGEDY: BRET TARVER'S LEGACY

INTRODUCTION

On March 14, 2001, 40-year-old Bret Tarver, a firefighter-paramedic in the Phoenix Fire Department, died in the line of duty while battling a fire at the Southwest Supermarket. Bret Tarver became separated from his crew, ran out of air in his SCBA, took off his facepiece, and breathed the toxic and poisonous gases that are present in smoke (see chap. 5). Bret Tarver died of carbon monoxide poisoning.

This chapter recounts the Southwest Supermarket fire and the tragedy of Bret Tarver's death in detail, based on the NIOSH Fire Fighter Fatality Investigation and the Phoenix Fire Department's own final report of the Southwest Supermarket fire.[1] This chapter reviews the details of the fire and discusses the important lessons that can be drawn from this line-of-duty death. Hopefully, every firefighter will learn something from Bret Tarver's death. These lessons represent Bret Tarver's gift to the U.S. fire service. This chapter is respectfully dedicated to the memory of Bret Tarver.

THE SOUTHWEST SUPERMARKET FIRE

The Southwest Supermarket was constructed in 1956 and was approximately 20,132 square feet. The foundation was concrete slab, with no sublevels. The walls of the supermarket were constructed of 12" × 4" × 16" concrete cinder blocks (unreinforced CMUs). The structural framework for the roof assemblies consisted of steel columns and beams with open-web steel trusses that were 6 ft on center. The roof system was composed of open-web steel trusses covered with ½" plywood decking, foam insulation, and tar asphalt roofing—a typical lightweight, panelized roof. The ceiling height was about 20 ft throughout the supermarket. There were second-floor offices and storage areas above the coolers and the first-floor storage areas. The building was L-shaped and was part of a strip mall (fig. 7–1). The supermarket did not have a sprinkler system.

Everyone should read the NIOSH report (Fire Fighter Fatality Investigation Report F2001-13) of Bret Tarver's death at the Southwest Supermarket fire. Therefore, the NIOSH report has been used as the source of the following details and the time line of events. Much of the actual text of the report has been included so that the facts are presented correctly and clearly.

On Wednesday, March 14, 2001, a caller telephoned 911 to report a fire in a pile of debris at the rear of an Ace Hardware store, near the back loading dock of the Southwest Supermarket.

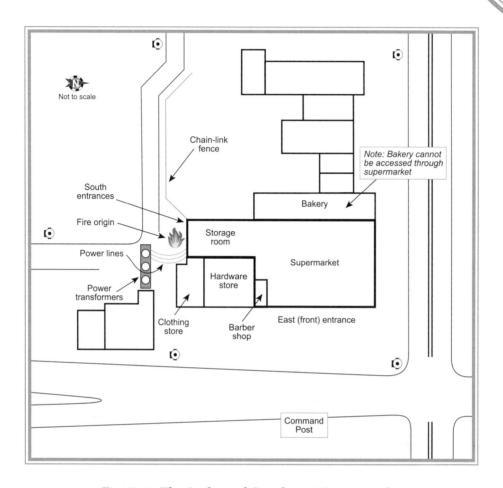

Fig. 7–1. The L-shaped Southwest Supermarket
(Source: NIOSH Fire Fighter Fatality Investigation F2001-13)

At 1645 hours, Engine 24 was dispatched to an exterior cardboard fire at the rear (the south side) of the supermarket (fig. 7–2). This fire is believed to have been arson, since during the course of the investigation, no accidental causes were discovered. The fire was exposing both the supermarket and the electrical service drop that ran into the supermarket. The Hazmat 4 engineer (who was in the area while returning from another call) self-dispatched to the alarm, saw the exposure problems, and requested two engines and a ladder company. Dispatch balanced the alarm with Engine 21, Ladder 24, Battalion Chief 3, and Rescue 25. Engine 14 and Rescue 21 self-dispatched in response to the alarm.

Fig. 7–2. The Southwest Supermarket fire, which originated as a rubbish fire but quickly moved into the structure (Source: NIOSH Fire Fighter Fatality Investigation F2001-13)

At 1702 hours, Engine 14—on which Bret Tarver was working that day—arrived. The crew from Engine 14 searched the main shopping area of the supermarket and encountered a light haze of smoke banked down about four feet below the ceiling. As they moved from the front toward the rear of the store, they found that the smoke was getting thicker and hotter. The officer from Engine 14 reported the conditions and told the IC that Engine 14 was going to stretch a hand line into the supermarket.

Nine minutes later, Tarver and the crew from Engine 14, along with an officer and two firefighters from Engine 3 and the crew from Rescue 3, advanced two hand lines through the front (the east side) entrance to the supermarket. Meanwhile, the crews from Rescue 25 and Engine 24 were advancing two hand lines through the roll-up door, on the south side, into the storage area, where they encountered heavy smoke all the way to the floor. The IC informed the south-sector officer that Engine 14 was operating on the opposite side of the storage area and that Rescue 25 and Engine 24 should be aware of opposing hose streams. (fig. 7–3)

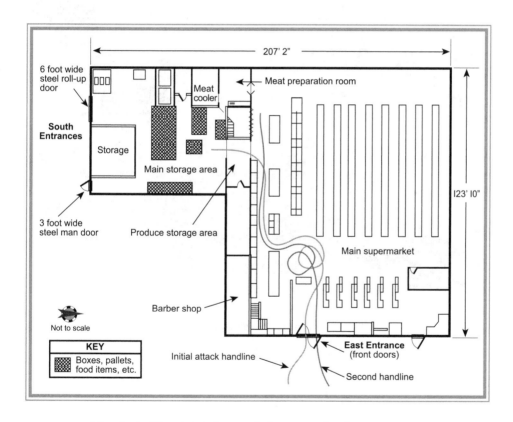

Fig. 7–3. The initial attack line inside the structure. While Engine 14 were met with light smoke initially, as they advanced farther into the structure, the smoke got thicker, and they encountered more heat. (Source: NIOSH Fire Fighter Fatality Investigation F2001-13)

At 1714 hours, Engine 14, Engine 3, and Rescue 3 advanced a hand line into the produce storage area, and Engine 3 moved the second hand line through the supermarket toward the meat preparation room (fig. 7–4). Conditions in the supermarket began to deteriorate, and the structure filled with thick, black smoke.

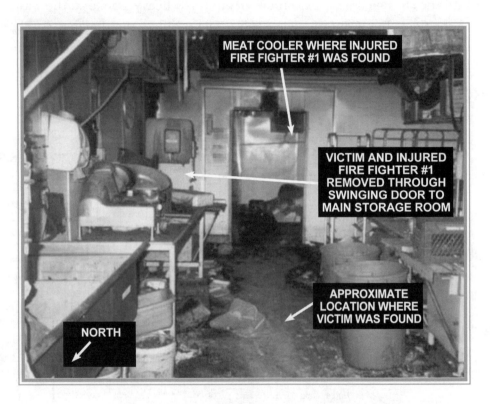

Fig. 7–4. The meat preparation room, where several firefighters encountered Bret Tarver (Source: NIOSH Fire Fighter Fatality Investigation F2001-13)

While the crews from Engine 14, Engine 3, and Rescue 3 were operating in the produce storage area, they began to encounter intensifying heat and thick, black smoke down to the floor level. The fire along the ceiling was growing. The crews from Engine 14, Engine 3, and Rescue 3 pulled the ceiling and searched for fire extension.

It was at this point in the incident that Bret Tarver told his officer that he was low on air. The officer pulled together the crew from Engine 14 and told them that they would follow the hand line out of the supermarket as a crew. At this time, another Engine 14 crew member reported that he was low on air as well.

The engineer, who was on the nozzle of the hoseline that Engine 14 had advanced into the supermarket, led the exiting crew, followed by Tarver, a fellow firefighter, and the officer. As the Engine 14 crew followed the hand line out of the produce storage area, Tarver and the firefighter behind him fell over some debris and became separated from the hand line. The officer from Engine 14, who was still in the storage area, fell backward over the debris and became disoriented; the officer from Engine 3 found him and put him back on the hand line and sent him in the direction of the exit.

Meanwhile, Tarver and the other firefighter from Engine 14 got to their feet and walked into a wall and then fell again. It is believed that Tarver and the Engine 14 firefighter had stumbled into the main supermarket shopping area and were heading toward the meat preparation room. As Tarver started to move quickly through the shopping area, the Engine 14 firefighter grabbed his coat to stay with him.

About the same time, as he was heading for the exit along the hoseline, the Engine 14 officer's low-air alarm activated, and he ran out of air near the front door (on the east side). As he left the building, he found the engineer who had led the crew of Engine 14 out of the supermarket.

Bret Tarver and the other firefighter from Engine 14 realized that they were lost and decided to radio a Mayday. The firefighter tried to radio the Mayday, but his transmission was unsuccessful. Tarver then successfully radioed the Mayday. The IC immediately told the crews of Engine 18 and Ladder 9 to assume rapid intervention duties. Hearing the Mayday over the radio, the officer from Engine 14 told the officer from Engine 21 to follow the hand line through the front door (on the east side) of the supermarket to the area where his crew had been. At this time, the Engine 14 officer mistakenly believed that only one member of his crew was missing, not two members.

Back inside the supermarket, the lost firefighter from Engine 14 became separated from Tarver near the meat preparation room. The lost firefighter reentered the meat preparation room just as he ran out of air in his SCBA. He heard the radios of other firefighters—an officer and a firefighter from Engine 34—and moved toward them in the main storage area. He became debilitated by the smoke and was assisted out through the door on the south side of the building by the officer and firefighter from Engine 34.

The officer and the firefighter from Engine 34 thought that they had rescued Bret Tarver. However, once they removed the Engine 14 firefighter from the building, they realized that Tarver was still inside, still missing.

At 1729 hours, Tarver radioed to request that crews not back out of the building because he needed help. The IC radioed back, asking for his location. Tarver replied that he was in the rear of the supermarket, behind something, out of air, and down on the ground, breathing in smoke. Bret Tarver was lost in heavy smoke, breathing in all the poisons that comprise today's smoke. The IC radioed back, telling Tarver to stay calm and informing him that crews were on their way.

About a minute later, the officer from Engine 21 heard someone yelling from the produce storage area. He followed the yelling and ran into Bret Tarver, who was standing near one of the initial attack lines in the main supermarket area. The officer from Engine 21 grabbed Tarver, asked him to identify himself, and tried to put him on the hand line. Tarver was out of air and had removed the regulator from his facepiece. He was disoriented and resisted getting to his knees. On a second attempt, the officer from Engine 21 got Tarver to his knees and on the hand line. However, Tarver soon stood up, turned, and quickly moved toward the rear of the supermarket in the thick smoke. It is believed that Tarver headed back toward the meat preparation room. Tarver was most likely suffering from hydrogen cyanide poisoning, which would explain his bizarre behavior. Hydrogen cyanide gas in smoke is notorious as an escape inhibitor.

A firefighter from Rescue 3 who was in the main storage area heard Tarver's yelling, moved toward the sound, and found Tarver just outside the swinging door that led from the main storage area into the meat preparation room. Tarver told the Rescue 3 firefighter that he was out of air, and the firefighter told him to stay calm and follow him. However, Tarver turned and moved away from the Rescue 3 firefighter, heading toward the meat preparation room. The Rescue 3 firefighter grabbed Tarver and told him that he was headed the wrong way. As Tarver turned, he knocked the Rescue 3 firefighter down, and the two became separated in the black smoke. Getting back onto his feet, the Rescue 3 firefighter tried to find Tarver but ran out of air and was forced to leave the supermarket.

During this time, the IC made numerous attempts to contact Tarver to tell him to activate his PASS device. However, Tarver did not respond to the IC's radio transmissions.

Another firefighter from Rescue 3, who was low on air, heard Tarver's voice coming from the direction of the meat preparation room. He followed the voice and found Tarver in the meat preparation area. The firefighter from Rescue 3 radioed the IC that he had found Tarver, that they were by themselves, and that they were both out of air. The Rescue 3 firefighter, after running out of air, removed his facepiece and pulled his hood over his face.

At 1736 hours, the IC thought that Bret Tarver had been rescued and removed from the supermarket. However, Tarver was still inside the building. The firefighter from Engine 14 who had been with Tarver had again been mistaken for Tarver. The IC ordered all firefighters out of the building, because he was changing firefighting operations from offensive to defensive.

After crews exited the building, the fire took over the front of the supermarket (the east side), preventing any further entry through the front door. From this point on, all rapid intervention crews entered and exited through the two doors on the south side of the building.

The officer from Engine 25, who was operating in the main storage area, followed a path to the meat cooler and found the firefighter from Rescue 3 who was out of air. The officer from Engine 25 radioed the IC that he had one firefighter down and that he needed assistance. He made another radio transmission, saying that he and his crew had found Bret Tarver and that they were bringing him out. (Note that when the Engine 25 officer referred to the Rescue 3 firefighter as "one firefighter down," his crew and the IC assumed that he was referring to Bret Tarver.) The Engine 25 officer passed the Rescue 3 firefighter to his crew in the main storage area, and they took him out of the building through one of the doors on the south side. This Rescue 3 firefighter was in full respiratory arrest; he was treated on the scene and transported to the hospital via medic unit.

At 1739 hours, the IC ordered all firefighters off the roof and out of the building, because they were going defensive as soon as the victim was rescued. Meanwhile, the officer from Engine 25 located Bret Tarver in the meat preparation room. Tarver was unresponsive, and his PASS device was sounding. The Engine 25 officer, who was alone with Tarver, made an emergency radio transmission that was not received by the IC. At 1740 hours, an officer from Engine 4 heard the transmission and radioed that a firefighter was down in the southwest corner. The Engine 25 officer, unable to move Tarver, ran out of air, removed his facepiece and regulator, and attempted to crawl out of the building.

At the same time, crews from Engine 6 and Engine 710 entered the south side of the supermarket and moved toward the produce storage area. As crew members from Engine 6 neared the produce storage area, they heard a PASS device sounding from the meat preparation room. The Engine 6 crew found the officer from Engine 25 (his PASS device was not sounding) and passed him to the Engine 710 crew, who assisted him out of the building. The Engine 25 officer was transported and treated for smoke inhalation.

Engine 6 continued toward the sounding PASS device and found Bret Tarver lying unconscious on his back with his facepiece partially removed. They checked him for a pulse but could not find one.

Many crews worked to remove Bret Tarver from the Southwest Supermarket. These firefighters were hindered in their removal efforts by Tarver's size (he was 6'4" tall and weighed more than 300 pounds with all his gear on). They were also hampered by the amount of debris blocking their path through the main storage area to the south-side doors (fig. 7–5). Approximately 19 minutes elapsed from the time when the Engine 6 crew found Tarver and when they finally removed him from the building.

Fig. 7–5. The main storage area of the supermarket, full of sacks, boxes, plastic containers, and pallets of food (Source: NIOSH Fire Fighter Fatality Investigation F2001-13)

When Bret Tarver was brought outside the building, firefighters immediately attempted cardiopulmonary resuscitation and initiated advanced life support. Unfortunately, these efforts did not work. Bret Tarver was dead.

The medical examiner listed the cause of Bret Tarver's death as thermal burns and smoke inhalation. Tarver's carboxyhemoglobin level was listed as 61% at the time of death (fig. 7–6).

Fig. 7–6. The Southwest Supermarket, during the rescue effort to find and extricate Bret Tarver (Source: NIOSH Fire Fighter Fatality Investigation F2001-13)

THE SOUTHWEST SUPERMARKET FIRE

The 2001 Southwest Supermarket fire was the most significant tactical challenge in the history of the Phoenix Fire Department (organized in 1887). The department had actively developed, practiced, and refined incident command, fireground safety, and effective tactical procedures since the mid-1970s. The profound lessons of the Southwest Supermarket incident show that we had previously applied these operational systems to the fires in single-family houses fires that we fought most often. Those systems were quickly outperformed by the size, fire loading, and complexity of the old, commercial store. By exceeding our house fire approach, the situation quickly became fatal for our firefighters.

We quickly found ourselves in offensive positions under defensive conditions. Bret Tarver (huge guy) was down and nailed to the floor. Our firefighters (huge risk takers when the chips are down) were not going to leave him inside. The rescuers ran out of air, the IC processed multiple Maydays, and things pretty much sucked on 35th Avenue. This was the worst day that a fire department could possibly have.

During the five-year recovery period after Bret's death, we started training in big buildings. Based on those drills and the development of our command training center, we revised the applications of our operational and command systems to more effectively deal with large fire areas.

Developing these new procedures instilled a respectful understanding of our actual tactical capabilities and limitations, particularly in large fire areas. This recovery response produced a more effective connection between water application and SCBA air management. We now appreciate that if our hose streams (in any building) do not quickly (and offensively) control the fire, then we are in for a longer firefight, which will typically exceed the air supply of the initial attackers. If the IC cannot continuously rotate firefighter teams through the hazard zone, then the operation should immediately become defensive and all firefighters should be accounted for and moved outside the collapse zone.

We must stay aware that any situation requiring SCBA is an inherently high-hazard operation. In fact, wearing SCBA in the products of combustion creates a definitive (and very accurate) prediction of the life expectancy of the wearer. Merely running out of air in a fire-based hazard zone is consistently fatal. Anytime there is any question about an ongoing and adequate air supply, the IC must automatically write off the entire (involved) fire area. Air management must become a major fireground factor in determining, managing, and maintaining the overall strategic mode (offensive vs. defensive).

Another important lesson is that we must always apply standard (commercial) firefighting procedures to routine house fire incidents. These frequent, smaller events provide both the opportunity for us to practice and refine safe practices and the focus to develop effective operational habits. We must use the little deals to get ready for the big deals—that is to say, if we run out of air 25 ft inside a house fire, then we will definitely run out of air 125 feet inside a commercial building.

—*Alan Brunacini (Chief, retired, Phoenix Fire Department)*

LESSONS LEARNED

Bret Tarver got lost in heavy smoke, ran out of air in his SCBA, took off his facepiece, and breathed in poisonous gases, including hydrogen cyanide and carbon monoxide. As stated in chapter 4, the five ways firefighters die in structure fires: smoke, getting lost or separated, thermal insult, structural collapse, and running out of air. Bret Tarver had three of the five working against him—he was in heavy smoke, got lost, and ran out of air. Consequently, he did not survive.

Firefighters must understand that the air in their SCBA is critically important to survival in structural firefighting. If firefighters run out of air in a structure fire, their chances of being injured or dying increases exponentially. Firefighters must manage their air so that they have an emergency reserve should they get lost, trapped, or find themselves in other unforeseen situations (See chapter 10 for a complete discussion of the solution to the air management problem—the ROAM).

Shortly after Bret Tarver's death, the Phoenix Fire Department and NIOSH recognized that firefighters need to understand air management and be trained on air management, both individually and as a crew. Recommendation 4 from NIOSH's investigation into Tarver's death states that, "fire departments should ensure that fire fighters manage their air supplies as warranted by the size of the structure involved."[2]

This was the first time that NIOSH had ever recommended that firefighters manage their air supplies. After investigating Bret Tarver's death, NIOSH recognized that firefighters need to change their behavior— or learn to manage the air on their backs—instead of relying on technology to solve the problem. (For a discussion of technology and the inherent problems associated with relying on technology, see chap. 8.)

The NIOSH investigators understood that, depending on the individual firefighter's air consumption rate and the amount of time required in order to exit a hostile environment, the low-air alarm might not activate soon enough to provide adequate time to exit. In other words, firefighters should not use the low-air alarm as a signal to leave IDLH environment, but rather should monitor and manage their air. NIOSH recognized that in larger structures—such as high-rise buildings, warehouses, supermarkets, or any larger commercial building—firefighters may need to exit the hostile environment prior to the activation of the low-air alarm. The investigators also concluded that when conditions deteriorate and visibility becomes limited, it may take firefighters longer to exit as compared to the time it took them to enter the structure.

In their final report of this incident, the Phoenix Fire Department came to several very important conclusions. The most significant of these is that the window of firefighter survivability is directly related to the firefighter's SCBA air supply. The Phoenix Fire Department learned the hard way that the air firefighters bring with them on their backs into hazardous environments equals time, and that air is critical to firefighter safety and survival. This holds true both for the firefighters fighting the fire and for the rapid intervention teams (RITs) going in to rescue firefighters in trouble. Everyone inside an IDLH environment is limited by the air supply in their SCBA, and this air supply must be monitored constantly. A firefighter's survivability profile in a structure fire is directly related to the available air supply.

The Phoenix Fire Department did not merely document the difficult lessons learned from Tarver's death. Rather, they designed hands-on training programs that addressed air management, along with firefighter self-survival, fireground communications, RIT search-and-rescue techniques, and RIT deployment and responsibilities.

THE PHOENIX FIRE DEPARTMENT DEVELOPS AN AIR MANAGEMENT PROGRAM

The Phoenix Fire Department's air management training class was divided into two distinct areas: air management as a crew and air management as an individual. The Phoenix Fire Department wanted to teach their firefighters two important concepts. The first was that all crew members are responsible for keeping track of each other's air consumption and reporting it on a regular basis. And secondly, that individual firefighters have a responsibility to their crew, as well as everyone at the incident, not to run out of air and create a crisis situation. The Phoenix Fire Department also wanted every firefighter, officer, and chief officer to understand:

- Air management as it relates to firefighting
- Scene size-up as it relates to air management
- The work vs. air consumption equation
- The company officer's responsibilities as they relate to air management
- Critical decision making and behaviors should the individual firefighter or crew run low or out of air
- Emergency SCBA procedures

These training programs emphasized the tough lessons learned at the Southwest Supermarket fire: air management and critical decisions concerning firefighters' air supplies are just as important as fireground tasks.

Under the progressive and inspired leadership of Chief Alan Brunacini, the Phoenix Fire Department recognized the importance of air management on the fireground. In light of the recovery effort that culminated in Bret Tarver's death, the Phoenix Fire Department came up with procedure and policy solutions addressing their air management needs. As a result, the Phoenix Fire Department became the first big-city fire department in the U.S. fire service to adopt an air management program for its firefighters. Phoenix firefighters are safer today because they now practice air management at every incident at which they put on their SCBA.

BRET TARVER'S GIFTS TO THE FIRE SERVICE

Air management is one of the many gifts that Bret Tarver gave the U.S. fire service. Because of Bret, we now understand that all tasks on the fireground that occur in the IDLH environment—pulling and advancing hand lines, putting water on the fire, search and rescue, raising ladders, ventilation, and rapid intervention—depend on the air supply of the firefighters assigned these tasks. The reality of interior structural firefighting is this: everything depends on firefighters' air supply. Air is the limiting factor in all interior structural firefighting endeavors.

Bret Tarver left the U.S. Fire Service many other gifts as well. Because of Tarver's death, we now know that rapid intervention is not rapid and cannot be conducted with just a few firefighters. In a landmark article, Steve Kreis, assistant chief of the Phoenix Fire Department, explains how long it takes for rapid intervention teams to deploy to a firefighter emergency.[3] In Bret Tarver's case, 53 minutes elapsed from Tarver's first Mayday broadcast until he was brought out of the structure. Kreis argues that the reality of rapid intervention is far from the well-intentioned concept. The expectation of quickly and efficiently rescuing a lost, downed firefighter in a structure is not realistic. In the Southwest Supermarket, the theory of rapid intervention was put to the test, and it failed.

Kreis explains that the rapid intervention crews trying to save Tarver experienced problems ranging from running out of air themselves, to losing the hoseline and getting lost in the thick, black smoke. Twelve additional Maydays were transmitted by the firefighters trying to rescue Tarver. These real-life obstacles were never anticipated in the theory of rapid intervention.

After Tarver's death, The Phoenix Fire Department conducted extensive scenario-based training evolutions to study rapid intervention. They discovered that it took more than 21 minutes to rescue one lost, downed firefighter in a commercial structure. They also found that rapid intervention firefighters had, on average, 16.5–18.5 minutes to operate in a structure before their low-air alarms activated. In combination, these findings show that firefighters attempting a rescue do not have enough air to locate and extricate a downed firefighter from a commercial structure without running out of air themselves. This explains why there were so many Mayday events at the Southwest Supermarket, and why one member of the rescue effort was brought out of the Southwest Supermarket in full respiratory arrest.

Scenario-based training demonstrated that 12 firefighters were required in order to find and rescue one downed firefighter. Sending in a two-firefighter rescue team is not realistic and sets that team up for failure—and, most likely, serious injury or death. This tragic lesson was also learned in 1999, in Worchester, Massachusetts, at the Cold Storage Warehouse fire. Four of the six firefighters who died were sent in as two-firefighter rescue teams. Both of these two-firefighter teams got lost in heavy smoke while searching for the two original victims. Four of the six firefighter fatalities in that fire were due to running out of air.

Another gift that Bret Tarver gave us is the knowledge that approximately one in five rescuers involved in a firefighter rescue will get into trouble in the structure.[4] ICs must preload their rescue efforts, bringing in more resources and more rapid intervention crews to rescue the rescuers. Additional alarms and rescue capacity must be forecast and staged as soon as the first rapid intervention crew is activated. ICs who do not heed this lesson are destined to have members participating in rescue efforts run into trouble and have to be rescued themselves. Remember, in the Southwest Supermarket fire, a number of firefighters trying to rescue Bret Tarver ran out of air and had to be rescued themselves. These firefighters running into trouble decreased the survivability profile of Bret Tarver, since they were rescued before Bret.

Kreis has also points out that fire departments need to practice calling a Mayday. Firefighters hate to lose and are used to helping others, not themselves. Because they never want to admit defeat, they hesitate to call the Mayday. Kreis explains that shortly after the Southwest Supermarket fire, most firefighters and officers in the Phoenix Fire Department admitted that they would call the Mayday only "on their last dying breath." And as Kreis points out, this is much too late to call for help.

SEATTLE FIRE DEPARTMENT AND MAYDAYS

Several years ago, in Seattle, we had a couple of our own close calls, with one firefighter running out of air in a large apartment building and another running out of air inside a large fish-processing ship. Neither of these firefighters called a Mayday. Why? Because both of these firefighters had bought into the belief that calling for help was a sign of weakness and defeat. Both of these incidents forced all of us in the Seattle Fire Department to take a good, hard look at ourselves and recognize that we also had a cultural bias against calling the Mayday.

Once we acknowledged our own shortcomings when it came to calling the Mayday, the Seattle Fire Department designed a department-wide drill to teach members of the department to call a Mayday. In the drill scenario, firefighters were placed in a very stressful situation and forced to call a Mayday. In addition, each member called the Mayday over the radio, reading the Mayday transmission from a script that was provided. In doing this, members learned to call a Mayday and practiced telling the IC what was wrong, who they were, and where they were in the structure. This was the first time in many of the firefighters' careers that they had ever practiced calling a Mayday.

As fate would have it, several years after this training, two members of the Seattle Fire Department became trapped in the basement of a single-family dwelling, with fire rolling over their heads, cutting off their exit up the basement stairs. They called a perfect Mayday, and the IC took steps to rescue them. After calling the Mayday, the two firefighters found their way to the only window in the basement and exited to safety. The Mayday called in that hot, fiery basement was not the first one that they had ever called over the radio. They had been trained in calling the Mayday and in what to do once they found themselves in trouble. Consequently, these two firefighters performed perfectly and survived.

Interestingly, every vessel in the U.S. Navy, big or small, practices the "abandon ship" drill over and over again while at sea. The captains of these ships know that during a true emergency requiring the ship to be abandoned, each sailor must know where to go and what to do beforehand, so that everyone gets off the ship safely and effectively. We, as firefighters, must train on the worst-case scenarios that we might run into inside a burning structure. Just like the U.S. Navy, we have to be prepared and know what to do when we get into trouble. This is another gift that Bret Tarver has given to us.

Bret Tarver did not die in vain. Thanks to the open-mindedness of Chief Alan Brunacini and the Phoenix Fire Department, Bret Tarver's legacy—his gifts—will live on in the U.S. fire service. The lessons that we have learned from Bret's tragic death will keep other firefighters safer and make them more prepared on the fireground—that is, as long as we study the lessons that Bret and the Phoenix Fire Department have given to us all.

MAKING IT HAPPEN

1. Does your city, town, district have a building that resembles the Southwest Supermarket? If so, have your company do a prefire of this building, noting where the exits are, how the building is constructed, and whether the building has a sprinkler system.

2. In a large acquired structure, such as a commercial warehouse, place a "victim" deep inside, in a zero-visibility environment. Then have your RIT enter, find the victim, and remove him or her to the outside of the structure. Record how long it takes from the activation of the RIT to their removal of the victim from the structure.

3. Design a drill focused on air management. For ideas and help, go to the Web site of the Seattle Fire Department (http://manageyourair.com; click on the "SMART Drills" tab).

STUDY QUESTIONS

1. What was the main factor contributing to Bret Tarver's death?

2. Of the five ways firefighters die in structure fires, how many were working against Bret Tarver at the Southwest Supermarket fire? Name them.

3. List three important lessons that the U.S. fire service has learned from Bret Tarver's death.

4. Why is practicing a Mayday transmission so important?

NOTES

[1]   NIOSH Fire Fighter Fatality Investigation Report F2001-13, pp. 1–24; Phoenix Fire Department. Final report on the Southwest Supermarket fire. 2002, pp. 1–98.

[2]   NIOSH Fire Fighter Fatality Investigation Report F2001-13, pp. 10–11.

[3]   Kreis, Steve. Rapid Intervention Isn't Rapid. *Fire Engineering.* December, 2003.

[4]   Ibid.

8

TECHNOLOGY AS THE ANSWER: NOTHING CAN GO WRONG

INTRODUCTION

This chapter examines the role that technology plays in the fire service and examines technology's effects on air management. Many companies claim to have technological air management solutions. Some of these products are beneficial. However, an air management policy based entirely on technology is not the answer to the air management problem.

EVER-ADVANCING TECHNOLOGY

Technology—it is everywhere we look. Telephones, televisions, automobiles, computers, medical diagnostic equipment, personal stereos, power tools, and even our toothbrushes are constantly undergoing change. Advances in technology are occurring at a faster and faster pace. For the most part, as technology continually changes our lives, it improves them.

The fire service is not immune to this tide of technological advances. Think back to when you first came into the fire service: What kind of turnout gear did you wear? Did you have thermal imaging cameras (TICs) back then (fig. 8–1)? Were they widely used? What kind of nozzles were on the hoselines? What was the first apparatus you rode on? Did it have enclosed cabs? Did it have a manual transmission? Were there onboard computers and locators? Was the pump easy to use? How much water could it deliver? Did you use power saws to ventilate roofs? Or did you use an ax? Were the power saws heavy and tough to handle? Did you carry a personal radio? Did you have an SCBA (fig. 8–2)? How much air did the cylinder carry? How heavy was it? Was there a PASS device on the SCBA? Did you have to turn it on to activate it? Was it reliable? Were there portable gas monitors on your rig?

Fig. 8–1. TICs—revolutionizing the way firefighters search for fire victims and hidden fire (Photograph by Mike Wernet, South Kitsap [WA] Fire and Rescue)

Fig. 8–2. An early SCBA. Note the low-pressure
hose running from the regulator on the waist
belt to the facepiece. (Photograph courtesy
Seattle Fire Fighters Union, IAFF Local 27)

The technological advances over the history of the fire service have
been amazing. Think about it. We have moved from throwing buckets
of water on fires from the outside of structures to aggressively putting
out fires from the inside, with hoses and nozzles that can deliver up to
300 gallons of water each minute, directly on the seat of the fire. Our
protective clothing has evolved from canvas-and-rubber jackets to space-
age materials that can withstand heat up to hundreds of degrees.

The modern SCBA is perhaps the best example of how technology has
helped the fire service (fig. 8–3). With SCBA, firefighters no longer need
to breathe in the toxic smoke and gases from fires. Instead, they can bring
fresh, clean air inside the fire on their backs. Because of the modern SCBA,
firefighters' health and safety has dramatically increased.

*Fig. 8–3. Modern SCBA. Firefighters no longer have
to breathe in toxic smoke and gases. (Photograph by
Mike Wernet, South Kitsap [WA] Fire and Rescue)*

As remarkable as these technological advances are, the truth is
that sometimes technology fails. Failure may result from mechanical
defects—for example, when a device breaks down at a critical moment.
Or failure may be due to human error—humans not using, monitoring, or
maintaining the technology correctly. Whatever the cause of the failure,
the result can be deadly to firefighters. Because it fails, firefighters cannot
rely solely on technology. They must use their experience and training to
make good decisions based on information that they see, hear, or feel.

TECHNOLOGY AND THE FIRE SERVICE

The fire service has benefited from technological advances ever since its inception. Ours is a profession that uses whatever tools and equipment are available at the time to protect lives and property. If a new tool or piece of equipment comes along that makes our job easier, then we use it until something better replaces it.

The origin of firefighting groups can be traced back to ancient Egypt and Rome. The first firefighters used bucket brigades, poles, and hooks to fight fires in structures. In Rome, the *Vigiles Urbani,* or watchers of the city, were organized in AD 6 by Augustus Caesar. They patrolled the streets of Rome, mostly at night, and watched for fires. They had their own quarters, pumps, buckets, hooks, and *mattocks,* the precursors of today's pickaxes.

Over the centuries, firefighting didn't change much, because the technology was slow to change. The bucket brigade gave way to fire engines in the 17th century, with the advent of the suction and force pump. In 1672, Dutch inventor Jan van der Heiden invented and put into service the first fire hose. This hose was constructed of flexible leather and was coupled every 50 feet with brass fittings. The 50 foot length and the brass connections remained the standard for fire hose for centuries.

Time and technology marched forward. A good example of how technology had a positive impact on the fire service is the evolution of the municipal fire hydrant system. Over time, pumps gave way to pumpers, or engines, which drew water first from water cisterns and later from wooden pipes under the streets of cities. When a fire occurred, a *fireplug* was pulled out of the top of the wooden pipe, and a suction hose was inserted to get the water to the pumper, or engine. If a fireplug wasn't available, a hole was drilled in the pipe to gain access to the water, and a plug was later inserted to fill the hole in the wooden pipe. Later, city water systems used pressurized fire hydrants, and the pressure to the system was increased when a fire alarm went out. As technology advanced, these pressurized fire hydrants, which were often unreliable and harmful to the water systems, were replaced by a valved hydrant system that was kept under pressure at all times. These valved systems are the water delivery systems that we use today in most cities.

MODERN TECHNOLOGY

Technological improvements have changed almost every aspect of our job. Just examine any piece of equipment in use today: The Navy fog nozzles developed in the early 20th century have evolved into the combination and automatic nozzles that we use today. Our turnout gear is now made of Nomex, Gore-Tex, and other laminates that allow firefighters to stay comfortable in environments that would have burned the firefighters of yesteryear in their canvas jackets.

Fire apparatus have undergone revolutionary technological advances. We now arrive on incident scenes in enclosed cabs, instead of riding on the back tailboard of the engine or the turntable of the ladder truck (fig. 8–4). Today's apparatus have rear-mounted cameras so that the chauffeur can see what is behind the rig from the driver's seat. Today's aerial ladders are strong and can support enormous tip loads as compared with the ladders of just a few decades ago. Whereas the engines at the turn of the century were cumbersome and inefficient, present-day engines have pumps that are rated to deliver thousands of gallons of water each minute.

Even the water that we use to put out fires is undergoing a technological change. Many of the newer fire engines are now carrying Class A foam systems, which mix water with foam concentrate and deliver wet-style foam through the hoselines, to extinguish structure fires. Class B foams have been used for years for flammable liquid spills and fires, but development of Class A foam applications is just beginning. Wildland firefighters have used foam in initial attack lines for decades; recently, however, wildland firefighters have begun employing compressed-air foam systems (CAFs) for static fire lines and to protect exposed structures in wildland/rural interface areas.

Fig. 8–4. A Modern fire engine. Note the enclosed cab and
the mid-mounted pump panel. These modern apparatus
are safer and easier to operate at emergency incidents.

Firefighting foam is one of the most interesting technological advances
to come along in a while. Will foam replace water as the standard
suppression agent of the 21st century? Only time will tell.

Consider how technology has defined the fields of fire detection and
prevention. Detection and prevention have come a long way from the
days when the *Vigiles Urbani* patrolled the streets of Rome. Forty years
ago, detection and prevention systems were primitive, unreliable, and
prone to malfunctions. Today, modern fire protection systems—sprinklers,
smoke detectors, heat detectors, and automatic fire alarm systems—have
dramatically decreased the number of structure fires and fire deaths over
the last several decades. The technological advances in how these systems
work and what they are capable of doing is nothing short of revolutionary.
Today, many of these systems are monitored 24 hours a day, 365 days

a year. Fire codes have been written to require these systems in many multifamily, institutional, and high-rise structures. And these systems are only getting better.

The technologies in firefighting and fire prevention are evolving at an ever-increasing rate. As technology gets better and better at a faster rate, our job as firefighters becomes more and more technology-based. As firefighters, we are expected to understand how to use all the technology that is being developed for us.

Examples of how our job has become more technology-based can be found in the escalating use of computers in the fire service. Several decades ago, there were no desktop computers at the fire stations, no touch-screen mobile data computers (MDCs) on the fire apparatus (fig. 8–5), no building-inspection databases, and no computer-aided dispatch (CAD). Computers have transformed how we do our job. Building-inspection file cabinets have been replaced with databases that are only a keystroke away (fig. 8–6). The data from the automatic external defibrillator (AED), used to save a patient's life, is downloaded over the Internet to the medical control division, who analyze it and review our performance. We start each shift behind the computer, logging the names of the firefighters assigned to our apparatus onto CAD. The computer in our air monitors tells us what percentages of oxygen, carbon monoxide, and hydrogen sulfide are present in the air we sample. The computer in our SCBA knows how much air we have used, knows how much air we have left, and can tell us how many minutes of air we have remaining at the current consumption rate. Using e-mail and the World Wide Web, we stay in touch with other firefighters around the world, and we learn the details of close calls and firefighter line-of-duty deaths from first-rate Web sites (e.g., see Billy Goldfeder's Firefighterclosecalls.com). Computers run our dispatch centers, our communication systems, and many of the systems on our fire engines and fire trucks. Honestly, where would today's fire service be without computers?

Fig. 8–5. MDCs on fire apparatus

Fig. 8–6. Desktop computers in the fire stations. Firefighters use computers for recording building inspections, logging on personnel to CAD, reading e-mail messages from officers and administrators, downloading information from AEDs, and keeping informed of national fire service news and events.

Since the time of the Roman *Vigiles Urbani,* technology has profoundly improved the fire service. As fire service professionals and volunteers, we are expected to learn how to use evolving technology to better serve the citizens in our cities, towns, and districts. New technologies are guaranteed to present us with innovative ways to better fulfill our mission.

TECHNOLOGY FAILURES

Have you ever been working your shift at the fire station when the department's computer system crashed? In our department, not much can get done without the station computers. Has your radio system ever gone down, making communications impossible? Have you ever carried a TIC into a zero-visibility fire (fig. 8–7) only to have the TIC suddenly stop working? Has your SCBA ever started leaking inside a hazardous environment? Has your fire apparatus ever failed to start after you received an alarm and jumped in the cab?

As wonderful as technology has been for the fire service, it does fail. Firefighters must understand the limitations of technology and not place their lives in jeopardy by relying solely on technology, particularly when they are inside a structure fire. Murphy's Law states, "Anything that can go wrong, will go wrong—and at the worst possible time." And Murphy, they say, was an optimist. Perhaps he was also a firefighter. If you have worked in the fire service long enough, you will understand that equipment breaks or fails exactly when you need it to work the most, especially on an emergency scene.

To offset the fact that technology will eventually fail—and fail at the worst possible time—firefighters must maintain a sound foundation of basic firefighting skills. These are behaviors learned through repetitive hands-on training sessions. These sessions should be provided by fire departments to every firefighter in their operations division. (For a complete discussion of training, see chap. 13.)

Fig. 8–7. A firefighter using a TIC in a training environment (Courtesy of John Lewis, Passaic [NJ] Fire Department, and Chief Rob Moran, Englewood [NJ] Fire Department)

Training in the basics must be the foundation of any fire department. Firefighters must be proficient in basic search techniques, hose handling, pump operations, ladder operations, fire behavior, vertical ventilation skills, PPV techniques, building construction, fireground decision-making skills, and much more (e.g., see fig. 8–8). These basic firefighting skills are imperative, because when the technology firefighters are using on the fireground fails, as it inevitably will, at least they can fall back on the basics to help them out of a potentially dangerous situation.

Fig. 8–8. Firefighters practicing trench cuts on an
acquired structure (Photograph by Joel Andrus)

Picture yourself searching for victims in a large residential structure, such as a high-rise apartment building. It is 0135 hours on a Monday night, and fire is reportedly coming from a two-bedroom apartment on floor 14, exposing the apartments on floor 15. Heavy smoke is on floor 15, and you and your ladder crew have been assigned to search that floor. You grab the TIC and head into the building with your crew. The firefighters who took lobby control put you on an elevator with a firefighter operator, and up you and your crew go, to floor 12, two floors below the fire. As the officer, you get off the elevator on floor 12 and quickly study the floor layout. In this case, a center hallway divides apartments on either side. You have one of your firefighters force open the door to one of the apartments, and you bring everyone inside to see the layout. After this short reconnaissance mission, you head over to the attack stairwell and climb up a flight, to floor 13. You have your crew put on their SCBA, and you pass an engine company about to make entry on floor 14 to attack the fire. You head up to 15, the floor you have been assigned to search.

When you arrive on floor 15, you open the stairwell door and are met with hot, black smoke from floor to ceiling. The smoke is so thick that none of you can see your hands in front of your facepieces.

"OK," you tell your crew, "let's stay together. Everyone, check your air." You find that everyone has a full cylinder.

"Good thing I have the TIC," you think to yourself, as you begin leading the crew down the smoke-filled hallway. You look down the hallway with the TIC and don't see any victims, so you move quickly to the apartment right over the fire, the exposure apartment. The door is open, and you can feel the heat. You use the TIC to scan the ceiling of the apartment from the hallway door. The temperature gauge on the TIC reads more than 200°F, but you don't see any fire moving above your head.

"It's not rolling over us, so we have some time," you tell yourself.

You have your crew check their air again. Being satisfied that they all have enough air to complete the search, you direct them inside the two-bedroom apartment. You stay oriented to the right wall, scanning with the TIC as you go. You move quickly to the apartment hallway, where you know the bedrooms to be, based on your earlier reconnaissance on floor 12. Your crew splits up and moves into the bedrooms, while you stay oriented in the hallway, checking overhead with the TIC. Soon one of your crew yells that he has found a victim. Sure enough, he drags an adult from one of the bedrooms. You radio to the IC that you have found a victim on floor 15, in apartment 1506, and that you need a hose team, another truck company to continue the search for victims, stairwell support personnel, and a medic team for the victim.

Meanwhile, the other firefighters in your crew have finished searching the other bedroom and the rest of the exposure apartment. You have everyone check their air once more and assign the firefighter with the most air to take the victim. You tell your crew to follow you, and you lift up the TIC to visualize your exit route. However, the TIC isn't working—the screen is black. You try to power up the TIC again, but it's no use; the TIC is dead. It is out of service (OS).

When you tell your crew that the TIC is OS, the younger members begin to worry out loud.

"Now what are we going to do?" one asks.

"How are we going to get out of here, Lieutenant? Where are we?"

Luckily, you have years of hands-on search training under your belt, and you have stayed oriented to the wall. You tell them to relax, that you know exactly where you are going, and you calmly and quickly lead everyone out of the apartment and into the hallway.

You stop to once again check your crew's air supply and to have the firefighter with the most air take over pulling the unconscious victim. Soon you are at the stairwell, where you meet two firefighters who tell you that they are there to take the victim to the medics waiting on floor 12. You pass off the victim and then make sure everyone on your crew gets into the stairwell before you close the door.

Down on floor 12, in staging, one of the younger crew members says, "Wow, Lieutenant, I thought we were in trouble when you told us that the TIC wasn't working."

"Yeah. How did you know where we were?"

"Remember those blacked-out search drills I make you guys do, where I turned off the TIC on you halfway through the drill?"

"I sure do. You tell us to make sure that we stay oriented to the wall. I guess I forgot to do that this time."

"Me too," says the other young crew member sheepishly.

"Today is a great lesson," you tell them. "Always stay oriented. The TIC is a great tool, but you can't always rely on it."

Your veteran driver nods in agreement and smiles with a knowing smile that tells you that he, too, had remained oriented on the fire floor.

"Hey, guys," one of the medics says as he walks over to you. "That was a great pull you guys made with that victim. We tubed him, got him on high-flow oxygen, and he's on his way to the ER. His oxygen saturation was down, but it was coming up. Can't say for sure, but it looks like he's going to make it. Nice work."

With that, you nod your head, smile, and tell the crew how proud you are of them. Then, you take a few moments to remember all those past officers and veteran firefighters who made it a point to train you in search basics.

"God bless those guys," you tell yourself. Lowering your head, you whisper, "Thanks."

AIR MANAGEMENT AND TECHNOLOGY

The SCBA is arguably the most important technological advance in the history of the fire service. The SCBA allows firefighters to breathe fresh air while inside a structure fire, instead of the toxic smoke and gases that these fires produce.

Air management is nothing more than the skill of managing the supply of air in your SCBA and keeping an air reserve in case you find yourself in trouble inside the IDLH atmosphere. Later in this book, you will be given the solution to the air management problem in the fire service; chapters 10 and 15 describe, in detail, how to manage your air and your crew's air inside a hazardous atmosphere.

In the first chapters of this book, the case was made that the fire service needs to have a simple and effective air management system that works. This is because firefighters continue to run out of air in structure fires and die from breathing in the products of combustion—that is, toxic smoke and gases. In fact, the majority of firefighters who die in structure fires died because they ran out of air and succumbed to such gases as carbon monoxide and hydrogen cyanide (For a detailed account of the modern smoke environment, see chap. 5.)

The modern SCBA has an audible low-air alarm that indicates there is one-quarter of the total air supply left in the SCBA air cylinder. This *end-of-service-time indicator* (EOSTI) tells the firefighter that they have used three-quarters of the air supply. SCBA manufacturers developed the low-air alarm so that firefighters would be reminded that their air supply was getting critically low. This is a good example of how the technology was developed to force firefighters to behave a certain way—the audible low-air alarm forces firefighters to acknowledge their low-air situation.

Most firefighters have been taught that they should begin leaving the hazardous atmosphere when their low-air alarm activates. In other words, firefighters do not think about their air supply until they hear the low-air alarm activate. For a majority of the fire service, this is exactly what we were taught in recruit school: wait until the low-air alarm activates and then leave the structure fire. Around the country, firefighters—some of whom have been in the fire service for 20 years or more—have attested that they have never, in their entire careers, looked at their SCBA air-pressure gauge while working inside a structure fire. They relate that they have always relied on the low-air alarm as a signal of when to think about leaving the fire.

Firefighters who use the low-air alarm as a signal of when to exit the fire are relying on the assumption that when the low-air alarm activates, they have enough air left in their cylinder to make it safely out of the structure fire. This assumption is completely wrong. Firefighters who wait until the low-air alarm activates to exit a structure fire are literally betting their lives that nothing bad will happen to them on their way out. This is a deadly wager to make.

TIME-TO-EXIT FACTORS

Firefighters should not depend on the technology of the low-air alarm to make the decision about when to leave a structure fire. The low-air alarm was never intended to be a time-to-exit alarm. The decision to exit must be made by firefighters, who are taking into account the many different factors that affect their air supply. These factors include:

- Rate of breathing
- Physical conditioning
- Stress level
- Location inside the structure (how far or deep)
- Visibility inside the structure
- Fire conditions
- Debris, furniture, or stock inside the structure
- Physical layout of the structure

By considering all of these factors, firefighters are practicing good situational awareness. Firefighters must remain aware of what is happening around them inside a structure fire at all times. Situational awareness is a skill that must be applied and reenforced in every drill and hands-on training scenario.

Rather than have firefighters rely on the technology of the low-air alarm, decision-making skills should be employed by firefighters to make their own determination of when to leave the hazardous atmosphere. Firefighters who are actively monitoring their surroundings and all the factors that affect their air supply are better able to decide when to safely exit the fire than relying on the technology of the SCBA low-air alarm. Proficiency in decision-making skills can be learned and reinforced through scenario-based training.

There are some who are lobbying to extend the low-air alarm time, so that the alarm will activate when half of the air cylinder has been used. The problems with this ill-advised idea is discussed in greater detail in chapter 10 and 21. Suffice it to say that extending the low-air alarm time creates more problems than it would solve. It is not a practical solution to the air management problem.

SCBA manufacturers are proposing technologies to resolve the air management problem. Fire departments are being offered a wide array of computer-driven solutions that monitor the real-time air consumption of firefighters on the fireground. Telemetry units exist that allow each SCBA to communicate with a central computer located at the command post. These systems track how much air firefighters have left in their SCBA cylinders. Some central computers are capable of displaying each firefighter's air consumption rate and can even be used to activate each firefighter's emergency alarm remotely, to warn firefighters to quickly exit the building. However, one glaring problem with this technology-driven solution is that it relies upon a person to actively monitor a computer screen at the command post. How many fire departments have the staffing to dedicate one person to watch a computer screen at every incident where firefighters are using their SCBA? Not many—especially in this era of decreased staffing levels in the fire service. With all the tasks ICs already have to do at a structure fire, would they really be able to keep an eye on every firefighters' air supply? More importantly, would you want to entrust such a sensitive task to anyone but the individual firefighters themselves? (For in-depth discussion of technology-driven solutions to the problem of air management, see chap. 21.)

Air management is the responsibility of the individual firefighters and crews on the fireground, not the responsibility of someone outside, at the command post, who may or may not be monitoring the firefighters' air supplies. Firefighters and crews possess the ability to manage their own air supplies. Air management is a personal responsibility, and chapters 10 and 15 explain exactly how firefighters and crews can manage their air supplies in hazardous environments.

SCBA AND BELOW-GRADE AREAS

Firefighters ordered to shut off utility control valves for gas or electric power must consider the possibility of carbon monoxide and smoke accumulation in the cellar. This is particularly important when a fire of long duration has been extinguished in a first-floor store directly above the cellar and the cellar is completely below grade and without windows. SCBA must be worn in the cellar, even if there is only a light haze of smoke. Carbon monoxide, a deadly gaseous by-product of combustion, is colorless, odorless, and explosive and quickly builds up in unventilated below-grade areas.

Notify your officer and wear SCBA before entering a cellar to shut off utilities. If there is no confirmation of the shutoff within a reasonable amount of time or if there is not radio contact, the officer must make an immediate effort to locate the firefighter sent below grade and ensure his or her safety.

Do not let the presence of an operating sprinkler give you a false sense of security. Wear your SCBA before entering a cellar, because carbon monoxide gas can be present even when a sprinkler is discharging and controlling a smoldering fire.

LESSONS LEARNED

The fire service needs a new SCBA.

The fire service is going into the 21st century with the same heavy mask and tank that we used in the 1950s. Firefighters in the high-rise district wear masks weighing approximately 35 pounds that supply air for 60 minutes. If you breathe deeply, you may get only 30 minutes of air. We search high-rise buildings for several hours and have to change air tanks several times during a fire or emergency.

If a mask weighs 30 pounds, our turnout gear weighs 30 pounds, and a length of hose weighs 30 pounds, then you may be burdened with 90 pounds of tools. Heavy masks also slow a firefighter's escape from a collapsing building. Firefighters all over the country need a 21st-century mask. We can do better. We need a lightweight mask that weighs only 10 or 15 pounds and can give a firefighter air for two or more hours.

—*Vincent Dunn (Deputy Chief, retired, FDNY)*
http://vincentdunn.com/

A BRIEF LOOK INTO THE FUTURE

The future of the fire service will be greatly influenced by emerging technologies—technologies that will certainly make our job easier and safer. The future of firefighting is wide open, limited only by our imaginations.

What will the future of firefighting look like? We can anticipate many exciting changes:

- Lighter air cylinders that hold more air

- Lighter SCBA backpacks that are easier on our backs

- One-piece bunking gear suits that integrate with snap-on helmets and internal air supplies, like the spacesuits that astronauts wear

- Foam/water systems, on the engines, that can cool overhead and knock down fire quickly

- Better nozzles that can penetrate farther with less pump pressure

- Stronger aerial ladders that can safely hold more people at the tip

- Better firefighter emergency location devices that employ Global Positioning System (GPS) and three-dimensional positioning technologies

- Personal communication equipment integrated into our helmets

- Heads-up displays on facepieces, allowing us to monitor the air in our cylinders; the temperature of our surroundings; the oxygen, carbon monoxide, and hydrogen cyanide concentration in the smoke; and the layouts of the building in which we are operating

- Personal cameras integrated into our helmets, allowing those at the command post to see what we are seeing

- Ground ladders made of stronger, lighter metal alloys, to spare our backs

- Fire apparatus with smart restraint systems that will not allow the apparatus to move until everyone is buckled into their seats

- Handheld devices that will help us track the building-inspection data for all the occupancies in our city or town

- Search ropes that can withstand temperatures well over 800°F

- Extrication tools that are lighter, stronger, and easier to deploy and use

- Positive-pressure fans that can move three or four times the amount of air that our current fans move

- AED units that not only shock the patient's heart back into a perfusing rhythm but also perform all of the chest compressions

Whatever lies ahead, we can rest assured that the future of the fire service looks bright and that it holds the promise of wonderful technological advances that will increase firefighter safety on the fireground.

The technological advances that have taken place over the history of the fire service have been nothing short of revolutionary. We have come a long way from the bucket brigades and the Roman *Vigiles Urbani*. As amazing as the technology has become, it sometimes fails—and usually in the worst possible way, at the worst possible time.

Since we know that technology fails, it makes sense that the air management problem cannot be addressed by technological solutions alone. Firefighters cannot rely on the technology of the low-air alarm, which was never intended as a time-to-exit alarm, to make the decision for them about when to leave a structure fire. Too many firefighters have run out of air in structure fires and lost their lives because they waited for the low-air alarm to activate before they began exiting (see chap. 7).

Firefighters must be individually responsible for managing their air supply. Firefighters possess the ability to make their own decisions about their air supply and about when to safely exit a structure fire. They must be taught, however, to manage their air and make decisions based on their remaining air supply. Good decision-making skills are much more reliable than technology, and these skills must be reinforced through effective, ongoing scenario-based training.

MAKING IT HAPPEN

1. In an acquired structure or a training facility, place several dummies in different locations throughout the structure, far from the entrance. Use a smoke machine to fill the structure with smoke. Send your crew into the smoke-filled structure to search for the dummy victims. Let them use a TIC to assist with their search. When they get deep inside the structure, tell them that the TIC has failed and watch how they react. The teaching point is that someone on the crew must remain oriented inside the structure. Practice staying oriented in a zero-visibility environment.

2. Open the compartments on your fire apparatus and initiate a discussion with your crew about the changes in technology that have taken place since they entered the fire service. Look at your tools, equipment, and PPE. Discuss with your crew how far technical advances have brought the fire service and how these advances have made their jobs easier and safer.

3. In a large acquired structure or a training facility, stretch 400 ft of charged hose to the back corner farthest from the entry to the structure/facility. Have the hose go into rooms or areas and loop back on itself. Either black the structure out with black plastic or use a smoke machine to add smoke. In this zero-visibility environment, have the crew bring in a charged 1 ¾" hoseline as a backup. Allow them to operate as they normally do. When the first SCBA low-air alarm activates, see if the entire crew can make it out of the structure before any of the members run out of air. The point of this drill is that from deep inside a large structure, most firefighters cannot make it outside before running out of air. When they exit the structure, explain that relying on the technology of the low-air alarm is not a safe practice. (For related training tips, see chaps. 10 and 15, which explain how to use the ROAM. Go to our website, manageyourair. com for other hands-on drills).

4. Ask your crew if they have ever had any piece of equipment on the apparatus fail when they were using it at the emergency scene. Listen to their stories and experiences. Keep the discussion focused on how equipment and technology does indeed fail. Relate this discussion to what happened to them in the preceding drill.

5. On the Internet, find the new, computer-driven solution that the manufacturer of your particular SCBA has developed to solve the air management problem. Print out this material and have your crew read and discuss it. Ask your crew whether your department or district has the staffing to dedicate one person to actively monitor the air supply of every firefighter in the IDLH environment. Discuss the pros and cons of having someone else monitor your personal air supply.

STUDY QUESTIONS

1. Who were the *Vigiles Urbani*? When was this group first organized?

2. Who invented the first fire hose and when?

3. What is *CAFS*? Who used it first?

4. Give several examples of how firefighting has become increasingly technology based.

6. What must be the foundation of any fire department?

7. What is arguably the most important technological advance in the history of the fire service? Why?

8. Briefly define air management.

9. Most firefighters wait until what occurence before they begin leaving the hazardous environment? Why is this practice dangerous?

CHANGE–RESISTANCE–CHANGE

INTRODUCTION

Newton's first law of motion states that an object at rest will remain at rest until acted upon by an outside force. This is also true of fire departments. Fire departments and firefighters like to perform their duties using the tried-and-true methods of their predecessors. Perhaps the strongest influence in fire service organizations is held by the training handed down from the veteran to the probationer, or probie: training in the techniques used to save and protect civilians and firefighters from serious injury or death; training that provides a safe and effective operating environment; training that defines the operational philosophy and actions to be taken by firefighters on the fireground. But what happens when this training—this operational philosophy, this way of doing business—is broken? In other words, what happens when the conventional wisdom is not wise?

Fire department conventional wisdom as it applies to air management is demonstrated in the common belief that it is safe to operate, before beginning to exit, until the low-air alarm activates. As explained already in this book, the current belief system as it applies to SCBA operations in IDLH environments is flawed. Chapters 1–5 demonstrate the *need* for air management. Chapters 6–9 demonstrate the *mandate* for air management. This and the preceding chapters establish that there is a compelling mandate to change how we are *implementing* air management, to prepare you for a change in the operational philosophy of your fire department. The recommended change is following the ROAM.

To answer the question "Why change?" the problem must first be clearly stated, as follows:

- Breathing smoke is not safe.
- Current fireground practice routinely results in firefighters breathing smoke.
- We must change fireground practice so that firefighters do not breathe smoke.

This book makes the case that the ROAM is the proper change that solves the problem of firefighters breathing smoke.

In this chapter, other fire service changes are investigated, along with a discussion of methods of resistance encoutered and methods to facilitate change in your organization. An individual examination will be made of the methods that people use to resist each change. Guidance, methods, and examples of how people in fire service organizations have been able to overcome resistance to change will be given. These examples are drawn from experiences teaching across the country and represent the ideas of class attendees, based on their own experiences in the fire service.

SCBA

The SCBA is widely considered to be the most significant improvement related to the safety of firefighters in modern times. The SCBA has demonstrated itself to be a useful tool, helping firefighters do their job in hot and smoky conditions. (Chap. 1 tells the story of the introduction of the SCBA and how this important tool was met with resistance by the fire service.) SCBA use is invaluable in preventing smoke exposure and its negative short- and long-term consequences. That said, the SCBA was—and still is—the recipient of some of the most vehement and consistent resistance from firefighters and fire departments.

Resistance to the SCBA has manifested itself in many ways since the very first day the equipment showed up on the rig or in the station. One of the classic methods of resistance to using the SCBA was simply leaving it on the apparatus. When the department purchased the equipment and placed it on the apparatus, there was often no expectation that the SCBA would be used. In many cases, use of the SCBA was passively, if not actively, discouraged by the department leadership. After all, if a firefighter were to use the SCBA, that meant the cylinder would have to be refilled. To refill a cylinder, the chief might have to drive to an adjacent district. Stories abound about the practice of "throwing the thing in the bushes" when the chief showed up. You don't want the chief thinking you are weak, do you?

With time—and fatalities—departments began to implement policy changes mandating that firefighters wear their SCBA. Resistance to this policy is a classic example of *malicious compliance*. Malicious compliance is when a person intentionally inflicts harm by strictly following the orders of management.

In regards to the policy dictating that firefighters wear SCBA, malicious compliance means following the letter of the policy—that is, firefighters wear but do not use the SCBA (fig. 9–1). Firefighters dutifully remove the SCBA from the compartment and place them on their backs. They have their facepieces ready for use and therefore are strictly following the policy that they wear the SCBA. While this is still allowed in many fire departments, the downside for firefighters is that their malicious compliance is not causing harm to the department. Most of the harm is done to firefighters and their families. Eventually, fire departments wanting to ensure that the SCBA did not just take up space on the backs of firefighters implemented policies mandating that firefighters wear and use the SCBA provided.

Fig. 9–1. A firefighter wearing but not using the SCBA during roof operations at a structure fire (Photograph courtesy Seattle Fire Fighters Union, IAFF Local 27)

Peer pressure

As departments began implementing mandatory SCBA use in recruit academies, new firefighters reported to the company already trained to wear and use their SCBA (fig. 9–2). The probies then got a lesson in another form of resistance—namely, peer pressure. *Peer pressure* is defined as the influence that friends have on one's own morals.[1] It is dubious to say that anyone who pressures a firefighter into not using the SCBA is a friend; unfortunately, though, the attitude that the SCBA is optional still exists in some fire departments. What kinds of morals does it take to pressure a new firefighter into breathing smoke? What example is provided when the senior firefighter refuses to properly use the SCBA? What precedent is set when the officer allows this behavior toward the probie? What message is given when the department leadership turns a blind eye to this practice?

Fig. 9–2. Recruits performing timed donning of SCBA during recruit training. New firefighters are trained to wear and use the SCBA when they report to the company. (Photograph by Lt. Tim Dungan, Seattle Fire Department)

Other methods of resistance can fall under the umrbella of *making excuses*:

- The SCBA is too heavy
- Having one more thing to do slows me down
- We are being "safety-ed" to death
- We fought fire for years without it
- The thing never works anyway

And the list goes on, to the point of absurdity. In spite of these excuses, SCBA is a great tool and is reliable. SCBA is a lightweight addition to the firefighting ensemble in exchange for the level of protection provided. There really is no acceptable excuse for not using SCBA any time there is smoke present. Breathing smoke is unacceptable. Breathing smoke kills firefighters.

OVERCOMING RESISTANCE

Now that the types of resistance to SCBA have been identified, let's take a look at the methods that have been used to overcome this resistance. Understanding these methods will makes it easier to devise ways of overcoming resistance to an air management policy such as the Rule Of Air Management (see chapter 10).

One of the first steps to instigate change in any fire service organization is to develop, to train, and to implement a policy. This is true of SCBA as well. Every fire department should have, and firefighters should be trained on, an SCBA policy. At a minimum, the SCBA policy should comply with all NFPA standards, NIOSH regulations, and any state or local rules or regulations specific to the jurisdiction.

Writing the policy is only part of the battle. Training must accompany any new policy. For an SCBA policy, this training should include all portions of NFPA 1404 (see chap. 6). The training should also encourage acceptance of the SCBA policy by making the compelling case that SCBA improves the fireground operation and the survivability of firefighters throughout their careers and into retirement.

Another method to overcome resistance is by attrition. Keep training the new people in your organization. Keep them focused on the good aspects of the new policy or procedure. Keep them in the loop by telling them how the change is making a positive impact on them and the organization. It is often said that progress comes one retirement at a time. While this method to overcome resistance may work, it takes a long time.

An additional way of overcoming resistance to SCBA use is to reverse the peer pressure. This works well when people begin to experience the improvements that the SCBA bring to the fireground:

- The ability to get to the seat of the fire quickly
- The ability to function without coughing or having snot run from your nose. (fig. 9–3)
- The ability to breathe easier, without coughing up a lung for a week after a working fire

When firefighters begin to wear and use SCBA, they can attest to how effective it is. These firefighters employ their own peer pressure to influence the remaining members of the team to wear and use SCBA. This is a great example of how firefighters can use peer pressure to the benefit of their coworkers and, in the big picture, to ensure incident scene safety.

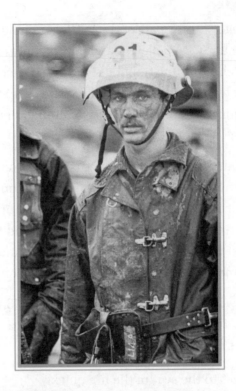

*Fig. 9–3. A firefighter covered in products
of combustion and insulation after a
house fire (Photograph courtesy Seattle
Fire Fighters Union, IAFF Local 27)*

PASS DEVICES

The PASS device as an addition to the SCBA was a prime candidate for resistance when it first arrived on the incident scene (fig. 9–4). The original add-on device required the user to activate the PASS device when donning the SCBA prior to entry into the IDLH environment. Resistance to the use of the PASS device came in many forms, some of which were similar to the resistance approaches used with the SCBA. After all, if something works, why not use it again?

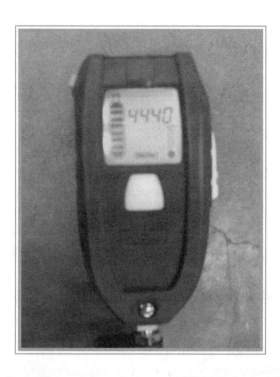

Fig. 9–4. The modern integrated PASS device. This unit turns on automatically when pressurized. This model from MSA has a large button in the middle of the unit to activate the device for a Mayday situation. This unit also gives the user information including air remaining (in psi), temperature, time left until empty at current rate, and battery life.

When the PASS device was introduced, many firefighters stated simply, "I don't need it." This form of resistance at least shows honesty and candor on the side of the resistor. Based on their individual experience in firefighting, these firefighters, many with decades of service over hundreds of fires, had never encountered a situation in which they or their teammates were in need of the PASS device. The downside of this line of reasoning is that the same holds true for most firefighters in most fire departments. The percentage of firefighters who will ever experience a situation that requires the PASS device is minimal. Most firefighters will never experience a Mayday situation or need a RIT activation. These are all good things that demonstrate how, by and large, the fire service is excellent at structural firefighting.

A problem exists, though, in our reliance on the rareness of the occurrence of Mayday situations or PASS activations as an indication that we do not need the PASS device. The reality is that no one knows when the Mayday or the PASS device will be necessary to save the life of a firefighter. No one can accurately predict which firefighter, at which fire, on which day, at which time, will get into trouble. We can't even predict when we will have a fire, much less when the fire will result in firefighter emergencies. This is why we need the PASS device. The PASS device clearly addresses a need on the fireground, providing a means for incapacitated firefighters to transmit a Mayday.

When you anticipate confrontation from the resistor who claims "I don't need it," effective training is in order. Teach firefighters the benefits of the PASS device during RIT training and during regular multiple-company operations exercises. In addition, the NIOSH firefighter fatality reports provide examples of situations in which multiple firefighters were lost, trapped, or went missing. In these case studies, firefighters were within feet of the exit and safety. In one particularly telling case, RITs searched for but were unable to locate a missing firefighter who had not activated the PASS device.[2]

Resistance to the PASS device comes in other forms too. Firefighters complain that the PASS device is too noisy or that the device is a work detector. These firefighters may not be actively resisting the use of the PASS device as much as they are actively persisting in maintaining the status quo. *Active persistence* describes the tendency of people to cling to the old way.[3] This is a tried-and-true method of resistance in the fire service—and for good reason: Firefighters rely on the techniques that have proven to be successful in the past.

With the PASS device, it is not acceptable for firefighters to wait until they are in trouble to begin using the device. By the time trouble comes their way, the PASS device must already have been activated to be effective. One simple way to overcome this resistance is to make PASS activation a regular part of the preentry *buddy check* during training exercises (fig. 9–5). (For a good buddy-check exercise, see chap. 14.)

Fig. 9–5. Firefighters from Poulsbo, WA, after receiving their assignment and completing a buddy check before deploying hose during a training exercise (Photograph courtesy Mike Wernet, South Kitsap [WA] Fire Department)

LARGE-DIAMETER SUPPLY HOSE

Large-diameter hose (LDH) seemed like a no-brainer addition to the fire service when it was first introduced. Who wouldn't want more water, delivered with less friction loss, on any fireground? While there are obvious advantages to the use of LDH, its implementation within a fire department does not happen without resistance. LDH provides a good example of a change to which resistance is, in part, based on legitimate concerns by firefighters, who are on the working end of the change.

LDH, in the form of 4" or 5" hose, is certainly heavier than 2½" hose (fig. 9–6). Thus, each piece represents more work for the firefighters during pickup. With the heavier LDH, the potential for injury is a serious concern for firefighters and company officers. Such concerns and the resistance they foster can be effectively dealt with through training exercises and changes to the practices used when hose is picked up after the incident. While an individual firefighter can drain and shoulder load a section of 2½" hose safely, the same cannot be said of 5" hose. The department must make the tools necessary to support a safe after-fire environment readily accessible.

A second line of resistance to LDH may come from the city council or the board of commissioners for the fire department: LDH, by section, is more expensive to purchase than smaller-diameter hose. Switching an entire department over to LDH can be an expensive proposition. Elected officials and administrative personnel have a responsibility to ensure that the cost of such a change can be justified to the voters. Fire chiefs and other senior promoted and elected personnel demonstrate many of the same resistance behaviors as are seen among the workers in the organization.

*Fig. 9–6 - Multiple hose sizes, from 1¾" to 4", hanging
from the tailboard of a Seattle Fire Department engine.
Various hose sizes are carried on every fire engine.
Each increase in size brings an increase in weight.*

THE CHANGE AGENT

The *change agent,* or advocate of a change, will go through many meetings to discuss changing to LDH with supervisors and decision makers. Each level of the chain of command will need to work through their resistance to the change. When higher levels finally come on board and approve the change, they will often announce the change to the organization with a united voice—without acknowledgment that they, as a group, had to overcome their own resistance. The same senior executives will then act amazed when resistance is encountered among the troops. The change agent in this case must be prepared to help the senior executives and decision makers work through their own resistance while preparing to deal with the resistance of the line workers.

SEAT BELTS

Seat belts are an interesting study in resistance in the fire service. While firefighters, company officers, chief officers, and the fire chief will all diligently enforce the use of the seat belt within the confines of their personal vehicle with their own spouse, children, and grandchildren, they fail to accept the need to wear a seat belt while responding in a fire apparatus. The need for seat belt use is obvious when you consider that since 1984, motor vehicle collisions have accounted for between 20% and 25% of firefighter fatalities annually.[4]

Several strategies have been employed by class attendees to gain department compliance with seat belt use:

- Make the seat belts easily visible. This can be done by specifying orange or bright red webbing when ordering new apparatus from the factory. Eventually, your entire fleet will have the bright seat belt material. In addition, consider changing the seat belt material in your current fleet, as a relatively low-cost change that improves firefighter safety (see fig. 9–7).

- Specify, in policy, training, and enforcement, the mandatory use of seat belts. Emphasize the specific responsibilities of the driver and the company officer to ensure that everyone is belted before the apparatus moves.

- Place sensors with indicator lights to determine whether everyone is buckled up. This action supports the driver and the company officer in their efforts to comply with policy.

- Change the rules for injury to place more responsibility—and liability—on individuals when they are injured as a result of the decision not to wear a seat belt.

- Encourage positive peer pressure. Firefighters should make sure that their partners wear their seat belts. Give firefighters small rewards when they or their teams demonstrate compliance with the policy.

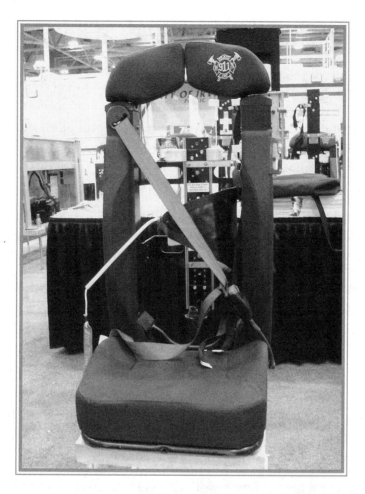

Fig. 9–7. A modern seat for a fire apparatus,
including a lap/shoulder seat belt and a device
that keeps the SCBA cylinder secured to the seat
in case of an accident. The seat belt material
is easily visible when it is properly fastened
during a response.

EMERGENCY MEDICAL SERVICES

For many of us, providing emergency medical services (EMS) is synonymous with being a firefighter. While EMS is provided by most fire departments today, it is not yet a universal standard, nor has EMS always been provided as an integral part of the fire service. Even if your organization has been providing EMS for decades, there was still a time when the service was new, a change to the operational and training requirements of firefighters. There are some organizations that are in the midst of this change today and others that will be changing to a fire/EMS model in the future.

When a change to fire-based EMS service is proposed, some firefighters and department leadership will respond by saying, "That's not my job and not what I signed up to do." While it may be true that they did not "sign up" to perform EMS service, the organization clearly expects them to perform EMS in the future. Resistance may boil down to a question of *why* instead of *what* and *how*.[5] In any organization, administrators and change advocates have a tendency to bypass the question of why, going directly to the *what* and *how* of the change. This may be even more true in fire service organizations because of the paramilitary structure.

While the management team has made the decision to accept new responsibilities and change to a fire/EMS department, the training will often focus on how this is to be done and what changes will be made to the operation. For example, there will be changes to training, dispatch, response, staffing, and station assignments. An abundance of information must be relayed to the firefighter and company officers. When developing the training package, it is easy to overlook or deemphasize the underlying reasons—the *why* portion. This is frequently the case because the change agent and the administrators have already been through their own change process, answered all of their own *why* questions, and made their decision. Training often downplays the importance of explaining why the change must be made, instead moving quickly to how and what must be changed.

No single question is more important than why the change is necessary. Firefighters and company officers are much more likely to accept any change in operation if a clear and compelling case can be made as to why the change must be made. As a firefighter, why should I support this change? As a company officer, why should I implement this change? As an organization, why are we moving in this direction? Why is this expected to improve service to the customer? By providing the answers to these and other *why* questions, the change agent will lay the foundation for compliance with the change and will facilitate a smoother transition to the new what and how.

Once firefighters and company officers accept the why of a change, such as adopting EMS into their service, they will want to know the how and what of the change. In addition, the change agent may find that the how and what questions and problems can be solved easily with the insight of the firefighters and officers. This can also provide a significant buy-in from the people doing the work, as they will have had a hand in developing the system that will be implemented.

INCIDENT COMMAND/MANAGEMENT SYSTEM

The introduction of a formalized fireground command system was—and is—a huge change that brought with it a multitude of resistance models. Evidence of this resistance and the powers that have an influence over the why, how, and what of incident command is provided by the sheer number of command systems that have been introduced to the fire service. Each system has its own cadre of supporters who insist their system is the one that will be able to keep a fireground organized, keep firefighters safe, and put out the fire the quickest.

Whatever system your fire department chooses to implement, behind the system was a change agent who believed that the system would improve the organization. Certainly the introduction of or change to an incident management system, by whatever name, is a major modification to the way a fire department operates. New terminology, new responsibility, new training, and a new operational philosophy each represent major changes; together, they can be rather daunting, challenging the tried-and-true methods that an organization has used for years. One of the first modes of resistance that a change agent should anticipate is falling back on tradition.

TRADITION VERSUS PRACTICE

A debate now rages in the fire service. One side of the argument states, "We must change. Our traditions are getting us killed." The other side counters, "Our traditions make us who we are." The fact of the matter is that both of these statements are true. The discussion needs to move forward by redefining *tradition* and *practice*.

Tradition, in essence, is the best and longest-lasting part of the U.S. fire service. From Ben Franklin to Tom Brennan, traditions remind us of our history and provide us with the necessary background to take pride in this noble profession (fig. 9–8). Tradition includes the awarding of medals, the company Christmas party, reasonable practical jokes, Class A dress uniforms, polishing brass on Friday, saving a bit of the bottle for the brothers and the sisters, the firehouse dinner, and engine/truck rivalries. When taken as a whole, these make us who we are.

Practice encompasses the way in which we do things. For example, practices include how we put up a ladder, how we ride the rig, how we attack a fire. how we ventilate, how we drive, who we hire, and how we deal with the public.

Tradition is rightfully slow to change. There is often no good reason to change a time-honored tradition when it is grounded in the desire to make the profession stronger and more able to provide service. Tradition must be abandoned, however, when it denigrates the individual or the service.

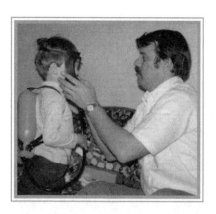

Fig. 9–8. A father and second-generation firefighter showing his son, now a deputy chief, how to use the SCBA. The tradition of teaching the next generation is one of the best aspects of the fire service. (Photograph courtesy of Assistant Chief Brian Schaeffer, Spokane [WA] Fire Department)

Practice, by contrast, changes daily. There is not a fire department in the country that does things in the same manner now as 200 years ago. There has been too much change in our practice to say, "This is the way we have always done it." This cannot be true unless you are still using hand-drilled wells and bucket brigades. Rather, the statement could be made, "This is the way I have always done it," or, "This is the way we did it when I came in." These statements have some merit.

There should not be change for the sake of change. But let's quit confusing tradition and practice. If your organization refuses to change and uses the word *tradition* to justify practices that are contrary to national standards and common sense, you might as well change your saying to "Tradition kills." The modern fire environment has changed, will change, and is changing. Advances in building materials, firefighting equipment, firefighter training, leadership, politics, and other outside influences will continue to mandate change in the fire service. Let's keep our traditions and change our practices, so that fewer of our firefighters have to be involved in our grandest tradition—*funerals* (fig. 9–9).

Fig. 9–9. The pomp and circumstance of the National Fallen Firefighter Memorial Ceremony in 2006. The Seattle Fire Department memorialized two fallen brothers in 2006, James Scott Barnard and Nate Ford. (Photograph courtesy Seattle Fire Fighters Union, IAFF Local 27)

RAPID INTERVENTION

The RIT represents another change to the way we practice our profession. During the mid-1990s, the RIT was the newest and best approach to firefighter safety on the fireground. Brutal experience has demonstrated that when firefighters are in trouble, it is necessary to have a well-trained, staffed, equipped, and ready firefighter rescue squad standing by. Hard lessons experienced by departments nationwide turned into firefighter rescue exercises (fig. 9–10).

Fig. 9–10. New firefighters being trained using the Nance drill techniques. This standard procedure of the Seattle Fire Department is a great example of learning from the tragedies experienced by other fire departments. (Photograph courtesy Seattle Fire Fighters Union, IAFF Local 27)

THE JOHN NANCE KNOT

On the night of July 25, 1987, Firefighter John Nance of the Columbus (OH) Division of Fire was killed in an arson fire in the downtown section of the city. John was riding in charge of Engine 3 when it responded to a commercial building at 151 North High Street. At the time of the fire, the Columbus Division of Fire used 30-minute–rated SCBA (i.e., 1200L cylinders) on all of their equipment.

Engine 3 was on the initial assignment to the report of a fire in a commercial building. On arrival, crews encountered heavy smoke coming from low on the first floor, indicating a probable basement fire. John and his crew advanced a charged line from the rear of the structure to search for the fire. Even though multiple companies were operating inside the structure, little progress was being made in finding the seat of the fire.

While working inside, on his second air bottle, John fell through a weakened area of the first floor into the basement below. Although John fell a distance equal to the height of a commercial basement, he was relatively unhurt. He communicated this to the firefighter who was with him, and a rescue effort was initiated.

When a firefighter falls through a floor or roof, one of two things usually happens. Either the firefighter will catch him- or herself with arms outstretched, to keep from falling all the way through into the fire, or his or her arms will be forced up overhead as he or she falls through. If the latter happens, as in John's case, this position represents the narrowest profile that the firefighter can assume while falling through a floor or roof. To pull someone back up this hole, unless you widen it with saws or other tools, you have to make the victim as small as he or she was when he or she fell through it.

In John's case, the opening that he fell through acted as a vent for the fire, which was in the basement area. When firefighters tried to widen the hole to rescue John, the smoke from the first floor and the fire below choked out the gas-powered saws that they were using, rendering them inoperable. Widening the hole in the thick commercial floor was a slow process, since the rescuers were forced to use axes and pry bars to accomplish the task.

Heat stress, fatigue, and exhaustion started to take their toll on John. A rope with a loop was dropped to him, but he fell out of it as he was pulled up. A narrow but long attic ladder was fed down to him through the hole, but he had great difficulty in climbing up it. Finally, two smaller-framed firefighters descended the attic ladder with a 1½" line, but John could not be located. (The Columbus Division of Fire did not have PASS devices at the time of this incident.)

At the same time that rescue efforts were being made from above the hole that John fell through, another crew was trying to breach a basement wall to get into the area where John fell. Despite multiple rescue attempts by some of the best firefighters in Columbus, the fire continued to grow, and John Nance was killed that night.

John's death is used as an example in countless firefighter rescue classes across the country. As a result of John's death and others like his, the *rescue handcuff knot*, or *John Nance knot*, was devised to pull an unconscious or conscious firefighter from a lower level and bring him or her back into the narrowest profile possible to enhance a rescue effort.

—*Dave Karn (Columbus [OH] Division of Fire)*

Throughout the fire service, *RIT* became part of the standard terminology, and fire departments worked to introduce the RIT as part of standard fireground operations. Federal and state agencies began to mandate the implementation of RIT, effectively forcing departments to change their practice, regardless of whether the department recognized the need. Again, part of the process of making any change, particularly one of this magnitude that is backed by political and statutory authority, is meeting resistance.

One argument against the implementation of RITs stems from concern about staffing levels. The mentality that "It's only for big departments" ignores one simple fact: When you are trapped, the fire doesn't care how big your department is or how many people you run on a rig. The fire does not concern itself with such factors but reacts only to the laws of thermodynamics. The laws of physics and chemistry dictate how the fire will respond in any situation.

Fire departments of any size must establish an effective strategy for firefighter rescue and RIT. The overall number of firefighters necessary to fight the fire, provide a safe and effective incident operation, and make the building behave should take into account the number of firefighters necessary for RIT. Whatever it takes to ensure adequate resources to commit to the chosen strategy must be done. It may be necessary to make changes in strategic decision making to reflect the reality of your staffing, instead of approaching the incident strategy as though your staffing were unlimited. Critical decision making that puts the strategy in line with the staffing level recognizes the need for an effective RIT function on every fireground. If you wait until you need the RIT to establish it, the likely result will be disappointment or disaster.

Although some big departments immediately embraced rapid intervention, the need for RIT was not universally appreciated. Many large department members believed that RIT, as an assigned function, was necessary only for small departments. After all, the argument went, the big departments have the resources necessary to implement a RIT immediately when a firefighter emergency happens.

SEATTLE'S RIT PROGRAM ESTABLISHED

Within the Seattle Fire Department, this thinking proved to be wrong at the Mary Pang warehouse fire in 1995. Seattle lost four firefighters at this fire when the floor on which they were operating collapsed.[6] No RIT had been established at the time of the floor collapse. Following the floor collapse, one of the division supervisors asked, "Where's my RIT? Are you my RIT?" Perhaps if Seattle had heeded the warning and changed in advance of the Mary Pang fire, the outcome would have been different—or perhaps not. Nevertheless, the outcome being what it was, there is no doubt that the makeshift rescue team put together after the floor collapsed did not have the tools, the training, the experience, or the focus of a committed RIT.

In response to this incident, it took Seattle only eight years to establish a committed, trained, equipped, and automatically dispatched rapid intervention group (RIG) for the fireground (fig. 9–11). What is the time needed for your fire department to change? How much resistance can you endure? Change provokes resistance, which precedes change. Ultimately, the change will happen.

A review of fire service changes, methods of resistance, and strategies of overcoming resistance can prepare change agents to implement an air management policy such as the Rule Of Air Management (ROAM, see chapter 10). The ROAM is mandated by the most basic of SCBA standards, NFPA 1404. The ROAM is recommended in NIOSH Fire Fighter Fatality Investigation 2001-13. The ROAM has been implemented in big departments as well as in small departments. The ROAM is a major change that is necessary for firefighter safety, health, and survival. The ROAM is not a change in tradition, but rather a change in practice.

Be the change agent for the ROAM. Improve your fire departments operations, safety, and the survivability profile of your firefighters. It can be done, and you can do it.

Prepare yourself to be the change agent. Study this book and the resources mentioned in the endnotes. Create a plan for implementation by taking a critical look at how your department operates and how changing to the ROAM will improve the effectiveness and safety of your department. If you are not a player in the power structure of your department, then attempt to enlist someone who is to assist you in implementing this change.

Fig. 9–11. A rapid intervention group (RIG) training prop, including collapsed roof sections and general debris. Annual training helps RIGs maintain their skill set should they be needed at an actual incident.

Be prepared to face resistance. Also, be prepared to change your plan or idea when good information and ideas are presented. Keep your attitude and your interactions with other members of the department positive and focused on the goal you have set. Don't let minor setbacks stop you; change takes time.

The practice of air management and the ROAM have been demonstrated to improve the safety and effectiveness of incident operations in departments of all sizes. Ideas like this will eventually catch on. You can do it!

RESISTANCE TO CHANGE: A CULTURAL BARRIER

The National Fallen Firefighters landmark Line of Duty Death summit in Tampa, Florida, in 2004, initiated a significant conversation about the need to change the culture of the U.S. fire service if any substantial reduction in line-of-duty deaths is to occur. This has generated a call for change in the culture from every level of the service. However, to successfully change the culture, a key barrier must be addressed. That barrier is resistance to change itself. Addressing this barrier requires an understanding of the concept of resistance.

Human beings resist change for a variety of reasons:

- *Habit.* We are creatures of habit. Humans develop a series of patterns of behavior, generally to assist in the satisfaction of their needs. Habits, also described as traditions or customs, are a method of satisfying needs in a consistent way. When one discovers a method of satisfying a need, it is highly likely that the method will be repeated until such time as it fails to meet the need on a consistent basis.

- *Fear of the unknown.* The success of our habits creates a fear of the unknown. In the constant search for stability in terms of satisfying our needs, there is a tendency to be fearful of things that appear to interrupt the safety that we feel in continuing our habits and traditions. This is complicated by the extent to which we believe that we have experienced loss as a result of surprises or situations in which control was lost.

- *Human emotion.* Our emotions are an extremely complex set of impulses that begin in our brains and manifest themselves in a variety of ways. Often, resistance to change is purely emotional, and the initial emotion is supported through rationalization.

- *Ignorance.* Resistance to change can be based on a misunderstanding of the change itself or the impact that it will have. The misunderstanding can be brought about by a lack of information on the subject matter and its effects.

Resistance is not necessarily based on sinister motives. A degree of resistance is healthy to the extent that it minimizes the exposure of the resister to harm or loss. When resistance is based on healthy skepticism or reasonable human fear, it can be a positive reaction that can benefit not only the person who feels the resistance but also those who react to it. Individual reaction to change in an organization is directly tied to one's perception of its personal impact.

In the final analysis, the motivation for and reaction to the impulse to resist determines whether the resistance is a good instinct. It is often incumbent on the change agent to educate and inform those affected by a change as to how it will affect others. Effective management of change can result in minimal resistance—and, thus, more effective change. The lower the resistance to change is, the quicker the change will occur. Although resistance to change in the fire service creates a barrier to changing our industry's long-standing trend of injuring and killing its members, carefully managed, change can become a way of life.

—I. David Daniels (Fire Chief/Emergency Services Administrator, City of Renton, WA)

MAKING IT HAPPEN

To identify the types of resistance that you may experience while implementing air management in your organization, perform this exercise:

- Get your change team together.

    - Include people with official and unofficial power.

    - Include people committed to making the change.

- Identify the goal: Getting the ROAM implemented in our department.

- Identify the top five types/methods of resistance your group thinks will be used to defeat the change to air management.

    - Hold a brainstorming session to generate ideas. Follow up by prioritizing and filtering the list.

    - Identify three methods to counteract each type/method of resistance anticipated.

    - Shoot from the hip and be prepared to research methods online or in the library, using the references in this chapter.

    - Identify which methods are proactive and which are reactive. Wherever possible, be proactive to defeat resistance.

- Document the outcome and make copies available to your change team.

- Follow up regularly to identify and deal with resistance early.

- Always keep an open mind and be positive in your approach to resisters. Their arguments may make your ideas for a change to air management better.

STUDY QUESTIONS

1. What piece of respiratory protection equipment is critical to firefighter safety but still meets resistance to its use?

2. What does the term *malicious compliance* mean?

3. What is one of the first steps for fire departments to overcome resistance to change?

4. How can peer pressure be used to overcome resistance to change?

5. How did firefighters react when the original PASS devices malfunctioned or needed to be serviced? How did this differ in comparison to other tools?

6. What is a good way to ensure that nonintegrated PASS devices are activated prior to entering an IDLH environment?

7. Are there times when resistance to change is based on legitimate concerns?

8. What is the second-leading cause of firefighter fatalities?

9. How can these fatalities be reduced?

10. Which are more important when planning to implement a change, the *what* and *how* questions or the *why* questions?

REMOVING A FIREFIGHTER FROM A TIGHT SPACE WITH A HIGH WINDOWSILL

On September 28, 1992, Engineer Mark Langvardt of the Denver Fire Department died in the line of duty. After getting separated from his crew owing to a floor collapse and becoming trapped in a small storage room on the second floor of a commercial occupancy, Langvardt was overcome. The storage room, measuring 6 ft wide by 11 ft deep, was filled with cabinets and business equipment on both sides, creating an aisle only 28" wide, with an exterior window at one end. The drop from the windowsill to the floor was 42". Firefighters entering the storage room through the second-floor window had to crawl over Langvardt, who was lying face down in the aisle in the fetal position, head pressed against the interior of the front wall, just under the window. Because of the restricted size of the aisle, there was room for only one rescuer to bend over the victim and attempt to lift him, thus making conventional windowsill lift and removal techniques next to impossible.

Following this incident, many fire departments and instructors developed techniques for removing a firefighter from a confined area with a high windowsill, most originating from the Denver drill. Special recognition goes to David M. McGrail and Jack A. Rogers from the Denver Fire Department for sharing this information in their April 1993 *Fire Engineering* article, "Confined Space Claims Denver Firefighter in a Tragic Building Fire," and to Chief Rick Laskey and the Illinois Fire Service Institute, from whom I had the opportunity to learn the Denver drill at the Fire Department Instructors Conference hands-on training session.

A variation of the Denver drill is taught by Tim Sendelbach, Chief of Training, Savannah (GA) Fire Department. Another variation is used by the Columbus (OH) Division of Fire. These are discussed on Rapidintervention.com.

This technique is practiced using a first-floor window to allow students the opportunity to get comfortable with removal techniques. In the event that a firefighter needs to be removed from a second- or third-story window, several ladder carries or lowering options are available.

The Denver drill simulator is fairly inexpensive to construct. Two sheets of plywood and a couple 2 × 4 × 8s are all that is necessary. Using bolts instead of nails allows the simulator to be easily set up and torn down and makes it portable by truck. The plywood surface, a full sheet on each side, should be placed on the inside of the framework, to re-create a 28" aisle. Braces are then placed on the exterior of the plywood walls, to add stability and keep them from falling over. An existing window from an acquired structure can be framed by 2 × 4s to create a 42"-high windowsill with a 20"-wide window, or a window of like dimensions can be framed on one end of the plywood walls.

—*Nick Sohyda (Mt. Lebanon [PA] Fire Department)*

from www.RapidIntervention.com

NOTES

1　*The New Webster's Lexicon Dictionary of the English Language.* 1990. Lexicon Publications Inc., New York, New York.

2　NIOSH Fatality Assessment and Control Evaluation (FACE) report 99F-02. Single-family dwelling fire claims the life of a volunteer fire fighter—Indiana.

3　Brenner, Rick. 2006. Is it really resistance? http://www.chacocanyon.com.

4　USFA. *Historical Overview of Firefighter Fatalities 1977–2006.* http://www.usfa.dhs.gov/fireservice/fatalities/statistics/history.shtm. If you need additional evidence, please visit Firefighterclosecalls.com, which has more than enough examples in the close-call and fatality sections to convince even the most ardent resistor.

5　Maurer, Rick. 2005. Making a case for change: Change without migraines. Maurer and Associates, http://www.beyondresistance.com.

6　Further information on the Mary Pang warehouse fire can be found in a USFA report, available online at http://www.usfa.dhs.gov/applications/publications/browse.cfm?sc=361.

PART **3**

THE SOLUTION

THE RULE OF AIR MANAGEMENT

INTRODUCTION

In this chapter, w give you our solution to the air management problem. This solution is the Rule Of Air Management (the ROAM). We will also discuss the single-family dwelling air-use mentality and how scuba divers and firefighters have the same air management issues. Finally, an outline of how to put the ROAM into practice is provided, as well as a brief introduction to other air management systems, which are explored more fully in chapter 21.

THE AIR MANAGEMENT SOLUTION

In Part 1 of this book, we made the case that there is a need for air management because firefighters are running out of air in structure fires and getting seriously injured or dying. We also explored the dangerous toxic gases and carcinogens that make up today's smoke, and explained how firefighters cannot expose themselves to these poisons by running out of air inside the IDLH environment of a structure fire.

In Part 2 of this book we discussed the mandate for air management—specifically the mandate of the NFPA 1404, *The Standard for Fire Service Respiratory Protection Training, 2007 edition*. NFPA 1404 states that fire departments must have an air management policy, that firefighters must be trained in air management, and that all firefighters must be out of the IDLH atmosphere before their low-air alarms activate (see chap. 6). In light of the NFPA 1404 mandate, we discussed the Southwest Supermarket fire in Phoenix as a case study. In this fire, Bret Tarver lost his life after he ran out of SCBA air and breathed in the deadly toxins that are found in today's smoke (see chap. 7). Bret Tarver's legacy is the mandate that we manage the air on our backs when we battle structure fires.

In Part 3 of this book, we provide our solution to the air management problem, which is the Rule Of Air Management, or the ROAM. We will give you a step-by-step method to manage both your personal SCBA air and your team's SCBA air, so that you can maximize your time inside a structure fire. We will also explain how you can implement an air management program in your fire department by describing how other departments, large and small, are adopting the ROAM as part of their Standard Operating Procedures.

Throughout this book, we have provided many case studies that describe situations in which firefighters ran out of air in structure fires and were seriously injured or killed. We have tried to make the point that most firefighters in the U.S. fire service have not been managing their air. Basically, firefighters have been using their low-air alarms as a signal to exit the the fire building. Firefighters have been waiting to leave the fire building until the low-air alarm activates because this is what they were taught in recruit school, when they first learned how to fight structure fires. Interestingly, in many of the air management classes that we teach around the U.S., veteran firefighters reveal that they have *never* checked their air inside a structure fire—never in their entire career! Instead, they have used their low-air alarm as a signal to begin leaving the IDLH atmosphere.

ROAM: The solution to a deadly practice

This practice of waiting until the low-air alarm activates to exit the hazardous environment must stop. This practice is not working. By waiting to leave the hazardous environment until the low-air alarm activates, firefighters are betting their lives that nothing will go wrong on their way out of the structure fire. Many firefighters have lost this all-too-costly bet.

So how can this deadly problem be solved? We advocate a simple solution—The Rule Of Air Mangement, or the ROAM. The ROAM states:

> *Know how much air you have in your SCBA, and manage that air so that you leave the hazardous environment* **before** *your low-air alarm activates.*

Or, put another way, know what you have and get out *before* your low-air alarm goes off.

The ROAM holds true for any fire building or hazardous atmosphere. Firefighters should follow it at all times—at the bread-and-butter room fire in a single-family dwelling, at the large apartment building fire, at the commercial fire, at the high-rise fire, at the confined-space rescue, and at any incident in which they are using their SCBA.

Why get out of the hazardous environment before the low-air alarm activates? By leaving the hazardous environment before the low-air alarm activates, firefighters achieve two goals. First, they leave themselves an emergency reserve of air in case something goes wrong on the way out of the IDLH atmosphere. In the Southwest Supermarket fire, Bret Tarver worked past the activation of his low-air alarm; then, on his way out of the supermarket, he got lost and separated from his team in zero visibility. Tarver soon ran out of air, removed his facepiece, and began breathing in all the toxins that are found in smoke—toxins such as hydrogen cyanide and carbon monoxide. Had Tarver and his team followed the ROAM, Tarver would have had his emergency reserve intact, which might have allowed the rapid intervention teams (RITs) to find him before he ran out of air. In other words, following the ROAM would have bought him some time in which to be rescued before he was forced to breathe in toxic smoke. Second, following the ROAM removes the noise of the low-air alarm from the fireground, thus making the low-air alarm a true emergency alarm.

In the U.S. fire service, firefighters routinely work into their emergency reserve, working in the hazardous environment as the low-air alarm sounds. These alarms have become nothing but a nuisance on the fireground— background noise to be ignored. But what happens when a firefighter's low-air alarm is sounding and they are truly in trouble? The answer is that because of the associated false-alarm mentality, the firefighter who is in trouble will most likely be ignored. For example, during an aggressive search for victims on the third floor of a large apartment building fire, a Seattle fire officer became lost in heavy smoke, and, with his low-air alarm ringing, breathed into his emergency reserve of SCBA air. Another team of firefighters rushed past in the smoke-filled hallway before the officer could reach out and grab them. They ignored the fire officer's low-air alarm and left him to completely run out of air. In the end, to escape, the fire officer was forced to bail out a window and fall three stories, which severely injured him and kept him off the job for well over a year.

By removing the low-air alarm from the fireground, we stop making it a nuisance alarm. Think about the positive impact on firefighter safety this would have on the fireground. If firefighters routinely were out of the IDLH atmosphere before their low-air alarms activated, then a low-air alarm ringing inside a structure fire would be a heads-up to everyone who could hear it on the fireground. The instance a low-air alarm sounded inside the fire building, following the ROAM, other firefighters in the area would go check on the firefighter whose alarm was ringing, providing assistance if necessary.

Older PASS devices were famous for sounding false alarms. They didn't work correctly; there were too many nuisance alarms. Because of these repeated nuisance alarms, firefighters refused to turn on their PASS devices. Consequently, engineers integrated the PASS device into the SCBA and made them work as they were intended. Today, no one hears a PASS device going off accidentally, because the engineers fixed that problem. When firefighters hear a PASS device going off inside a structure now, they radio out that someone's PASS device has been activated and then investigate. Thus, there are no more nuisance alarms—and no more ignoring a PASS alarm inside a structure fire. Remove the low-air alarm from the fireground, and it is no longer a nuisance alarm but rather a true emergency alarm that calls for immediate action.

THE EMERGENCY AIR RESERVE

The emergency air reserve comprises the last quarter of the SCBA cylinder. On the SCBA cylinder gauge, which is found at the neck of the cylinder, this emergency reserve is usually signified by the color red (fig. 10–1). This is literally the last air in the cylinder, and as such, it is given the international warning color.

Fig. 10–1. The emergency air reserve, comprising the last quarter of the air cylinder and signified by the color red on the cylinder gauge

The emergency air reserve was never intended to be used as operational air. In fact, the manufacturers refer to the low-air alarm, which signifies that a firefighter is breathing into the emergency air reserve, as the *end-of-service-time indicator* (EOSTI). The original intent of the EOSTI air was that it should be used by firefighters only in the event of an emergency—when things go bad, as when a firefighter gets lost in heavy-smoke conditions or falls into the basement. This air is the firefighters' air, not air for the fireground operation.

When the SCBA was first introduced to the U.S. fire service, misinformation and bad practices were incorporated into our culture (see chap. 1). The idea that firefighters could use their emergency reserve of SCBA air for firefighting activities is a bad habit that we are only now beginning to rectify.

How did this happen? How did the U.S. fire service get into the dangerous habit of routinely using the emergency reserve for fireground operations?

Conversations with people in both the U.S. fire service and the SCBA industry revealed that this practice started when the modern SCBA were first introduced. When fire departments started purchasing the breathing apparatus that we know today—modern SCBA with a separate air cylinder and a separate facepiece—many departments simply handed out these new SCBA to their firefighters without providing any training whatsoever. Firefighters used the SCBA only when necessary, at first because of the limited air volumes of those early SCBA. This practice, coupled with the glorification of the "smoke-eater" image of the veteran firefighter, provided yet another excuse by firefighters to use the SCBA only intermittently. As a result, the U.S. fire service adopted an intermittent-use mind-set that has been carried over to the modern SCBA and persists today (fig. 10–2).

The intermittent-use mind-set caused many fire companies to leave their new SCBA on the rigs. Many companies neither placed them into service nor tried to understand how they worked. Unfortunately, those firefighters who initially wore or wanted to wear these new SCBA were derided and called derogatory names by veteran firefighters in their companies and firehouses. This provided yet another incentive not to wear the new SCBA.

Fig. 10–2. Firefighters not using their SCBA because of the intermittent-use mind-set. This attitude has been part of the culture of the fire service for decades and persists today, although it is getting better. (Photograph by Seattle Fire Fighters Union, IAFF Local 27)

With the passage of time and the advent of NFPA regulations, firefighters were mandated by their fire departments to wear their SCBA into hazardous atmospheres. With time, firefighters began to wear the SCBA and accept them (fig. 10–3). The firefighters who used the SCBA quickly realized that they could breathe fresh air, stay inside the fire building longer, and fight more fire—a huge plus to any aggressive firefighter or crew. Furthermore, since staying in longer and fighting fire was good, the firefighters who used SCBA began staying in the fire as long as they could. These firefighters began staying in the hazard area until their low-air alarms activated, mistakenly believing the myth that this alarm was a signal to exit the fire building. Because the U.S. fire service had not been trained correctly on the use of these modern SCBA when they were first introduced, the idea and practice of working into the SCBA's emergency air reserve became universally accepted.

Fig. 10–3. Firefighters using their SCBA. Department policies and the realization that structure fires could be more aggressively fought prompted firefighters to use their SCBA. (Photograph by Mike Heaton)

How many of us were taught in drill school that the signal to leave the fire building was the low-air alarm? This piece of misinformation has been propagated throughout the U.S. fire service, passed down to generations of new firefighters. And it is still being taught today.

For almost half a century, the American firefighter has been remaining in the hazard area until the low-air alarm activates. Why? Because the majority of fires occur in single-family dwellings and apartment houses, and these structures have several characteristics in common. Both single-family dwellings and apartment houses have many exits (windows and doors), and they have common layouts that do not vary much from structure to structure. For example, we all know that the bedrooms in a two-story house are located upstairs. We know that the staircase in a two-story house is usually found near the front door. We also know that there are only a limited number of apartment building layouts—for example, center hallway, garden style, H type, and E type. Firefighters fighting fires in single-family dwellings and apartment houses know that there is usually an exit nearby, and they typically know where they are in these structures (fig. 10–4). Because we live in these structures, we are very familiar with their layouts and features.

Fig. 10–4. Firefighters from a ladder company working to ventilate and dig out fire from a single-family dwelling fire. (Photograph courtesy Seattle Fire Fighters Union, IAFF Local 27)

For the most part, firefighters have been able to work far into their emergency reserve at these single-family dwelling and apartment building fires and still exit the structure before their air has run out. These single-family dwelling and apartment building fires have dictated how the U.S. fire service uses its SCBA and have given us a false sense of security in how we use our air in structure fires (fig. 10–5).

Fig. 10–5. Firefighters at a single-family dwelling fire, preparing to vent, enter, and search a second-floor bedroom for a confirmed victim. These types of structures have dictated how the U.S. fire service uses SCBA. (Photograph by John Lewis, Passaic [NJ] Fire Department, and Chief Rob Moran, Englewood [NJ] Fire Department)

Do not misunderstand us. The majority of firefighters who die in structure fires die in single-family dwellings. Firefighters have run out of air in single-family dwellings and have died by breathing in the deadly gases found in today's smoke. Chapter 11 presents the story of Bart Bradberry, a firefighter with the Fort Worth (TX) Fire Department. A veteran firefighter, Bradberry worked into his emergency reserve in a single-family dwelling fire, became disoriented on the second floor, ran completely out of air, and almost perished.

Firefighters who work into their emergency reserve can quickly get into trouble when fighting fires in large commercial structures, in structures with odd layouts, in high-rises, or in below-grade compartments, such as cellars, basements, or belowdecks on ship fires (fig. 10–6). Several examples of firefighters getting seriously injured or dying in these types of structures have already been provided, including the Southwest Supermarket fire, where Phoenix firefighter-paramedic Bret Tarver lost his life.

Fig. 10–6. Firefighters at a commercial structure fire. Firefighters should never work into the emergency reserve, especially in large structure fires. Before this photograph was taken, firefighters attempted to aggressively fight this fire from inside. They were unsuccessful and had to resort to defensive operations. (Photograph by Mike Heaton)

Unlike single-family dwellings and apartment houses, most commercial structures have limited egress and do not have common layouts. In other words, firefighters working deep inside these structures in heavy-smoke conditions cannot get out easily and don't know where they are. Consider the commercial buildings in your city or district. How many times have these occupancies been remodeled? Would you know where you were inside one of these buildings if there was heavy smoke? How many exits do these buildings have?

Commercial occupancies are not like single-family dwellings. However, firefighters who work into their emergency reserve in these commercial structures are using a mind-set geared toward single-family dwellings when it comes to SCBA air use. This is referred to as the *single-family dwelling air-use mentality*. The single-family dwelling air-use mentality is commonly used by most firefighters in the U.S. fire service today, and it is literally killing us. A tragic example in which the single-family dwelling air use mentality played a role was a fire in 1999 in Worcester, Massachusetts, at a cold-storage warehouse. Four of the six firefighters who died in this large commercial structure ran out of air and died "of inhalation of smoke, soot, and hot gases" (fig. 10–7).[1]

Fig. 10–7. The cold-storage warehouse building shortly after first alarm (Photograph by Roger B. Conant; source: NIOSH Fire Fighter Fatality Investigation 99-F47)

SCUBA AND SCBA

What do scuba divers who are underwater and firefighters in an IDLH atmosphere have in common? Both breathe the air they bring with them on their backs, and both know that if they run out of air, there is a high likelihood that they will become seriously injured or die.

The commercial and recreational scuba industries learned long ago that divers need to practice air management and need to retain an emergency reserve in their air tanks in case something goes wrong. How did they learn these lessons? They learned the hard way early on, by having divers die. It didn't take the dive industry long to recognize the problem and come up with a workable solution to this deadly problem. The dive industry decided to require that all divers be certified. As part of the certification process, the dive industry mandated that divers check their air before they go underwater, that they check their air at intervals while they are underwater, and that all divers return to the surface with 500 psi of air (for recreational divers) remaining in their tanks. This 500 psi is the diver's emergency reserve, only to be used if something unexpected happens while the driver is underwater.

Unfortunately, the U.S. fire service did not follow the lead of the dive industry with regard to dedicated emergency air reserve and air management practices. When the modern SCBA were first introduced, the U.S. fire service should have heeded the hard lessons in air management learned by the dive industry and should have passed these lessons onto its firefighters. Instead, because of the intermittent-use mind-set (see chap. 1), because of the universally practiced single-family dwelling air-use mentality, and because SCBA were introduced with little or no associated training, the U.S. fire service never adopted the ideas of a dedicated emergency air reserve or air management. This is why, to this day, firefighters continue to run out of air in structure fires and die.

The goal of this book is to bring the same standards regarding the emergency air reserve and air management that have been accepted by the dive industry to the fireground. The ROAM does this. The ROAM is nothing more complicated than the air management standard of the scuba-diving industry. Let's stop killing firefighters by allowing them to routinely use their emergency reserve as operational air. Let's reserve our emergency air for emergencies. Let's have firefighters look at their air gauges before they enter a fire building. Let's have them check their air at intervals while they are working inside the structure fire (fig. 10–8). Finally, let's have them leave the hazardous atmosphere before their low-air alarms activate.

*Fig. 10–8. Firefighters checking their air gauges
in an IDLH atmosphere. The ROAM mandates
that firefighter check their air at intervals while
working inside structure fires.*

The ROAM is a new, basic firefighting skill—a skill that should be taught to all firefighters. The ROAM costs fire departments and fire districts nothing to implement. There is no new equipment to buy, no upgrade to purchase. The ROAM is a change in behavior—nothing more.

Firefighters must be trained to use the ROAM correctly. Training is the key to being successful at air management. Just as the dive industry trains all new divers to manage their air, so should the fire service. Firefighters should consider themselves smoke divers, and should be as serious about air management and a dedicated emergency air reserve as the dive industry is.

SPECIALTY TEAMS

Air management is already being used in the fire service by its specialty teams—specifically, hazardous materials (hazmat) and confined-space teams.

Hazmat technicians have been using air management for many years. The last place a firefighter wants to run out of air is in the hazmat hot zone. In the Seattle Fire Department, one member of a hazmat team's only responsibility is to ensure that the entry team makes it through decontamination and back into the cold zone before their low-air alarms activate. This air monitor position is critical to the safety of our members and to the operation as a whole.

Hazmat technicians understand that they have only a limited amount of air. They also know that working in the hot zone and having to go through decontamination in the warm zone takes time. As with any operation in an IDLH environment, work time is limited by SCBA air supply. In the case of the hazmat entry team, not only do they have to watch their air supply in the hot zone, but they must also ensure that they have enough air to make it through the decontamination corridor in the warm zone (fig. 10–9).

Like hazmat teams, confined-space teams have been practicing air management for years. Confined spaces typically have noxious and deadly gases associated with them, and wearing a SCBA or a direct-supply air system is typically not an option. The members of the confined-space team watch their air constantly. In the Seattle Fire Department, confined-space technicians follow the ROAM, ensuring that they are out of the confined space before their low-air alarms activate. Following the ROAM is particularly important when the confined space is down a vertical shaft and the members have to be brought up to the surface by way of a rope system.

If hazmat and confined-space teams in the fire service are doing air management, then why aren't operations firefighters using air management in structure fires? This is an important question that needs to be answered.

Fig. 10–9. A hazmat entry team. Hazmat entry teams check their air constantly. They must manage their air supply so that enough air remains to complete decontamination in the warm zone. (Photograph by Charles Cordova)

No one is dying on the specialty teams

Firefighters are not dying on the hazmat fireground or the confined-space fireground. Across the United States, not a single technician on a specialty team has died on a true hazmat incident or confined-space incident in many years. Over the past 20 years, these specialty technicians have learned to slow down, take their time, and follow best practices. One of these best practices is the implementation and use of air management.

Contrast these specialty teams with structural firefighters. Each year, firefighters die in structure fires because they run out of air and breathe in the poisonous and deadly products of combustion. Yet structural firefighters in operations divisions across the country do not practice air management. If any part of the U.S. fire service needs air management, it is operations, where firefighters are dying each year. These firefighters do not have the luxury of slowing down and taking their time like the specialty teams do.

When a fire engine pulls up to a single-family dwelling with fire showing out of the first floor and smoke showing from the second floor and a mother tells the officer that her two children are still on the second floor, the firefighters have an obligation to try to save those children. There is no time for a formal safety briefing, no time to set up zones, no time to prepare equipment, no time to pull out air monitors and gas analyzers, no time to initiate lock-out/tag-out, and no time to discuss tactical options. This company of firefighters must act aggressively and quickly because lives are in the balance in the critical first minutes of the operation (fig. 10–10).

Firefighters are dying from running out of air in structure fires. The U.S. fire service needs to bring air management practices from specialty teams to operations division firefighters.

*Fig. 10–10. Firefighters masking up before attacking a fire
in an occupied residential structure. These firefighters
do not have the option of slowing down. (Photograph
by John Lewis, Passaic [NJ] Fire Department, and
Chief Rob Moran, Englewood [NJ] Fire Department)*

HOW TO PUT THE ROAM INTO PRACTICE

How exactly does a firefighter and a fire company follow the ROAM? What are the nuts and bolts of checking your air supply? When and where do you check your air once you are inside a structure fire?

The ROAM starts with knowing your air supply before you enter the fire. You must check your air the minute you don the SCBA and turn on the air supply, and do this outside the hazardous environment (fig. 10–11).

Fig. 10–11. A firefighter checking his air supply outside the hazardous environment

There are several reasons to check your air supply before entering the IDLH environment. First, firefighters must know that they have a full air cylinder before they enter the IDLH environment. If their cylinders are not full, this is the time to take care of the problem by changing out the cylinder with a full one (see chap. 14). Firefighters have perished in structure fires because they have entered structure fires with cylinders that were not full. Second, firefighters must know that their SCBA are functioning properly.

The only way to know this is to put the SCBA into service outside the hazard area and take a few breaths. Perhaps there is a regulator malfunction and air is escaping, or maybe there is a facepiece problem, or maybe the batteries that operate the digital pressure gauge and the PASS device are dead. The place to discover these problems is outside the fire building, not inside. Just as scuba divers check that their tanks are full and that the entire air supply system is working correctly before going underwater, firefighters must check that their SCBA are functioning properly and that the air cylinders are full before they enter a structure fire.

Once a company (or team) of firefighters have entered the fire building, how do they use the ROAM? The firefighters must check their air at intervals and let the officer know their pressures. The company is always limited by the firefighter who has the lowest pressure. The officer should request air checks about every five minutes, depending on the task that the company has been assigned to complete. If the company's task is

ROAM Rules

The concepts in this book take the science of air management to the next level. In the fire service, we wear seat belts as a precaution to ensure that our firefighters arrive safely at the fire scene. Incorporating an air management policy helps ensure that firefighters return home safely.

This chapter describes an approach to the fireground that will better prepare firefighters in your community. In the Seattle Fire Department, we follow the ROAM and have successfully incorporated it into our regular training. Even with new technologies, the lack of a standard policy for air management can place firefighters at risk. If we manage our air properly, alarms from bottles low on air will become a thing of the past.

It is my hope that you will integrate this very practical approach to air management into your fire department safety procedures.

—*Gregory M. Dean (Fire Chief, Seattle Fire Department)*

physically demanding, like moving a 2½" hoseline down a hallway, the officer should have the team check their air more frequently than if they were advancing a 1¾" hoseline (fig. 10–12).

Fig. 10–12. A firefighter checking the air gauge while helping to advance a hoseline during an air management drill. Note the instructor behind the firefighter, making sure that the firefighter is working safely and effectively.

NATURAL BREAKS

Companies operating in the fire building usually experience natural breaks in the performance of their tasks. These natural breaks are a good time to have the company check their air. Companies advancing hose down hallways or up stairways usually stop every so often, to let the firefighter on the nozzle put water on the fire or to allow the officer to monitor conditions overhead. Firefighters should check their air immediately before or after changing levels in the structure, before entering a room, after moving down a hallway, before or after searching a room, and before or after completing a physically taxing task, such as moving furniture or debris on the way toward the seat of the fire. These natural breaks present themselves throughout a firefight, and firefighters can quickly and easily check their air and report the pressure to the company officer.

Air checks can also take place on an individual level. Firefighters are presented with many opportunities during a fire when they are not working. Thus, individual firefighters can check their own air pressure whenever they have a few seconds. By doing this, firefighters obtain a clear sense of their air supply and can make good decisions based on remaining air. These checks take only seconds to accomplish and do not interrupt the firefight at all.

REPORTING AIR PRESSURE

The most fundamental skill of air management is deciding how your department, company, or team is going to report air pressure. Everyone must be speaking the same language when talking about air pressure, so that there is no confusion. This important and basic skill takes training and practice and should be used any time members are using their SCBA, whether in training or on the fireground.

Does your department report the actual air pressure readings in psi? Or do you measure air pressure by quarters, for example, ¾, ½, and ¼? Whatever method of talking about air pressure your department chooses, everyone on the fireground must know it and use it. A common language when discussing air pressure is particularly important for fire departments and fire districts that routinely respond to other cities, towns, or districts for mutual aid. While not everyone on the fireground will be using the same type of SCBA, everyone should be using the same language when it comes to their air supplies.

DRAFTING OR BUMPING UP

The officer (or team leader) must take the lead in using the ROAM. The officer must make sure that the company gets out of the hazard area before any member of the team's low-air alarm activates. The officer must also maximize the company's time in the hazard area. By monitoring the workload and air use of the firefighters, the officer can redistribute the workload to other firefighters in the company, so that the work is shared. We call this "*drafting*" or "*bumping up*" within the team or company. Think of professional bicyclists in a road race. Each rider in a team takes a turn in front of the line so that the other members of the team do not have to expend so much energy and can recover a bit. By drafting, the team can conserve its energy and allow individuals to stay as fresh as possible. Professional race car drivers use the same technique, drafting off the lead car to conserve fuel.

If one member of the firefighting team is doing all the work, then it stands to reason that this member is probably using the most air. The officer can replace this firefighter, who has been working and breathing hard, with a firefighter who has been drafting, or not working as hard. Drafting allows the company to maximize its time in the fire, because time inside is directly related to air supply. The more air the team members have, the longer they can work inside the fire building.

The technique of drafting requires training and practice. However, companies who have never practiced this technique before have become proficient at it over the course of an eight-hour training day. This effort is well spent, since drafting extends the company's work time in the fire.

PRACTICING THE ROAM INCREASES SITUATIONAL AWARENESS

Because air checks take only seconds to accomplish, they do not have any negative effect on the firefight. In fact, we have discovered that when officers (or team leaders) call for an air check at a natural break, they also take a few seconds to observe their surroundings. The officers use this time to become more aware of where they are, what the fire and smoke conditions are like, and what is going on around them. In other words, following the ROAM increases situational awareness. This is a positive result of practicing the ROAM that was never predicted, and an additional benefit to the safety and effectiveness of the firefighters.

If firefighters are following the ROAM, then they should be out of the hazard area before their low-air alarm activates. But how does the officer/team leader know when it is time to leave the fire building? This is a skill that takes practice. The officer must make a decision when to leave, and this decision should be based on a number of factors:

- How far is the company from the entry point?
- How long did it take to get where they are now?
- Have conditions worsened?
- Will it take longer to get back to the entry point owing to a change in smoke or fire conditions?
- How tired are the members of the company?
- Who has the lowest air pressure in the company, and how fast is that firefighter consuming air?
- What are the stress levels of the firefighters in the company?

All of these factors affect the duration of the air supply and must be taken into account when making the decision to leave the hazard area. When firefighters are first learning to use the ROAM, it usually takes them a few rotations through the hands-on props to recognize when they should turn around and leave so that they are out before their low-air alarms activate. Again, this is a training issue. Firefighters must repeatedly practice ROAM to become proficient.

For some good hands-on drills that you can download for free, visit our website, Manageyourair.com. Under the "SMART Drills" tab, you will find a number of proven, time-tested drills that will help you train your firefighters to follow the ROAM. These hands-on drills range from the very basic to the advanced. Everything is there for you to make your firefighters successful in using the ROAM—from stated training objectives, to materials needed, to diagrams, to debriefing points. Again, training on the ROAM, which is a new basic firefighting skill, is a must.

OTHER METHODS OF AIR MANAGEMENT

While the ROAM is the simplest, easiest solution to air management, there are different methods of air management that are used effectively in the fire services of other countries and in other fire departments in the United States. We will discuss these options in much more detail in chapter 21.

After the death of Bret Tarver, the Phoenix Fire Department realized that the single-family dwelling air-use mentality was deadly and had to be rethought. Under the progressive leadership of Chief Alan Brunacini, the Phoenix Fire Department became the first U.S. fire department to adopt an air management policy.

Phoenix Fire Department uses an air management system that is based on work cycles and having companies *on deck*—that is, ready to replace the companies that have been working inside the fire building. The IC, the sector chiefs, and the fire officers know that their firefighting teams can work only for a certain amount of time inside a structure. Company officers inside the fire building are responsible for having their firefighters check their air pressure periodically. This method ensures that air checks are being performed by all members inside the hazardous atmosphere. This system works for Phoenix. However, Phoenix's system is resource dependent. Not many fire departments in the U. S. have the necessary staffing to be able to use the Phoenix Fire Department's system.

Some in the U.S. fire service advocate extending the low-air alarm time to solve the air management problem. They have proposed that the SCBA's low-air alarm should activate when the air cylinder is half empty. This, they argue, would signal the firefighters inside the hazard area to begin leaving the structure.

This idea of extending the low-air alarm time does nothing to deal with the single-family dwelling air-use mentality. If firefighters are reluctant to leave the fire building with a quarter cylinder of air left, as they are today, then imagine the trouble firefighters would have leaving the fire with half a cylinder of air left. Aggressive firefighters might say to themselves, "We have half our air left—so we can stay, fight some more fire, and still get out with plenty of air to spare."

Another problem with the idea of extending the low-air alarm time is that it would not remove the low-air alarm from the fireground. Thus, the dangerous practice of treating the low-air alarm as just another nuisance alarm—just another background noise to be ignored—would continue. How will firefighters distinguish whether the low-air alarm that they are hearing in the fire building is a firefighter leaving, or a firefighter who is in trouble? They cannot tell the difference. Therefore, the idea of extending the low-air alarm time does nothing to solve the dangerous practice of ignoring the low-air alarm.

When the low-air alarm is removed from the fireground by following the ROAM, several positive results are achieved—the nuisance alarms are eliminated, and the low-air alarm becomes a true emergency alarm. We almost lost two of our friends in the Seattle Fire Department because our firefighters simply ignored the low-air alarm on the fireground. This is why we are so passionate about having firefighters leave the hazard area before their low-air alarms activate.

Extending the low-air alarm time would create more problems than it would solve. It certainly is not a realistic solution to the fire service air management problem. The time to exit the fire building is before the low-air alarm activates.

The ROAM is the solution to the deadly problem of firefighters running out of air in structure fires. It is simple, easy to learn, and makes sense. The scuba diving industry has been using a similar air management standard for decades, with positive results.

The first three-quarters of the SCBA air cylinder is for the operation, for the citizens we have sworn to protect, and for our fire department or fire district. The air in the last quarter of the cylinder is for us. It is our emergency reserve, to be used only if something bad happens to us inside the hazardous environment of a fire. This emergency reserve is your family's air, your wife or husband or partner's air, your children's air, your loved ones' air; therefore, it should *never* be used for operational firefighting.

If you choose to use your emergency reserve of SCBA air for operational firefighting, understand the bet that you are making. You are literally betting your life that nothing bad will happen to you on the way out of the fire building. Consider all the case studies we have discussed throughout this book. Is breathing into your emergency air reserve in order to pull a bit more ceiling or put out some hot spots really worth the gamble? Remember, the time to exit the fire building is before the low-air alarm activates; your life depends on it.

MAKING IT HAPPEN

1. Have your firefighters wear their PPE and don their SCBA. Place them in a zero-visibility environment and ask them to read their air pressure gauges. See if they can read the pressure gauge. Note how they talk about pressure. This is a good drill to discover any problems with reading the pressures from the gauges and to discover whether your firefighters use the same language when discussing air pressures within the company. Afterward, hold a discussion on the gauge and the common language that should be used to discuss air pressure when inside an IDLH atmosphere.

2. At the Web site Manageyourair.com, click on the "SMART Drills" tab, and choose SMART Drill 60-1. Starting with this SMART Drill, complete as many SMART Drills as you can over the course of several months. Some of these drills require a large acquired structure.

3. Get several engine/truck companies together and ask the firefighters if they routinely check their air pressure when they are inside a fire. Initiate a discussion of air management.

4. Go to the Web site of the NIOSH Fire Fighter Fatality Investigation and Prevention Program (http://www.cdc.gov/niosh/fire/). From the list of all reports, select and print report 2001-33 for your crew(s). Discuss this report with the crew(s) in the context of the ROAM and the emergency air reserve.

STUDY QUESTIONS

1. What is the solution to the air management problem in the fire service?

2. If the low-air alarm is a nuisance alarm in your fire department, what actions can you take to make the low-air alarm a true emergency alarm?

3. Explain the single-family dwelling air-use mentality.

4. How are scuba divers and firefighters alike? What have scuba divers been doing for decades in regards to air management?

5. Which specialty teams in the fire service are currently following the ROAM? Why are they following the ROAM?

6. How could following the ROAM help the structural firefighters in operations?

7. When following the ROAM, how does the officer or team leader decide when it is time to leave the fire building? What are some of the factors that affect this decision?

8. When is the time to exit the fire building?

NOTES

[1] NIOSH Fire Fighter Fatality Investigation 99-F47, p. 5.

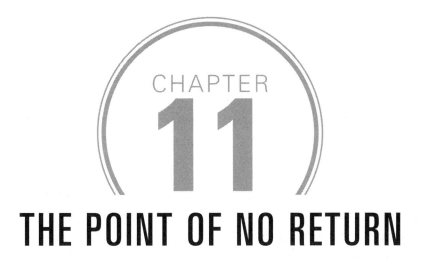

THE POINT OF NO RETURN

INTRODUCTION

The unfortunate reality of fireground operations is that most tasks will involve entering hazardous conditions for varying amounts of time. A suppression assignment usually requires that the hose team enter the structure and advance to wherever the fire is causing damage. Search teams, likewise, are likely to proceed to as many areas of the burning or damaged building as possible to provide help for those who couldn't otherwise make it out. This is the true nature of our operations, and it's the bread and butter of fire departments everywhere.

It is important for firefighters to recognize that the initial entry into the structure and the requisite tasks performed subsequent to that entry are not the only actions needed; there is also the important matter of exiting. A comprehensive operation will also take into account the following concept: at some point, a decision to cease our activities and/or turn around and depart must be made. This point varies with every operation, but the time-to-exit decision is one of the most critical choices a company officer will ever make. It usually makes the difference between a smooth and successful operation and one that is problematic.

Making an incorrect time-to-exit decision often occurs without any dramatic repercussions. We usually call these situations *close calls,* most of which are recognized only by the few individuals involved. Unfortunately, these continued close calls represent the tremors before the disaster, and continued bad practices will eventually result in a tragedy.

This chapter explores the concept of the *Point of NO Return* for firefighters and emphasizes a simple principle: Firefighters need to be part of the solution—not part of the problem. By bad practices, inadequate training, and/or simple refusal to operate safely, firefighters are routinely putting themselves in the position of becoming part of the problem. And, sadly, some are losing their lives.

WHAT IS THE POINT OF NO RETURN?

To the imagination, the concept of the Point of NO Return often suggests a situation in which there is no going back. In the case of airplane pilots, this occurs when they have used enough fuel such that a return to the initial departure point is not possible because they would run out before getting there. Another common image is a ship moving too close to the edge of the earth or a waterfall. At some point, the current becomes too strong to allow a return. At that point, there is no going back: the ship will be pulled over the edge. These examples and many others like them are what we typically think of when talking about the concept of the Point of NO Return. In essence, the end result that comes to mind is that you are dead.

The problem with this kind of imagery for the fire service is that it just isn't true. While we need to do everything within our power to avoid these types of situations, it is irresponsible to describe them as "game over." As a matter of fact, doing so increases the potential danger on the fireground.

For firefighters, the unexpected is an unfortunate part of the circumstances into which we are often thrust. Our planning, training, and best efforts enable us to minimize the risks of many variables on the emergency scene but will never eliminate all of them. There is no point at which firefighters should give up, quit trying, or roll over and die should things go bad for them. Firefighters never quit, never relinquish, and never give up.

This is an extremely important concept to fix in your mind, because mental imagery can come back to haunt you when everything has gone wrong. In police weapons training, instructors have learned to be adamant about the fact that getting shot is not fatal in most instances and any officer hit should not give up. This stands in sharp contrast to prior weapons training in which any student who sustained a hit, no matter where it was on his or her body, was instantly pronounced dead. The incredible reality of training the latter way was that officers in real shootouts gave up—and even died—on receiving only superficial wounds to such places as their arms and legs. The imagery of the training stuck.

A watershed incident for law enforcement officers occurred in Newhall, California, in 1970. Four officers died in a brutal gunfight, prompting many changes to weapons training. Of significance was the following:[1]

> *At least one of the officers was found clenching fired and spent .38 caliber casings in his cold, dead hand, the result of an instinctive reversion to range training during the firefight. During range training, cadets were required by firearms instructors to "pick up their brass" before moving on to the next stage, where they would fire at stationary targets, not ones that simulated movement as exhibited by Twining and Davis. The California Highway Patrol demanded a clean, static range at the expense of training its officers to utilize cover, movement, and quick reloading of weapons. Officers Gore, Alleyn, Frago, and Pence paid the ultimate price for the California Highway Patrol's inadequate firearms training practices.*

The practice of collecting brass was what the officer reverted to when things got tense. As firefighters, we cannot allow faulty mental imagery and training to be the default mechanism that kicks in when things go wrong. This is especially true when things go dramatically wrong.

WHAT IT'S NOT

The point of NO return is *not*

- The point at which you give up
- The point at which you lose hope
- The point at which you die

Firefighters should never give up, no matter how many mistakes they've made that have brought them to a position of being in trouble. We should do everything in our power to put ourselves in a position of being rescued. By the time a firefighter is at the point of being finished, he or she will not be aware of it. That is the mental imagery we want to instill.

WHAT IT IS

We spent quite a bit of time in discussion with colleagues about an appropriate definition of the Point of NO Return—one that would accurately apply to firefighters. The following fits perfectly: The Point of NO Return is the point at which you stop being part of the solution and start becoming part of the problem.

By your actions and/or the circumstances of the incident, you have moved to the point where you are no longer having a positive impact on the operation and instead are having a negative impact. This is significant because your assignment affects the entire operation. The IC has entrusted you with a portion of the operation that is expected to be accomplished. Now, not only is that element of the plan not being taken care of, but the IC must also allocate resources to take care of you (fig. 11–1).

Fig. 11–1. Firefighters being forced out the second-floor
window because they have run out of air. Fortunately,
the tower is able to rescue them. (Photograph by
Dennis Leger, Springfield [MA] Fire Department)

By allowing yourself to become part of the problem, you set in motion a series of events that hopefully will end with a good lesson learned and some good-natured ribbing. What cannot be overstated is that your mission was not accomplished, and valuable time and resources were wasted. This is in the best-case scenario. The reality of close calls is that they are a step away from injury and death.

FACTORS THAT AFFECT THE POINT OF NO RETURN

Many factors contribute to and affect the point of NO return. Some of these have been derived from NFPA 1404, and others are from research and experience. Each factor contributes, both in isolation and in tandem, causing firefighters to become part of the problem.

HAVING THE APPROPRIATE EQUIPMENT

Having the equipment needed to get the job done is a staple of effective fireground operations. When the expertise and training of a firefighter is combined with the proper equipment, operations usually proceed well on the emergency scene.

The opposite is equally true. Showing up to the controlled chaos that is the emergency scene with the wrong equipment is a recipe for disaster. Failing to utilize the equipment that you have in the proper way is also dangerous. Furthermore, lack of training in the use of your equipment is the icing on a professionally irresponsible cake. It is the responsibility of every firefighter to come equipped and ready to do the job.

The first area that needs our attention is PPE. The correct donning of bunking gear, including the protective hood, sets the tone for the operation. It is very difficult to recover from sloppy practices inside the hazardous atmosphere. Using bunking gear that is in disrepair is an invitation to injuries or burns that could lead to bigger problems should the firefighter panic (fig. 11–2).

*Fig. 11–2. Firefighters entering with the necessary
equipment to accomplish their assigned task
(Photograph by Lt. Tim Dungan, Seattle Fire Department)*

Respiratory protection is a second area that needs to be addressed prior to entry. Your mask should be checked every shift and after every use. As, arguably, the single-most important piece of equipment in the firefighter's arsenal, it should be treated that way. Going to battle with an SCBA that is unchecked, questionable, or damaged is indefensible behavior. Not training on your SCBA to a level of high proficiency is a dereliction of duty.

Finally, bring the tools necessary and appropriate for the task you are assigned. Don't bring a pick-head ax and a battle lantern with you when your assignment is to pull ceilings. Conversely, grabbing an eight-foot pike pole for a search team will make you look stupid. Firefighters should anticipate the needs of the job before they make entry and should bring appropriate equipment to meet those demands. Not doing so moves you one step closer to being part of the problem.

BEING PREPARED—PHYSICALLY, MENTALLY, AND EMOTIONALLY

Preparation is the key to performing at high levels in any endeavor. This is true of the fireground and any emergency scene. The point of NO return loves nothing more than an ill-prepared firefighter who is not ready to do battle.

Physical fitness is of primary importance owing to the intense work required and the adverse conditions encountered. It is common to face extreme temperatures, limited visibility, and confined areas all while doing labor-intensive work. To meet the demands of these situations, firefighters must train in the areas of strength, endurance, and flexibility. There can be no faking these traits when the job requires it. Your fitness has a direct impact on the amount of work that you can accomplish, your impact on the team, and the safety of firefighters on the fireground (fig. 11–3).

Fig. 11–3. Firefighters working to ensure that they are fit for duty (Photograph by Kim Favorite, Seattle Fire Department)

Mental readiness is another area that needs attention. All the physical preparation in the world will not be of much use if you can't use your head during emergencies. The ability to process information, make quality decisions, and react to changing variables are necessary elements for the successful mitigation of any emergency. Avoiding tunnel vision and minor distractions requires that firefighters think, as well as act. It is essential that training involve decision-making elements in its objectives and that firefighters discipline themselves to be mentally sharp at all emergencies.

Emotional preparation is often overlooked and minimized because of the macho culture of firefighting. This is both unfortunate and unnecessary. Emergency operations are usually centered around very emotional situations, and firefighters are in the middle of it all. Therefore, it is important to consider our own emotional status and ensure that it is sufficient for the job that we must do. Emotional state should be addressed before we are in the middle of a mass-casualty incident or a rescue of trapped firefighters.

Company officers should strive to stay aware of what is happening with the members of their team and provide help if it is needed. Follow up—as through *critical stress debriefing,* after tough incidents—should be used to address problems before they get out of hand. Many cities have employee assistance programs (EAPs) available to assist with family and personal problems. It is foolish to think that firefighters' personal lives will have no effect on them once they put on the uniform. Not all problems can be addressed by the company officer, but ensuring that our members are emotionally prepared to respond is essential.

KNOWING YOUR AIR SUPPLY ON ENTRY

Getting off to a good start is the best bet for effective operations. Beginning with a full bottle is essential if you expect to get the job done in interior operations. While firefighters routinely forget to verify their air supply prior to beginning work, this is not smart. The preentry air check verifies that nothing has changed since your initial SCBA check, and it gets the operation off on a firm foundation. Depleted air supply is an essential variable to eliminate and takes only seconds to check. Training makes this second nature, and firefighters will be safer as a result (fig. 11–4).

Fig. 11–4. Firefighters utilizing the ROAM by checking their air prior to entry

CONTINUALLY EVALUATING AIR SUPPLY AND TEAM STATUS

Starting off right is only half the battle; continuing best practices is essential, to avoid becoming part of the problem. Because of varied workloads and air use rates, the only way to know your team's air status is to check it. This doesn't happen much in current fireground practice because firefighters have come to rely on the bell to prompt an exit. As discussed earlier, this is a dangerous practice.

An ongoing evaluation of the team's air supply is quick, easy, and makes too much sense to ignore. It allows for better utilization of team resources since members with higher air levels can be assigned more strenuous tasks, thereby expanding the team's ability to get work done. Problems with team members' air status will be more readily identified if routine checks are part of your plan. Better time-to-exit decisions can be made while keeping the emergency reserve intact for true emergencies. As important as the air supply is to the safety of your team, there is no good excuse to not monitor it as you proceed.

Likewise, firefighters must be aware of both physical conditions of their team and what is happening around them. Maintaining good situational awareness is critical to accomplishing your task and not becoming part of the problem. Changes in smoke, heat, or structural stability are critical factors that need to be known by the team and relayed to the IC. Any changes in location of the structure or alteration of the assignment also apply. Likewise, team leaders should monitor fatigue levels and be prepared to give members less taxing roles if their strength or air levels become diminished (fig. 11–5).

Fig. 11–5. Firefighters discussing their next actions
(Photograph by Lt. Tim Dungan, Seattle Fire Department)

An interesting fact surfaced when drilling companies checked their air while committed to interior operations. The air check caused a pause in the action that greatly enhanced situational awareness. Instead of blindly rushing onward or getting tunnel vision, team leaders stopped and assessed their members. Directions thus became clearer and more focused. Instead of causing significant delays in the operation, as was feared by some, the opposite occurred. This proved to be one more reason for good air management.

KNOWING THE PHYSICAL LAYOUT OF THE STRUCTURE AND ANY VARIABLES AS YOU PROCEED

Most professional firefighters know that size-up does not end with the initial officer's report on arrival at an emergency. Size-up continues throughout the incident and is essential to a safe and effective attack. All firefighters should be involved in the ongoing size-up, no matter what their assignment or role in that task.

Recognition of the basic layout of the structure prior to entry is critical. Features that should be noted include the following:

- Type of construction. How will this building fail in fire?
- Type of occupancy. What is it used for? Where are the people likely to be?
- Means of egress. How do we get in—and, more important, how do we get out?
- Stories. Are there belowground levels (e.g., a basement)?
- General condition

This initial evaluation can prove invaluable when things do not go as planned. When the initial entry point becomes unusable as an exit, for whatever reason, having already determined alternative means of egress can save your life (figs. 11–6 and 11–7). Determining from the outset that the building is of wood-frame construction should trigger caution when it is determined that structural elements are well involved in fire. A few seconds of evaluation can be of enormous significance when the incident tries to make you its victim.

Fig. 11–6. A large commercial structure, facing the entrance. Note the lack of exits toward the back of this building. In your size-up, would you assume that crews could exit from the rear of the building?

Fig. 11–7. The same large commercial structure, viewed from the rear. Assuming that there is an exit in the back could prove to be a deadly miscalculation.

A continuous size-up should be performed after entry is made. Maintaining good situational awareness is essential and, at a minimum, should involve the following:

- Is our plan working?
- Do I have team integrity? How are team members doing?
- Where are we in relation to our exit?
- What are my current conditions? Are they changing?
- Have I given timely/accurate progress reports to the IC?
- What is the air status of the team?

Team leaders who are on top of these items should have great success, conducting interior operations in a safe and effective manner. They will also be prepared for the unexpected and able to adapt more readily when the plan goes awry. Remember that if things go wrong at the incident, it does not automatically mean that your team will become part of the problem. Good situational awareness can aid you in continuing to be part of the solution.

WHAT DRAWS YOU TOWARD THE POINT OF NO RETURN

From the moment you arrive at the fire station for your day's duty, the Point of NO Return is actively trying to ruin your day. Make no mistake, the Point of NO Return is a nasty, vicious animal that seeks to break your heart and the hearts of your family, crew, and department. While we are never going to completely eliminate the times when we become part of the problem in our careers, we can at least recognize what pulls us toward that end. Knowing the areas that cause problems allows us to minimize the Point of NO Return's incessant draw toward tragedy.

STAYING IN THE HAZARD AREA
UNTIL YOUR LOW-AIR ALARM ACTIVATES

Firefighters across the country routinely remain in the hazard area until their low air warning alarm activates, giving little thought to the danger to which they are exposing themselves and their crew on a shift-by-shift basis. While this is detailed in other chapters, it is well worth revisiting in this context. You are rolling the dice and betting that nothing goes wrong on the way to the exit when you consume your emergency reserve. However, things do go wrong, and there is no way to totally prevent them from occurring. Every firefighter worth his or her salt knows that. So why not consistently give yourself the time you need to deal with these occurrences?

The Point of NO Return loves the sound of low-air alarms activating inside a smoke-filled structure. Every alarm that sounds is one more firefighter who is becoming part of the problem.

In Fort Worth, Texas, a veteran firefighter experienced the point of NO return in a routine, bread-and-butter house fire that left a lasting impact on the way he views emergency operations. Bart Bradberry was fighting a routine house fire on the second floor when a very simple thing went wrong: He got disoriented. This led to his having to go out a second-floor window and almost ended his career. What should get your attention is how small the margin for error was that led to that plunge from the second-floor window; it was a five-inch wall that separated the stairway from just another second-floor room (fig. 11–8). That margin for error—and you face this at every fire—made the difference between a routine close call that no one ever hears about and the dramatic incident Firefighter Bart Bradberry shares with us in this book. Figures 11–9 through 11–14 detail Bradberry's plunge out a window after running out of air at a house fire.

*Fig. 11–8. The five-inch wall that made the
difference between safety and a desperate
plunge out a second-floor window*

*Fig. 11–9. Firefighter Bradberry attempting egress
through a window after running out of air at a house fire
(Photograph by Glen Ellman, Fortworthfire.com)*

Fig. 11–10. Firefighter Bradberry plunging out a
window after running out of air (Photograph by
Glen Ellman, Fortworthfire.com)

Fig. 11–11. Firefighters attempting to catch Bradberry
(Photograph by Glen Ellman, Fortworthfire.com)

*Fig. 11–12. Firefighters accompanying the injured
Bradberry from the house fire after his fall
(Photograph by Glen Ellman, Fortworthfire.com)*

Fig. 11–13. This is the face of the Point of NO Return. From the moment you arrive on duty it is trying to make this a reality in your department. (Photograph by Glen Ellman, Fortworthfire.com)

IN TOGETHER, OUT TOGETHER

On December 23, 2003, I was part of a response to a reported structure fire around 8:30 in the morning. My unit, Engine 8, was dispatched as the second due unit. Heavy smoke could be seen in the distance as we left the fire station, so we knew that we had a working fire.

As the first engine arrived on scene, they reported fire in the rear of the house, a house like you could find in any part of the country. From the front curb, the home appeared to be a single story, but as you looked down the side, you realized that there was a partial second story. Owing to some traffic and access issues, we actually arrived fourth on scene. Two hand lines were already being operated on the first floor, and one hand line was being operated between the house and an adjacent house because of an exposure fire. By this time, the fire had extended to the second floor.

On our arrival, the IC directed us to stretch a 1¾" hand line to the second floor and begin fire suppression. With my lieutenant positioned behind me, we advanced a hand line up a set of stairs to a second-floor landing. There was heavy fire venting from a room to the right of the stairs, as well as from a room to the left of the stairs. A hallway was barely visible past the room on the right. I began to attack the room to the left, which we later found out was the attic and had a normal-sized door as an entrance, and the other two firefighters with my company began advancing a 2" hand line to the room on the right. Simultaneously, a truck company cut two vent holes in the attic, which by now was well involved.

We were experiencing extreme heat and poor visibility and having difficulty knocking the fire down when the low-air alarm began sounding on my lieutenant's SCBA. Seconds later, my alarm began sounding, and I shut the line down in preparation for us to exit the building. Suddenly, the fire came roaring right back at us. To protect our exit path down the stairwell, I opened the line back up and kept flowing water while my lieutenant went down to the base of the stairs, where he directed a relief crew to my position.

When the two-person crew arrived to relieve me, I turned to my right and somehow one of my gloves became entangled in the air pack frame of one of the firefighters. My lieutenant, who was still at the base of the stairs, became concerned when I did not join him after being relieved and followed the hoseline back up to my position. I informed him that my glove had gotten stuck but that I would be right behind him as we exited. As he turned and began his way down the stairs again, I jerked my hand free of the SCBA frame, but in the process, I lost contact with the hoseline—our lifeline to the outside.

Even at that point, I was not concerned because I knew that I was only a step or two from the stairwell. However, when I turned to exit, I hit a wall where I thought the stairs were supposed to be. Apparently, as I pulled my hand free, I had also turned my body and was now oriented differently on the landing. The wall that I hit happened to be the area between the stairwell and the room to the right that was involved earlier. I then felt my way to the right but could not locate the stairs, and I hit another wall.

It was at this point that I realized I did not have enough air to find the stairs and would need to look for an alternative exit. I glanced back to the left and saw lighter smoke, which I determined must be due to light coming through a window. When I got to the window, I took my regulator off so that I could call to the firefighters outside and make them aware of my situation. The smoke was so hot that as I attempted to speak, it literally took my breath away. I had no choice but to hook my regulator back up.

Knowing that it was still not safe and I had just a few more breaths of air, I crawled out of the window headfirst onto a very small second-floor awning, with the intention of swinging my legs around and dropping to the ground feetfirst. The firefighters operating on the stairway landing saw me crawl out of the window and, not knowing that there was a ledge and fearing that I would fall, grabbed my legs and held on to me. At the same time, personnel on the ground— including my lieutenant, who had exited the building and, not seeing me behind him, was calling a Mayday—saw me and reached up to assist me to the ground. Even though my helmet, face shield, and hood were severely damaged owing to the high heat, I was not injured.

Looking back at this fire, I realize how important it is to conserve your air, stay with the hoseline (no matter how small an area you are operating in), and have a backup plan ready. The ability to create that backup plan comes from one thing alone: training. The Fort Worth (TX) Fire Department places an extremely high value on all areas of training. Our training division does a wonderful job of training us on firefighter survival, among many other topics.

I have no doubt that repetitive training played a major role in the outcome of the event. The area in which we were operating was small with the exit nearby. However, having high-heat conditions with near-zero visibility, becoming momentarily entangled in my gear, and then being separated from the hoseline in the process of freeing myself are all things that could as easily happen to anyone at any fire on any day.

Just as preplanning fires is important for a safe outcome, air management preplanning should begin prior to arrival on scene. Recognize that unforeseen events can and will happen; always allow enough time to safely exit the structure. We should rotate crews out sooner and not place ourselves in low-air situations. It is essential that firefighters exit before their low-air alarms begin sounding. Once the alarm sounds, we are literally on borrowed time.

As we say in Fort Worth, every firefighter's goal should be to go in together, come out together, and then go home safe and sound.

—Bart Bradberry (Fort Worth [TX] Fire Department)

TUNNEL VISION

Tunnel vision can impair any of us. Most firefighters are go-getters who want to get the job done and love being a part of the solution. There comes a point, however, when the determination to accomplish the task can narrow the focus so much that obvious danger signs are missed. Tunnel vision can prevent even the most seasoned veterans from using their experience to recognize changing conditions or alter their plan based on what is working—or, perhaps more important, based on what is not working.

Tunnel vision can develop because of many factors, especially in situations that require quick decisions and difficult working conditions. Fighting fires is the perfect arena for it to rear its ugly head.

There are two primary ways in which tunnel vision comes into play, causing you to become part of the problem. The first takes advantage of our fierce desire to get the job done. As stated earlier, firefighters are in the business of providing solutions to problems. Our practice of hitting hard, fast, and effectively can move us into a situation where we allow our focus to become too narrow. In pushing forward to accomplish our task, we can get so locked in that we are not maintaining the situational awareness that is critical to our job. The desire to get the hoseline to the seat of the fire may overcome our sensory recognition that the lightweight I beams of the floor below are sagging. Or the search-and-rescue assignment that we've been given may become too important to wait for a hoseline to protect the stairs that we are about to ascend. There must be a balance, which comes only through training and experience, of gung ho aggressiveness and situational awareness.

The second way in which tunnel vision can strike is in making assumptions and/or committing to preconceived notions that end up being wrong. An initial plan of attack needs to be made when we arrive at an emergency. We attack the problem based on what we know at the time. If our strategy and tactics do not evolve as the situation plays out, we become part of the problem.

This happens all the time when firefighters attempt to make aggressive attacks on quick-moving incidents. The need for ventilation cannot override the need to evaluate the roof before venting; search and rescue should not be performed in compartments that obviously cannot sustain life; and the list goes on. Make your initial size-up and go to work. Don't allow yourself to become fixated on and/or so enamored with your plan that it doesn't change to meet new conditions or information. That is the kind of tunnel vision on which the point of NO return thrives.

Ego

Many well-intentioned firefighters have succumbed because of the devastating effects that ego can have on a fireground operation. Ego takes many forms and is not always a bad thing. Pride in personal preparedness, in our company, and in our departments is essential and should be encouraged (fig. 11–14). Ego's positive manifestation is feeling good about working hard, taking the job seriously, and being among those who can be counted on to get the job done.

Fig. 11–14. Firefighters are proud of the job they do and the people they work with (Photograph by Lt. Tim Dungan, Seattle Fire Department)

Ego leads us astray, though, when it allows us to develop the attitude that we're the only ones on the fireground who can get the job done. This usually interferes with good teamwork, resulting in freelancing and distrust among the companies involved. These are the perfect ingredients for the point of NO return to do its damage. With firefighters working against each other, instead of pulling together, the operation is much less than what it could be.

Further problems occur when ego will not let you admit that the task cannot be completed. This is particularly true as it pertains to air levels because running out of air is an unforgiving tutor. If you are pushing into your emergency reserve simply because you don't want someone else to finish the job, you are moving toward becoming part of the problem. How can we speak glowingly of the great brother/sisterhood of the fire service, yet be too arrogant to work as a team to accomplish our jobs? This is the negative manifestation of ego that is injuring and killing firefighters every year.

POOR COMMAND STRUCTURE

As stated earlier in this chapter, getting off to a good start is essential. So it goes with incident command, and starting poorly usually leads to less-than-stellar operations. Poor strategic and tactical decisions are obvious factors that make us part of the problem. Setting up a good command structure and communicating clearly are imperative.

Two additional factors, while more subtle than a bad size-up, contribute to many close calls and fireground injuries/fatalities: lack of accountability and poor crew rotation (fig. 11–15). The fire service is doing much better as it pertains to accountability on the fireground. There are different systems, from Velcro name tags on the passport system to metal tags dropped in a bucket at the fire. Regardless of how it's done, the ultimate result must be the same: knowing who is in there and where they are supposed to be. This should be a major priority of ICs and should be tracked throughout the incident.

Figure 11-15. The Incident Commander must ensure
proper crew rotation and accountability. (Photograph
by Lt. Tim Dungan, Seattle Fire Department)

Imagine the stress level when a Mayday is called from the structure or someone turns up missing as crews are relieved. That is not the time to find out that accountability is lacking and there is really no clear idea of where to start the search. This is also true for team leaders who need to stay apprised of where their members are and not actively involved in the task at hand. If a member of a team goes missing or is having difficulty, the team leader should be aware of it instantly and take appropriate action.

It is also important that teams be able to rely on a timely replacement crew to relieve them when their air dictates that they should leave. ICs put crews in a difficult position when they do not have a crew ready to go and must rely on a delayed reaction from staging to cover relief. The Seattle Fire Department has addressed this by making it mandatory for crews to radio out when they have reached 50% of their bottles. This notifies the IC that the crew will be exiting soon and that a crew should be called up and ready to replace them. Rotation keeps crews from feeling like they have to stay in longer than their air allows and provides a smooth transition of teams to combat the emergency.

LACK OF RIT TRAINING

As you will read in chapter 18, the emphasis in RIT should be on prevention. We'd rather see the RIT never have to be used because we are not getting firefighters into trouble. Training hard to not need rapid intervention is one thing, but not being prepared when it is needed is deadly.

Interestingly, this facet of the point of NO return doubles the problem. Obviously, someone in the operation has reached the point of NO return when a Mayday has been called. The lack of an RIT or one that is poorly trained compounds the problem because now the rescue effort will be unsafe and will very likely lead to further complications. Remember that RITs are usually deployed under very difficult conditions. Even well-trained RITs face a difficult job. The Southwest Supermarket fire is a clear example of this, as the Phoenix Fire Department had 12 additional Maydays called during their efforts to rescue Bret Tarver—and this is a department that trains diligently on effective RIT operations and has sufficient personnel to staff them. Make no mistake, tempting fate with faulty RIT procedures is a fool's gamble (fig. 11–16).

*Fig. 11–16. Crews gathering the necessary equipment for RIT
(Photograph by Lt. Tim Dungan, Seattle Fire Department)*

INEFFECTIVE BASIC FIREFIGHTING TECHNIQUES

It always comes down to the basics. The point of NO return thrives in an environment where the basics are forgotten. Many fire departments have found themselves devoting the majority of their training budgets to WMD and technical rescue training. While these areas certainly merit attention, they should never take precedence over training for the events that we must expect. The priority should always be to train in the actions that you are likely to perform day in and day out, as that is where the injuries occur. There are very few injuries and fatalities at technical rescue incidents. The reason for this is that those operations are done in a methodical manner permitted by the type of response. Those incidents are low-frequency events.

You will go to structure fires on a regular basis no matter where you are in the country. At those fires, certain operations must happen, not only to end the emergency but also to keep firefighters safe on the fireground. Firefighters need to be consistently trained on the following skills (figs. 11–17 through 11–21):

- Hose
- Ladder
- Search and rescue
- Forcible entry
- Ventilation
- Overhaul
- Apparatus

This training gives the biggest bang for your training dollar and provides the best service for your customers. Letting this training slide allows the point of NO return an open door to your operations.

Fig. 11–17. Ladder training (Photograph by Lt.
Tim Dungan, Seattle Fire Department)

Fig. 11–18. Search and rescue training
(Photograph by Lt. Tim Dungan,
Seattle Fire Department)

Fig. 11–19. Ventilation training
(Photograph by Lt. Tim Dungan, Seattle Fire Department)

Fig. 11–20. Hose skill training (Photograph by
Lt. Tim Dungan, Seattle Fire Department)

*Fig. 11–21. Forcible-entry training (Photograph
by Lt. Tim Dungan, Seattle Fire Department)*

LACK OF TRAINING IN AIR MANAGEMENT TECHNIQUES

The thrust of this book is that training in air management is essential and will save the lives of firefighters. The point of NO return seeks to make you part of the problem, cause injury or death, and break the hearts of your family and your department. The quickest way to get to that point is to neglect training in air management.

Not being skilled in air management techniques means that you will be more likely to run out of air at emergencies on a routine basis. Consequently, you'll be exposed to the toxic products of combustion, which can kill you on the spot or down the road, in a distant hospital. It also means that your chances of getting lost or disoriented are greater. The RIT skills described previously will now have to come into play, putting even more firefighters at risk. This is the perfect breeding ground for the point of NO return, whose goal is that well-intentioned firefighters become part of the problem.

AIR MANAGEMENT DURING ENCLOSED STRUCTURE FIRES

In 1997, a three-alarm thrift store fire resulted in the disorientation and near death of a veteran San Antonio firefighter. Subsequently, a 3-year firefighter disorientation study, focusing on a 22-year time span, was conducted to determine the specific causes of and means to prevent *firefighter disorientation*, defined as loss of direction due to the lack of vision in a structure fire.

Seventeen national disorientation structure fires resulting in 23 firefighter fatalities were closely examined. The study's findings underscored several operational and structural similarities, including the existence of opened and extremely dangerous enclosed structures. Enclosed structures have been determined to be the type of structure that takes the lives of firefighters at the greatest rate.

Of particular interest, a disorientation sequence was uncovered that was followed in 100% of the enclosed structure fires studied. In general, the sequence is as follows:

- Fire in an enclosed structure
- Aggressive interior attack
- Deteriorating conditions
- Separated or entangled hand lines
- Disorientation

In the first step, firefighters found light, moderate, or heavy smoke showing 94% of the time. This has since been determined to be an indication of extremely dangerous conditions when associated with an enclosed structure, not the traditional sign to quickly search for and attack the fire.

An interesting aspect of the disorientation sequence was the multiple ways it unfolded on the fireground. Following an aggressive interior attack, conditions would deteriorate, at times suddenly and violently in conjunction with flashovers and backdrafts or collapse of floors or roofs. During one scenario, light-smoke conditions found on arrival slowly deteriorated into zero-visibility conditions 52 minutes after arrival at a Kansas City, Missouri, paper warehouse fire, taking the life of a veteran chief officer who was first to arrive on the scene.

Hand-line separation would similarly result from a variety of causes. In some cases, the blast of a flashover or backdraft literally knocked bunkered firefighters completely off the line as heavy fire or smoke filled the structure's interior, causing instantaneous disorientation. Firefighters also became separated from hand lines as they fought the fire or collided with other firefighters in the interior who were simply not visible. The most unlikely way in which hand-line separation occurred was by a conscious decision of the firefighter not to use a hand line, possibly because of the firefighter's perception that one was not needed for the situation.

The solutions proposed to prevent firefighter disorientation include the use of a *cautious interior assessment*, which is a risk management process to determine the safest tactic during an enclosed structure fire. The ultimate choices available to firefighters may include interior attack either from the initial point of entry or from a different side of the structure, closer to the seat of the fire. An integral component of this risk management process involves an understanding of the correct use of air management.

Regardless of the tactic used, every firefighter entering an environment with life-threatening, prolonged zero-visibility conditions, which are common in enclosed structure fires, must exercise safe air management procedures. This involves exiting the structure as a cohesive crew, with adequate breathing time to reach the safety of the exterior in the event of an unforeseen delay.

Such delays are quite real and can be fatal, as reported in the U.S. Firefighter Disorientation Study. Extremely dangerous enclosed structure fires presented evacuating firefighters with the following delays in their efforts to reach the safety of the exterior: Of 17 cases, in 11 (65%), firefighters exceeded their air supply while attempting to evacuate the structure; in 10 cases (51%), companies lost integrity during evacuation from the building; and in 10 cases (51%), firefighters became confused when encountering tangled hand lines in zero visibility as they attempted to evacuate the structure. These findings seriously underscore the requirement for safe air management, to provide any firefighter in trouble with added time to receive needed assistance for a safe evacuation. When coupled with a safe and calculated selection of the type of interior attack to be implemented, the instinctive use of air management during enclosed structure fires provides one of the most valuable layers of safety readily available for the prevention of line-of-duty deaths in the fire service (fig. 11–22).

Fig. 11–22. An enclosed structure. During a fire, enclosed structures contain and conceal heat and smoke only to expose advancing firefighters to life-threatening hazards such as prolonged zero visibility, flashover, backdraft, or partial or complete collapse of the roof. Without regimented practice of air management, the air supply may not provide firefighters with the ability to safely exit the structure during an emergency evacuation should a delay occur. (Photograph by Captain William R. Mora, San Antonio Fire Department).

—*William R. Mora (Captain, San Antonio Fire Department)*

MAKING IT HAPPEN

1. Complete SMART Drill 60-4: The Point of No Return (available at Manageyourair.com).

2. Create drills that include typical operations a firefighter would perform and evaluate each member's personal air use and turnaround times.

3. Hold discussions in which groups of firefighters share situations where they have become part of the problem. Evaluate those situations against the factors you have learned about in this chapter and determine areas where you can strengthen your approach to fireground operations.

STUDY QUESTIONS

1. True or false: The Point of NO Return is the point at which the firefighter cannot recover and will likely die.

2. What is the definition of the point of NO return?

3. Mental and physical preparation are critical to avoid the point of NO return. What other related factor is often overlooked in emergency events?

4. What three critical changes of the interior fire conditions must be noted and relayed to the IC?

5. True or false: Making assumptions and committing to preconceived notions about the incident are two elements that contribute to tunnel vision.

6. True or false: The pride that we take in our personal preparedness, our company, and our departments is a leading contributor to the ego that has a negative impact on the fireground.

7. What are two factors that subtly contribute to many close calls and fireground injuries/fatalities?

8. How many additional Maydays did the Phoenix Fire Department have during the Southwest Supermarket fire while attempting to rescue Bret Tarver?

9. Why are there very few injuries or fatalities during technical rescue incidents?

10. Not being well trained in Air Management skills will cause what to happen on a routine basis?

NOTES

1 The New Paradigm: Police Trends Toward More Powerful Handguns
   and the Mental Aspects of Combat Survival and Training, Chris
   B. Hankins Criminal Justice Institute School of Law Enforcement
   Supervision, Session XXIV, Dr. Kleine, November 8, 2004. http://www.
   cji.net/CJI/CenterInfo/lemc/papers/HankinsChris.pdf#search=%22F
   BI%20NEWHALL%20BRASS%22.

THE ART OF NOT
BREATHING SMOKE

INTRODUCTION

It's time to put it all together and draft your department's air management plan. While this chapter provides a basic outline, it is up to you to fill in the gaps based on how your department operates, on staffing levels, and on the leadership philosophy of your administration. The following plan has proven to be successful in departments throughout the country, dramatically reducing firefighter close calls and line-of-duty deaths. While the concepts that make up this outline have been discussed in detail earlier, this chapter relates it all together.

Every good change has been met with varying degrees of resistance in the fire service. It is worth reiterating that you will meet challenges from members who are comfortable doing things the way they've always done them. During the meetings to change the NFPA 1404 standard to reflect our core concepts, there were members who disagreed strongly. In some instances, that disagreement had more to do with the resistance that they would meet on returning to their own departments than with the proposed changes.

If the goal of your efforts is truly to keep your members from routine exposure to smoke and the residual products of the IDLH atmosphere, this is the single-most effective way to do it. As mentioned already, you do not have to buy expensive equipment, add more bodies to the fireground, or go through lengthy specialty training to adopt this program. It can be added to your operations in a short amount of time and confers the needed benefits immediately. The following is a blueprint for how we can all *not* breathe smoke.

IT SHOULD BEGIN AT THE TOP, BUT IT DOESN'T HAVE TO

There should not be a department in the country that is led by individuals who aren't committed to a comprehensive air management program and philosophy. No intelligent firefighter can deny the dangerous and deadly outcomes that result from the way we've been applying air management in the fire service. For chief officers to stand before their firefighters and say it's acceptable to disregard effective air management procedures is a dereliction of duty.

Every firefighter who has promoted up to rank of chief or officer agreed to assume responsibility for the people on their crew. This is not a token acceptance of some minor duty. This is life and death, and it had better be treated that way. The first—and best—step toward getting your air management program rolling is to have a firm commitment from the top that this is what will be done. When leaders demonstrate that firefighters are the top priority and all operations reflect that, acceptance and implementation of air management is a snap (fig. 12–1).

Fig. 12–1. Command staff for the Seattle Fire
Department addressing new recruits (Photograph
by Lt. Tim Dungan, Seattle Fire Department)

Admittedly, expecting all chief officers to buy in is unrealistic. Traditions, fear of not being "one of the guys," and plain old laziness are hurdles many of you will have to overcome to do what is right. Nevertheless, we are confident that you can do it, and it is happening across the country.

The ROAM is so simple and effective that it can be practiced at the lowest levels of the fireground, regardless of whether there is a buy-in from the top. The ROAM can be modeled by company officers who have weak-willed chief officers in control, and the results will speak for themselves. Whatever situation you find yourself in, it comes down to what you're willing to live with. You will not regret the commitment it takes to implement and practice air management. By contrast, not doing so has and will continue to produce results from which many individuals never recover.

DEVELOPING AN AIR MANAGEMENT PHILOSOPHY

Development of a philosophy of air management goes hand in hand with buy-in from the top, but it is done on an individual level as well. The philosophy boils down to determining in your mind that neither you nor your crew will accept that routine exposure to smoke is a necessary part of your procedures. It's that simple. Setting this as the standard makes determining how you train and how you perform on the fireground less confusing. When discussing your tactics or developing a training module, you simply have to ask yourself, does this meet our belief that we will not breathe smoke? By answering this, you address both the confusion and the emotion that often accompany IDLH situations, and you build a solid foundation on which to establish clearly that you will not unnecessarily expose your firefighters to smoke.

Air management needs to be comprehensive in nature and consistent in application. Your air management philosophy needs to include elements that happen long before the fire starts or potential exposure begins. As with most everything in the fire service, that begins with training.

PRACTICE MAKES POSSIBLE

As simple as the ROAM is, don't expect to show up at your next fire and do it well if you have not trained in advance. The ROAM is the new basic firefighting skill, and like every other skill you have, it takes practice to master. This is especially true since most of us will have to overcome years of faulty conditioning in our air management practices. It is an established fact that you default to your basic training when in the midst of the battle, and that is what you will do with air management as well.

The Seattle Fire Department incorporates air management into every drill that involves potential for interior operations. This reflects a commitment from the top, a philosophy of safe air management at the core, and a desire to train in the way we want to perform. As you can see, the successful incorporation of air management into the Seattle Fire Department follows the basic pattern outlined in this chapter.

It is essential that your training involve three basic components prior to arrival on the fireground:

- SCBA mastery
- Realistic operational drills
- Communications training

There isn't any more critical piece of equipment for firefighters than the SCBA. It is a shame that many firefighters check their mask at the beginning of the shift and don't think about it again until it is needed to keep them alive. You must know your SCBA inside and out. You must know how to make corrections on the fly when things go wrong and detect potential failures before they fully occur. With integrated PASS devices, heads-up displays, and better monitoring devices now part of the SCBA, it is essential that you train constantly on your mask. This includes drills that put the firefighter in low/zero-visibility scenarios, confined areas, and situations that simulate the stress and exertion of the fireground (fig. 12–2).

Fig. 12–2. Firefighters conducting timed SCBA drills
(Photograph by Lt. Tim Dungan, Seattle Fire Department)

There can be no substitute for well-run drills that involve realistic props, actual fireground scenarios, and real-time decision-making opportunities. Members learn far more in these drills than in multiple classroom sessions. You need to practice following your procedures in environments that come as close as possible to what your firefighters will encounter when the dangers are real. This includes heavy workloads, since air will be more rapidly consumed and exertion is needed to accomplish the task. Through realistic operational drills, firefighters practice how to conserve air, determine the point when time-to-exit decisions should be made, and learn how to utilize the air they have effectively. You must know how to check your air gauge, quickly and in difficult situations. These drills take time and creativity, but they are a necessary component of safe air management operations (fig. 12–3).

Fig. 12–3. A realistic drill reflecting the rigors of actual firefighting
(Photograph by Lt. Tim Dungan, Seattle Fire Department)

Communications training needs to be focused on two key elements that greatly affect your ability to safely manage your air. The first is communication between team members as they complete their tasks. The team leader must ensure, through proper communication with his or her team, that all members have sufficient air remaining for a safe exit from the structure. This is difficult in the hectic atmosphere of the emergency scene. The place to get proficient at this is in pre-incident training, not on the fireground.

The second area of emphasis is communication between the team leader and the command post. A coordinated effort is essential for accountability, crew rotation, and the safety of the crews. (This will be discussed in more detail later; see "Share the load.") Importantly, command personnel need to train just as diligently as the troops in the field for effective air management to succeed.

You cannot fake good training. While you can never prepare for all of the variables the fireground will throw your way, proactive training eliminates most of them. It is imperative that we not be in a continuous mode of reactionary training based on what we did wrong at fires. Take the time to prepare on the front end—and be prepared to be delighted with the results.

FOLLOWING THE ROAM

The foundational role that the ROAM plays in air management has already been discussed at length. The ROAM is the rock on which all your efforts should build and will make a major impact on your operations. Once again, the concept is to know how much air you have in your SCBA and manage that air so that you leave the hazardous environment *before* your low-air alarm activates. In simple terms,

- Know what you've got before you enter the IDLH
- Manage your air as you proceed
- Exit before your low air alarm activates

Exotic technology and complicated command configurations pale in importance compared to this simple concept. It puts the responsibility for your air right back on you—where it belonged all along.

READY CHECKS

Just as you have committed to training before entry, the READY check takes care of many of the problems that occur frequently in interior operations. This is your chance to do a quick, systematic buddy check, to start your operation off correctly and possibly keep you from trouble once committed to the interior attack. The READY check is:

- *Radio.* Is the radio on, turned to the correct channel, and does everyone on the team know to whom they report?

- *Equipment.* Does each member of the team have the appropriate equipment (including appropriate PPE) and know how to use that equipment?

- *Air.* Has the air level been checked prior to entry and assessed to be adequate for the assignment?

- *Duties.* Is the assignment known, are roles within that assignment clear, and is the team trained to do it?

- *Yes!* If you answer yes to all of the above questions, you are ready to proceed safely into the hazardous area. If you answer no to any of these questions, then fix the problem before you proceed (fig. 12–4).

The concept of READY-Checks is discussed in much more detail in chapter 14.

Fig. 12–4. Firefighters conducting a READY check
prior to interior operations (Photograph by
Lt. Tim Dungan, Seattle Fire Department)

It is extremely difficult to rectify problems on the interior of the fireground. Difficult visibility, tight spaces, and firefighting gloves make minor corrections that are easily completed on the outside extremely difficult on the inside. The READY check eliminates many of these problems and is essential to a comprehensive air management strategy.

How's your air?

You've done your READY check and now are on the inside. At this point, the second phase of the ROAM kicks in. Regardless of your assigned task, at various intervals, you must take a quick glance at your air gauge to remain constantly aware of your air situation. While this sounds like common sense and a simple task, ask yourself how many times you've actually done it. Doing this makes the difference between managing your air and having your air manage you.

In training sessions conducted across the country, many firefighters have had difficulties in checking the air gauge. Whether the problem is bad SCBA design, poor eyesight, or not knowing how to read the air levels, it is not automatic for many firefighters. Simply stated, it needs to be. The routine checking of your air not only makes you safer by virtue of doing appropriate air management but also assists in maintaining situational awareness and eliminating tunnel vision. Firefighters who practice this skill find it easy to incorporate into their routine and are able to continue aggressive interior operations at equal or better levels with greater safety.

One question that most firefighters have is when they should check their air. It's an important point to consider, as the goal is to meld effective air management with aggressive operations. The following are a few opportunities to check your air:

- *After periods of heavy workload.* Checking at this time is critical, because you may deplete your air quicker than you think during strenuous work periods. Fatigue is an issue as well, making a safe exit even more difficult should things go wrong with the operation (fig. 12–5).

- *At each change of rooms or area.* This serves two very important purposes: awareness of air status and situational awareness. These two factors are critical in making your time-to-exit decision.

Fig. 12–5. A firefighter who has used a lot of air. On completion of a difficult task, an air check is advised. (Photograph by Lt. Tim Dungan, Seattle Fire Department)

- *At a change of levels.* This is another critical time to verify your air status, because a change in levels usually increases the difficulty of egress and will dramatically affect your time-to-exit decision. At any change of level, you should be sending a radio report to the IC; therefore, an air check is an obvious task to complete as well (fig. 12–6).

- *During progress reports.* This is an obvious time to check air. The team leader can verify air status of the team, make decisions on how they will proceed, and let the IC know via the radio. (See also "Do you speak CARA?")

- *At rest intervals.* A great time to check your air is when there is a pause in the action. This doesn't happen with every task you'll be assigned at fires, but it does occur frequently. Think of it as the pause that refreshes.

- *Before starting a new assignment.* This is an important time to check your air status, because your previous task will have depleted some amount of your air. You need to know how much that is, and the IC needs to know as well, to determine whether the new assignment is doable given your current air status. It is unprofessional and dangerous to accept or ask for an assignment without having enough air to reasonably accomplish the job.

Fig. 12–6. A firefighter who has descended a flight of stairs. On moving to a different level of the structure, an air check is advised. (Photograph by Lt. Tim Dungan, Seattle Fire Department)

Performing a periodic air check on the interior of fireground operations is the simplest, most effective way to practice good air management. It costs nothing, requires no elaborate training, and puts you in a great position to safely do your job.

SHARE THE LOAD

Good air management begins with the time-honored firefighting principle of teamwork: Everyone pulls their weight. This concept works well in the world of air management, as rotating members based on their air consumption allows you to accomplish more work with your allotted air supply. The concept entails letting the member with the least air assume the least strenuous job. Implicit within this concept is the personal integrity of firefighters to not always have to be the hero. There has to be a willingness to work as a team and accomplish the task together.

DO YOU SPEAK CARA?

A concise progress report is an important component of air management. Giving progress reports makes the whole operation run better and enhances decision making at the command post, owing to the constant feed of information from inside. (For more on communications and progress reports, see chap. 16.)

The CARA report gives the IC the following information (fig. 12–7):

- *Conditions*. What do you have?
- *Actions*. What are you doing?
- *Resources*. What do you need?
- *Air*. What is the current status of your team's air?

Fig. 12–7. A firefighter giving a CARA report to the IC.
(Photograph by Lt. Tim Dungan, Seattle Fire Department)

The relaying of air levels, especially at the 50% mark, provides crucial notification to the IC that your crew will imminently need relief. Currently, in most operations, relief is called for when the crew comes out of the structure. Think how much better crew rotation will be when the IC is aware of your status and is already getting a crew ready while you are still inside. Good communication—among crews on the interior and to the command post on the exterior—greatly enhances the overall air management effort.

INCIDENT COMMAND

The command-level responsibility for air management varies depending on the system. Some programs, such as on deck (the Phoenix system; see chap. 21), put a heavier emphasis on work cycles and monitoring air from the outside. While this is a perfectly reasonable approach as part of a comprehensive air management program, it does not supersede the simple truth that your air is your responsibility. Reliance on someone else, on technology, or on fate is a risky proposition and should never be your first choice.

A well-developed incident command structure will enhance all elements of the air management plan by ensuring proper accountability, timely crew rotation, and effective tactical decisions. Maintaining a proper supervisory ratio is the key to making this successful, so that work is divided among supervisors and no one in the command structure gets overwhelmed. Thus, supervisors can actually lead their teams, preventing freelancing and poor operational tactics.

The IC needs to be in command from the start of the incident through its conclusion. This includes ensuring that safety in rehabilitation and salvage/overhaul is not an afterthought. ICs who take air management seriously on the fireground show that they truly care about the people they lead and the safety of their operations.

READY RAPID INTERVENTION

The RIT needs to be ready. In particular, they need to do everything possible to get air to firefighters who are in trouble. It shouldn't come as a surprise that many firefighters who have died in structures did not perish because of the initial event that went wrong. Many survived after getting lost or being in a collapse but died because they ran out of air. The RIT needs to be prepared in every possible way to get air to their fallen comrades (fig. 12–8).

Fig. 12–8. A RIT setting up equipment for entry
(Photograph by Lt. Tim Dungan, Seattle Fire Department)

The Seattle Fire Department uses a modified version of *AWARE*, developed by Jay Olson of Portland (OR) Fire and Rescue.[1] We train our RITs to provide the following when a Mayday is called:

- *Access.* You need to get to them before you can help.
- *Water.* Get between the fire and firefighter.
- *Air.* They need air to survive, and they need it fast
- *Radio.* As always, communication of needs and progress is critical.
- *Extrication.* Getting them out of the hazard area is the goal.

An effective air management program includes an IC who establishes an RIT early in the operation, an RIT that takes its job seriously and is prepared when the unthinkable occurs, and firefighters who will call the Mayday instead of hoping that things work out in the end.

REHABILITATION: THE PAUSE THAT REFRESHES

Rehabilitation may not be an obvious component of air management, but it should be. Your ability to do work and utilize your air to its fullest capacity is directly related to your level of fatigue. Well-hydrated, rested firefighters are able to perform at a higher level and use their air more effectively. Replacing fluids and electrolytes, as well as cooling our core temperatures, is vital for further work cycles (fig. 12–9).

Fig. 12–9. Firefighters resting, replenishing fluids, and checking vital signs in rehabilitation (Photograph by Lt. Tim Dungan, Seattle Fire Department)

Rehabilitation is also the place to catch any medical problems that might have developed. It is much better to discover an issue that needs addressing on the outside, with paramedics present, than on the inside, in the hazard area.

SALVAGE AND OVERHAUL

Most firefighters do not view salvage and overhaul as dangerous assignments. They have already battled the fire and are now just mopping up, so what's the big deal? Importantly, the air that firefighters are breathing while pulling those final ceilings or hitting those last hot spots contains the same contaminants as did the fire (fig. 12–10). As detailed earlier, the levels may even be higher, making the environment more, rather than less, hazardous.

Your air management program must demand that all firefighters performing salvage and overhaul maintain respiratory protection at all times. Performing salvage and overhaul is more difficult when wearing a mask. Moreover, more personnel will be required, because of the need for frequent work rotations. However, the benefit is that team members will not be inhaling stirred-up asbestos fibers and will not be absorbing carbon monoxide or hydrogen cyanide from smoldering remnants of the fire. Air management during salvage and overhaul will greatly reduce firefighters' chances of falling victim to cancer or respiratory disease.

While other actions, such as heavy PPV prior to entry, can assist the process, the one factor that should never be compromised is respiratory protection. How tragic is the reality that we safely attack a raging fire only to succumb to its byproducts when the fire has been extinguished? Good air management programs don't allow for unnecessary risks.

*Fig. 12–10. Firefighters pulling ceilings during overhaul.
Proper respiratory protection is essential because of the
presence of gases and particulates during this phase.
(Photograph by Lt. Tim Dungan, Seattle Fire Department)*

How did we do? How can we get better?

The final piece of the puzzle can prove the most problematic and difficult to implement in your organization. Unfortunately, many fire departments do not take advantage of the many lessons learned at the emergencies they go to every day. Whether the resistance stems from a false sense of pride, fear of legal/political ramifications, or laziness, there is no excuse for not training on postemergency lessons.

Many good fire departments have had to learn from mistakes. In addition, the rest of the fire service has benefited from their examples. Chief Brunacini of the Phoenix Fire Department has set a tremendous example by establishing a culture that celebrates learning from the successes and failures on the fireground. There are numerous close calls happening every day that would provide invaluable training if the message was allowed out.

For every firefighter, it is critical to improve on mistakes and work to enhance successes. You will have both, so get over it. With the implementation of your air management program, there will be things that you do well right from the start; other areas will be more difficult to assimilate into your routine. A commitment to training and sharing your experiences will speed up the process dramatically and will spare team members from repeating the same old mistakes.

Is that all there is to it?

The simple answer is *yes*. If you put all these pieces together into a comprehensive program, your fireground safety will improve dramatically, and close calls will diminish. There are other more complex elements that can be considered, and you can always find more ways to spend money to assist your efforts. Nevertheless, this plan works without making those elements necessary.

The air management program outlined here puts the majority of the responsibility on the individual, where it belongs. The professionalism, training, and integrity of the firefighter should prevail over any piece of equipment made. It boils down to this: Your air is your responsibility (fig. 12–11).

Fig. 12–11. A crew checking their air prior to entry

NEVER BREATHE SMOKE MEANS
NEVER BREATHE SMOKE

Early in 2003 "Air Management" was only a buzzword I was throwing around during lectures. I say that because sometimes all your awareness and understanding are useless until your experience confirms your education. I say it was a buzzword because although I truly believed it to be critical, I had not yet developed the personal understanding of what can happen when it is lacking or absent. When you are involved in an incident where a firefighter sustains an injury on your watch, there is a level of clarity that event brings that is like no other. I was very aware of the toxicity of smoke. I knew the cancer threat was growing and I had ample training in the dangerous effects it has on lost and disoriented firefighters. But I had no understanding of how very personal smoke is to each and every one of us.

On the surface, it seems as if smoke should affect us all in pretty much the same way and that is usually the case. What is not so well understood or widely known is how even small amounts of certain toxins in smoke can have devastating, even fatal, effects on individuals who are, for a wide variety reasons, more susceptible. I learned this the hard way and it was a horrific experience for the firefighter who was injured that night.

We responded to a routine residential structure fire and had the fire contained after a short but intense firefight. The fire had a considerable head start on us and it had spread to practically every room of the structure and even melted a hot tub in the back yard. This made all of us very curious about how it grew so quickly. Crews were checking for hot spots and I had joined several firefighters who were examining the garage, which seemed to be the area of origin. The CO levels had been measured and our safety protocols deemed it safe to remove face pieces so we were all looking at char patterns and debating theories about the possible origins.

Crossing the floor we noticed a small flare up under a car in garage, a firefighter with a handline directed a short burst on the seat of the fire. This extinguished it immediately, and a small cloud rose quickly. We all continued our inspection of the electrical panel and the area around it—all of us but the nozzleman, Shane Briggs, who said he needed to step outside. The next report was for a firefighter down in one of he medical units. It was Briggs.

As the paramedics worked on assisting Firefighter Briggs, they explained to me that he was experiencing a bronco spasm. The insult from the gases he inhaled was closing his airway and he was suffocating. The situation went from bad to worse quickly and he was rushed to a local hospital. The emergency continued over the next five to seven hours with Briggs struggling to breathe and all sorts of advanced methods being used. The room was full of firefighters for several hours as we watched our brother fight to breath on his own. After several hours he was stabilized, but he spent several months in respiratory rehabilitation (fig 12–12). He was able to return to full duty but was it largely due to his exceptional physical conditioning and extraordinary work ethic in therapy.

While in the hospital the doctors explained that we all have different levels of susceptibility or vulnerability to different types of respiratory irritants and exposures. It is impossible to determine what these are, often until it is too late. Now I understand why never breathe smoke means NEVER. We do not know what we inhaled that night; we only know it was almost fatal for one of us. It was a mistake I know no one on that fireground will ever repeat.

—Chief Bobby Halton retired
Editor in Chief, Fire Engineering
Education Director, FDIC

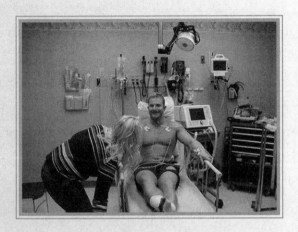

Fig. 12-12. Firefighter Shane Briggs needed lengthy respiratory rehab to recover from his "Breath from Hell" experience.

MAKING IT HAPPEN

1. Conduct a drill that investigates the significant close calls experienced by your department over the past year. Discuss how implementation of an air management program might have had an impact on those operations.

2. Check the calendar at Manageyourair.com and arrange for members of your department to attend a training session.

3. Assign a section of this air management program to individual companies within your department. Have them brainstorm ways to implement that segment.

4. Review the DVD *Fire Fighter Survival* and discuss how an effective air management program might have altered the outcomes for the firefighters involved. (To obtain the DVD, contact the Seattle Fire Department, at 206-386-1401.)

STUDY QUESTIONS

1. What are three hurdles to overcome in implementing an air management program?

2. True or false: It is an established fact that you default to your basic training.

3. What are three key components of preincident training?

4. True or false: It has proven to be easy to correct equipment problems once committed to interior operations.

5. Routinely checking your air makes your operations safer. Name two other benefits.

6. Checking your air at a change of level is advised. What two things are affected by changing levels in a structure?

7. What percentage of air is crucial to relay to the IC?

8. How can ICs ensure that no one within the structure is overwhelmed by having too many people reporting to them?

9. In the acronym *AWARE*, what does the first *A* stand for?

10. True or false: Fluid replacement and maintaining high core temperature are key components of rehabilitation.

11. True or false: All firefighters performing salvage and overhaul must maintain respiratory protection at all times.

NOTES

[1] Olson, Jay B. 1998. A.W.A.R.E.: A lifesaving plan for rescuing firefighters. *Fire Engineering.* December: 52–58.

CHAPTER 13

WHERE SHOULD OUR TRAINING EMPHASIS BE?

INTRODUCTION

In this chapter we focus on training and how it must be the foundation of any fire department. A fire department that is not training its members does a grave disservice to both the citizens it is sworn to protect and its firefighters. Our training philosophy relies heavily on hands-on training in the basic skills that operational firefighters use at structure fires: basic engine skills, truck skills, and critical decision-making skills. These basics are the core of structural firefighting. These basic skills must be practiced repeatedly, in interesting ways, so that firefighters stay sharp.

Air management is the new basic firefighting skill. The ROAM must be practiced in hands-on settings so that firefighters use it whenever they use their SCBA.

THE BASICS

Fire departments should be training their firefighters on how to handle emergencies effectively and safely. The best way to accomplish this is to focus on basic firefighting skills. These basic skills should not be taught exclusively in a classroom setting, with a PowerPoint slide show. While lectures and audiovisual presentations are important and useful parts of any basic skills training program, they should never be the sole components.

Basic firefighting skills must be taught in a hands-on training environment. Firefighters must use their motor skills and their senses to learn how to pull hose or ventilate a roof, and they must use these motor skills over and over again. You can show pictures and lecture all you want, but the only way to teach firefighters how to perform a fireground skill is to have them physically do it themselves and have them do it repeatedly (fig. 13–1).

Basic fireground skills are at the core of the fire service. Ours is a hands-on profession that must perform when lives are at stake. Firefighters cannot talk a fire out, nor can they rescue a trapped civilian with a grease pencil and a white board—no matter what some self-proclaimed incident management wizard might tell you.

The ICS is important and should be taught to every firefighter, but not at the expense of basic, hands-on training. Setting up the ICS doesn't automatically guarantee that a fire will go out or that a victim will be removed with a ladder before he or she jumps from a window. Firefighters must get water on the fire and throw the ground ladder to the upper-story window, and they must perform these fundamental skills aggressively and safely. The citizens, who are both our bosses and our customers, expect nothing less.

Fig. 13–1. Firefighters performing vertical ventilation on an acquired structure. This type of training is invaluable.

Effective operational fire training should emphasize the basics and include a large amount of hands-on, scenario-based training. Firefighters must flow water, throw and extend ground ladders, conduct rooftop ventilation operations, practice forcible entry with irons and rescue saws, pull hoselines, conduct interior searches in single-family dwellings and apartment houses, conduct rescue operations from the outside and inside, practice vehicle extrication, perform CPR, use the AED, and complete many other tasks expected of today's firefighters (figs. 13–2 through 13–5).

Fig. 13–2. Recruit firefighters practicing raising ladders during a drill at an acquired structure (Photograph by Lt. Tim Dungan, Seattle Fire Department)

Fig. 13–3. Firefighters from a ladder company cutting a heat hole in the rib-arch truss roof of an acquired structure during training (Photograph by Joel Andrus)

*Fig. 13–4. Firefighters pulling a 1¾" attack line at a
training exercise at an acquired structure (Photograph
by Lt. Tim Dungan, Seattle Fire Department)*

*Fig. 13–5. Recruit firefighters learning how to breach an interior
stud wall during a training exercise at an acquired structure
(Photograph by Lt. Tim Dungan, Seattle Fire Department)*

TRAINING VERSUS LEARNING OPPORTUNITIES

According to Mark J. Cotter—a 30 year veteran paramedic, fire officer,
and noted author—classes, drills, and seminars should not be called
training but are rather *learning opportunities*, which are the beginning of the
process of becoming a competent firefighter. In distinguishing training from
learning opportunities, Cotter argues that training "refers to the repetitive
performance of skills and application of knowledge in an effect to achieve
and maintain mastery."[1] He explains that to maintain proficiency:

> *Training must be ongoing and consistent, periodically
> introducing new concepts and skills, but continuously
> reinforcing fundamental knowledge and abilities already
> learned. Addressing deteriorations in skills and behavior,
> just like we do for our equipment and vehicles, is a*

vital component of this process. Still, despite the wide breadth of potential subject matter and techniques, the actual topics requiring such attention are basic and, sometimes even, mundane.[2]

Continually reinforcing fundamental knowledge, skills, and abilities already learned forms the foundation of training in the basics.

Recognizing that training must be ongoing and must be hands-on, Cotter explains that basic structural firefighting training is like training for a marathon or another sporting event. Marathon runners must run to stay in aerobic condition, practice the basics of stride and breathing while running, vary their training mileage during each week and leading up to the marathon. In other words, a marathon runner must physically run. Marathoners cannot just learn about increased aerobic capacity, stride efficiency, and breathing basics from a lecture, a slide show, or a video. Instead, they have to physically do it and practice it, over and over again. And this is true for firefighters and basic firefighting skills. Firefighters must physically practice firefighting skills over and over again.

Scenario-based training

The best basic, hands-on training is scenario based. In this type of training, firefighters are given a scenario similar to one they might encounter in the real world and must perform the firefighting tasks necessary to mitigate the problems. These tasks might be forcing entry into the occupancy, pulling attack lines, performing ventilation, and rescuing victims from upper floors.

Using acquired structures for scenario-based training can enhance the hands-on training experience for everyone involved. Acquired structures allow firefighters to practice their basic skills on the types of structures they are likely to find in their districts. Commercial, single-family residential, multifamily residential and high-rise structures bring a sense of reality to any hands-on training scenario.

Firefighters love to train in acquired structures. They immediately recognize the benefit of training in the same types of structures in which they might be fighting a fire later that night. This is called the *reality factor.* Hands-on training must include the reality factor. Above all else, firefighters appreciate reality in their hands-on training. They crave training that is genuine and authentic. If the hands-on training lacks the reality factor, then firefighters will either discount it entirely or will minimize its importance. Truly, there can be no substitute for reality-based, hands-on training in the basics.

Air is a basic resource and the ROAM is a basic firefighting skill

The ROAM is the new, basic firefighting skills. Firefighters must learn to manage their air supply just as they learn to deploy hose and throw ground ladders. Why? Because firefighters are running out of air in structure fires and getting seriously injured or killed (see chap. 4 and 7). The U.S. fire service can no longer afford to ignore this fact. Therefore, firefighters must be trained in how to manage one of the most important and basic resources they bring with them into structure fires—the air they carry on their backs.

Interestingly, firefighters and fire officers are managing other crucial resources at structure fires already. Firefighters manage the water that they bring with them to a structure fire. Fire officers manage the units and personnel at a structure fire by implementing and using an ICS. Fire officers manage their crews' actions and tasks at structure fires. Fire officers also manage the risk to their personnel by conducting an ongoing risk-benefit analysis. Why, then, are firefighters and fire officers not also managing their air supplies?

Air is a basic firefighting resource that, up to now, has not been managed. Until recently, the only form of air management practiced on firegrounds in the United States has been firefighters' reacting to the SCBA low-air alarm. As we have demonstrated throughout this book, using the low-air alarm as a signal to exit the hazardous environment of a structure fire can be deadly.

Since the ROAM is a new fundamental skill, it needs to be practiced over and over again so that firefighters can become proficient at it. Officers, training officers, and firefighters must integrate the ROAM into every training evolution in which firefighters are using their SCBA. As a result, firefighters will internalize air management and the ROAM as just another essential skill automatically employed during a fire. Recruits in drill school should learn about the ROAM when they first learn about the SCBA. Thus, the SCBA and the importance of managing the air in the SCBA are learned together—integrated and inseparable, as they should be (fig. 13–6).

Fig. 13–6. Firefighters navigating a confidence course. During this training, the team of firefighters were expected to manage their air using the ROAM. The course was blacked out to be a zero-visibility environment. The flash from the camera has illuminated the stairwell. Note the missing treads on the stairs and the instructor following the team of firefighters.

After the death of Bret Tarver in Phoenix, the Seattle Fire Department recognized the importance of air management and adopted the ROAM as policy. The Seattle Fire Department trained all operations firefighters in the ROAM and mandated that the training division begin teaching the ROAM to all our recruit firefighters. Probationary firefighters in Seattle now come out of recruit school knowing how to manage their air (fig. 13–7). For recruits in drill school and probationary firefighters in the companies, the ROAM is just another essential firefighting skill to master.

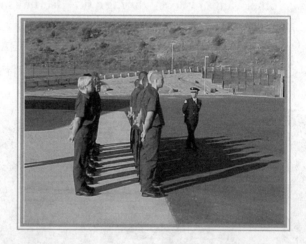

Fig. 13–7. Probationary firefighters during a roll-call inspection. Seattle Fire Department probationary firefighters are taught the ROAM. (Photograph by Lt. Tim Dungan, Seattle Fire Department)

OPERATIONAL SKILLS ENHANCEMENT TRAINING: BACK TO THE BASICS—A CASE STUDY

In 2000, several proactive members of the Seattle Fire Department realized that department-wide operations training needed to get back to basic fire operations. Since the mid-1990s, department-wide training efforts and budget had been concentrated on special operations—hazmat, confined spaces, trench rescue, and high-angle rescue. While special operations are indeed important and have their place in the fire service, they should never be the only form of operational training that a fire department provides its members. Providing special operations training at the expense of basic fire operations training is myopic at best, and a skilled attorney could easily argue that this represents willful negligence.

A bit of historical perspective and honesty is appropriate at this point. In the Seattle Fire Department, like other big-city fire departments, training firefighters in the business of fighting structure fires had not changed since the department's inception. When new firefighters were hired, they were sent to recruit school for four months, where they learned basic hose and ladder operations, along with everything else a probationary firefighter, or probie, needs to know. After graduation from this intense recruit school, probies were assigned to specific companies in the city, where for the remainder of their probationary year, they were taught everything they were expected to know by their officer and the firefighters at their assigned station.

From the beginnings of the Seattle Fire Department up to the 1960s, firefighters received almost all of their fire operations training on the job. That is to say, up to the 1960s, Seattle firefighters were going to jobs, or working fires, almost every shift throughout the city. This was the time before fire prevention and smoke detectors. Probationary firefighters learned how to fight structure fires by doing it on the job, under the guidance and direction of veteran officers and veteran firefighters, who had gained their firefighting experience on the job as well (figs. 13–8 and 13–9). Training in basic fire attack wasn't necessary, since members were going to structure fires almost every shift, doing the job and learning as they went.

Fig. 13–8. A working fire at a single-family dwelling.
Up to the 1960s, before smoke detectors and fire alarm
systems, firefighters saw quite a bit of fire. Firefighters
learned their craft through on-the-job-training.
(Photograph by Mike Heaton)

All this changed during the 1960s, with the advent of fire prevention.
A major paradigm shift was occurring in Seattle and across the United
States. Fire departments, city governments, and municipalities began to
preach and practice fire prevention. With new fire prevention systems like
sprinklers, fire alarm systems, smoke detectors, and building-inspection
programs, it was only a matter of time until the number of fires began to
decrease. And decrease they did, which was wonderful for the citizens we
protect. Fires have continued to steadily decline each year since.

Fig. 13–9. A firefighter climbing an aerial ladder to gain access to the roof and perform vertical ventilation. Firefighters went to many structure fires up to the 1960s. (Photograph by Mike Heaton)

The only problem with the decreasing number of fires is that the opportunity for on-the-job training in structural fire attack has all but disappeared, except in the most impoverished and dense areas. Although the number of fires in Seattle has steadily declined since the 1960s, the leadership of the Seattle Fire Department never thought to change the way it trained its firefighters.

We didn't modify or alter the way we trained our firefighters in basic structural fire attack because no one noticed the monumental change that was occurring around us. Poor planning, a lack of vision, the comfort of ignorance, complacency, a deficiency of critical analysis, overconfidence, incompetence—call it what you will, but the fundamentals of fire attack were not being taught to our newer people, and they were not getting the on-the-job training that previous generations of firefighters had received. In fact, newer members began to experience close calls at bread-and-butter house fires—the type of fire that kills the most firefighters each year. Keep in mind that the Seattle Fire Department was still operating under the old paradigm, which held that all our firefighters were going to one or two good structure fires every shift or two.

The point of this case study is not to cast blame or find fault. Rather, the hope is that by presenting the truth, other fire departments might learn from our experience. In fairness, the Seattle Fire Department was like many other fire departments when it came to its failure to train firefighters in basic fire attack. We were ignorant of what was happening around us, and we failed to keep up with change.

OPERATIONAL SKILLS ENHANCEMENT TRAINING

Fortunately, a dedicated group of individuals in the Seattle Fire Department recognized that operations firefighters were not being trained in basic engine and ladder company fire operations. They crafted a hands-on training program that brought a back-to-basics approach to structural firefighting. The fire chief saw the value of this program and gave it his full support. The philosophy underlying this training program is as follows: Operations firefighters need to learn and continually practice the fundamentals of fire behavior, fire attack, rescue, ventilation, water supply, hose streams, ladders, and air management, among other basic skills. The firefighters who developed this program understood the need for hands-on training in acquired structures, where firefighters could pull and advance hoselines, flow water, pull ceilings, practice rooftop ventilation, use air management, place portable ladders (fig. 13–10), and implement the ICS, among other important fireground tasks. This training package was ultimately named *Operational Skills Enhancement Training* (OSET) and has provided some of the most valuable training our members have received over the past 10 years.

Fig. 13–10. Firefighters raising a portable ladder for rooftop operations at an acquired structure

The OSET model is simple: provide basic, realistic, hands-on training in individual skills, company-level skills, and multiple-company–level skills. This three-part program was designed for three half-day (3½ hour) training sessions—one half-day for individual skills, one half-day for company-level skills, and one half-day for multiple-company evolutions. These training days are spread out over the year—one in spring, another in late summer, and the final one in early winter.

Individual skills are those that every individual firefighter is expected to know and perform on the fireground. These skills include, but are not limited to:

- Selecting, deploying, and operating the proper hoselines
- Raising ground ladders
- Positioning and climbing ground ladders
- Basic nozzle operations
- Operating and cleaning a chainsaw
- Operating and cleaning a rescue saw

- Using an SCBA
- Performing an emergency SCBA transfill in zero visibility
- Managing the SCBA air supply
- Operating PPV fans
- Using irons in various forcible-entry scenarios (e.g., see fig. 13–11)
- Using a rescue saw on various metal person doors and roll-up doors
- Opening a pitched roof from a roof ladder
- Opening a flat roof

*Fig. 13–11. A firefighter practicing
forcible entry using a rescue saw*

COMPANY-LEVEL SKILLS

Company-level skills are those that members of an engine or ladder company should know and be able to perform on the fireground. These skills include, but are not limited to:

- Ventilating flat roofs (figs. 13–12 and 13–13)
- Ventilating pitched roofs
- Deploying 2½" hose for commercial occupancy fires
- Deploying 1¾" hose for residential fires
- Standpipe operations
- Elevator operations
- Ground ladder rescue operations
- Aerial ladder rescue operations
- PPV for high-rise operations
- Foam operations
- Ship fire operations
- Water supply operations
- Pump operations (fig. 13–14)
- Aerial operations
- Spotting and positioning the aerial apparatus
- Forcible entry for larger commercial structures
- Company-level air management

Fig. 13–12. Firefighters from several ladder companies being taught how to cut an offensive heat hole in an acquired structure (Photograph by Joel Andrus)

Fig. 13–13. Firefighters being taught how to open up a flat roof during hands-on training at an acquired structure (Photograph by Joel Andrus)

Fig. 13–14. A chauffeur practicing pump operations during company-level skills training (Photograph by Mike Wernet, South Kitsap [WA] Fire Department)

At the end of these company-level scenarios, a critical debriefing session is held. During this meeting, members are asked what they did and why they made the decisions they made.

The last OSET session of the year is made up of multiple-company evolutions that bring both company-level and individual skills into play. This half-day is composed of several full-blown, multiple-company scenarios that take place at acquired structures. These can be residential, commercial, or high-rise scenarios, and they should simulate real events as closely as possible. Instructors are present to observe the actions of the companies and add as much realism as possible. At the end of these scenarios, everyone involved in the drill is again brought together for a critical debriefing.

Hands-on training scenarios provide firefighters with valuable experiences on which they can draw when they go to structure fires. This type of training is called *experiential training*. This type of training is essential in these days of declining structure fires. We must provide our firefighters with real-life experiences from which to learn and base their actions when they do go to a working fire.

Decision-making skills are key components of the OSET model. We want our fire officers and firefighters to make the right decisions at structure fires. Often, critical decisions must be made in a fraction of a second. Sometimes, these are life-and-death decisions. In the course of a structure fire, company officers and firefighters might have to decide:

- Where to place the fire apparatus
- Whether the fire dictates an offensive or defensive attack
- What size hoselines should be pulled and where they should be placed
- What type of ventilation technique to use
- Whether the hose team should keep trying to advance or back out, based on smoke and fire conditions
- Where to search first
- When the team should leave, based on the individual members' remaining air supply
- Whether the conditions in the structure are getting better or worse

The debriefing sessions at the end of company-level and multiple-company scenarios are essential to the learning that takes place. Decision-making skills are the focus of debriefing sessions (fig. 13–15). During debriefings, the firefighters discuss what happened during the training scenario. The debriefing facilitator uses the Socratic method, a format based on questions, to spur discussion. By asking questions, the facilitator gets the firefighters involved, leading them to understand the reasons behind the choices that they made. This discussion identifies the difficult decisions and prompts officers and firefighters to analyze why they made those decisions.

Fig. 13–15. A veteran engine company lieutenant debriefing members after an OSET hands-on exercise at an acquired structure. Debriefing sessions are where the real learning takes place.

These debriefing sessions should last no longer than 15–20 minutes. The debriefing facilitator should just ask questions and listen. A successful debriefing is one in which firefighters discover the lessons on their own, from their answers to the facilitator's questions and the ensuing discussion.

At the end of each year, Seattle's OSET development team meets and comes up with three or four core individual skills that will be the basis for the individual skill stations for the following year. These individual skills lead into the company-level skills that will be emphasized for the coming year. As an example, the OSET development team might have chainsaw operations be the focus of one of the individual skills stations. This skill station might be composed of several steps, such as having the firefighter put together a broken-down chainsaw, put the chainsaw into operation, and use the chainsaw on a flat roof prop to open up several ventilation holes. Each individual skill station would build to a company-level skill scenario later in the year, in which the company would be taught how to safely work on and successfully ventilate a flat roof. Thus, individual skills build toward company-level skills and culminate in multiple-company scenarios. Through this step-by-step approach to hands-on fire training, members of the Seattle Fire Department become successful on the fireground.

PREVENTION VERSUS INTERVENTION

The following question is posed at the beginning of our air management presentations: "What would you rather do if you were the IC at a structure fire: activate an RIT into a structure fire, to search for and attempt rescue of a downed firefighter, or prevent the firefighter from getting into trouble in the first place?" The answer to this question is always unanimous. Everyone would rather prevent the firefighter from getting into trouble in the first place.

Prevention is always a better option than intervention. Intervention is never a sure thing—and intervention is not always successful, particularly on the fireground. Don't misunderstand: RITs are an important and essential part of today's fireground, and there have been some wonderful saves recently by RITs around the country. However, the cold, hard truth of present-day rapid intervention is that most of the firefighters RITs are sent in to get are already dead or do not survive once they are found and brought out of the structure.

There is also the inescapable *reflex time* with rapid intervention. Reflex time is the time it takes for the RIT to get inside the building and begin the rescue efforts. This depends on where the RIT has been staged relative to the entry point, whether they already have all the equipment and personnel necessary to affect the rescue, and how well they have been trained in RIT operations, among other factors. In Seattle, we found that even in the best of circumstances, our rapid intervention groups (RIGs) had a reflex time of about two minutes. These tests were conducted during training exercises, with equipment already staged and personnel ready to go. In a real RIT event, more time would be added to the reflex time.

Training firefighters in the ROAM is practicing prevention. If we can prevent a firefighter from running out of air in a structure fire, then we do not have to launch the RIT. Remember, the RIT's success is far from guaranteed. Recall the 1999 cold-storage warehouse fire in Worcester, Massachusetts, in which six firefighters perished.[3] All of them got lost in the heavy-smoke conditions of the fire. Four of the six firefighters were part of the two RITs that went in to search for the two original lost firefighters. Four of the six firefighters died by breathing the products of combustion. In brief, they ran out of air and died from breathing in deadly smoke.

Training in the ROAM must be ongoing, must be realistic, and must be incorporated into every training exercise in which firefighters use their SCBA. As with basic fire attack skills, the ROAM needs to be reinforced using continuing education and training. After firefighters are introduced to the ROAM and understand what it is and how to use it on the fireground, they need to maintain proficiency through ongoing training.

Whenever firefighters use their SCBA in training, remind them that they must follow the ROAM. Make sure they understand that they are expected to be out of the hazard area *before* their low-air alarm activates. Regardless of whether firefighters are training in residential search techniques, RIT operations, or hoseline advancement during OSET evolutions, they are required to abide by the ROAM when using their SCBA (fig. 13–16). If they do not follow the ROAM, they should be reminded by instructors when the training scenario has ended.

Fig. 13–16. Firefighters negotiating overhead obstructions during training. These firefighters are following the ROAM during this training exercise and thus must be out of the hazard area before the low-air alarm activates.

COMMITMENT TO PREVENTION

Prevention is hard work. Prevention is proactive, not reactive. Prevention allows us to avoid potential problems before they arise. Prevention takes time, energy, and commitment.

The commitment to prevention seems to be the most difficult obstacle to overcome for many fire departments. Part of this commitment to prevention is the commitment of purpose—namely, to prevent firefighters from running out of air in structure fires. Another element is the allocation of money and resources, which are needed to develop and run a first-rate training program. The final part of the commitment to prevention is the commitment by the fire department administration to stay informed and be proactive in their training. Fire department administrators must get out from behind their desks and take an honest look at their firefighters' successes and failures on the fireground.

OPERATIONAL SKILLS ENHANCEMENT TRAINING

Operational skills enhancement training (OSET) provides hands-on training to operations personnel that is intended to fulfill the requirements of the Washington Administrative Code. More importantly, the training program has a dramatic impact on how the firefighting team performs during emergency responses. It is well documented that under stress, firefighters will perform based on their training and experience. With a continuing decrease in structure fire responses and the experience they provide, the need to increase the frequency of training and to maximize the realism of the training environment continues to grow.

Based on this premise, the learning environment should recreate the context and content of the actual emergency environment. Acquired structures, both large commercial structures and apartment houses, are required since this type of training cannot be replicated in a firehouse setting.

OSET focuses on development of decision-making skills and the role of leadership in developing and implementing action plans. Post-exercise debriefing allows evaluation of

- Critical decisions made during the exercise and how those decisions were made
- Outcomes and how they could be improved given similar conditions at an actual emergency

This is the basis for OSET—to provide our members with the knowledge, skills, and abilities to perform their assigned duties in a safe and competent manner.

In Seattle's OSET program, the training staff consists of instructors from all ranks. The daily training staff consists of one lead instructor and four instructors. Instructors are divided into four groups by platoon. This allowed for the same instructors to work together each day, providing a higher degree of consistency, enhancing safety, and improving the overall training experience for the students. Four companies (engine or truck), one chief officer, and one medic unit attend each day. The normal class size is 18–22 students.

There are 25 days of training. Each training day consists of two 3½-hour sessions. Approximately 900 members receive this training, at an average cost of $126 per student for a morning or afternoon of training.

The most important aspects of OSET are as follows:

- Firefighter skill levels. Throughout the past seven years, the Seattle Fire Department has focused on individual and company skills that have a direct impact on the safety of our members. Instructors continue to observe that individuals and companies experience difficulty performing basic skills, such as donning SCBA in a timely manner and search-and-rescue techniques. Still, instructors did witness an overall improvement of the majority of our members. It is believed this improvement is a direct result of seven years of the OSET program.

- Required participation. Instructors and the majority of participants support mandatory participation of all members. Throughout the past seven years, the Seattle Fire Department has focused on two performance standards:
  - Category I skills: The skills that every Firefighter 2 should know and be able to successfully perform 100% of the time. These skills comprise the minimum acceptable standards for firefighters.
  - Category II skills: The skills that firefighters learn at OSET. Category II skills are taught and demonstrated by the instructors during the skill stations. Members are then given the opportunity to demonstrate their proficiency at these skills.

During the seven years in which the program has been used by the Seattle Fire Department, OSET has continued to be the most effective way to train our members on new skills and provide them with the opportunity to develop and build on previous training. OSET allows us to become quickly familiar with new equipment, procedures, and training concepts. It also allows us to informally evaluate our members and get a feel for what direction our training is taking us.

Instructor feedback supports the need for this training. Instructors cite numerous observations showing a wide range of ability for each element of training. These elements range from simple (e.g., a firefighter's ability to don a mask, enter the prop, and perform routine tasks) to complex (e.g., decision making based on accurate situational awareness).

Feedback/critique forms from companies show that a large majority of our firefighters believe that this is valuable training and should be more frequent than annually. Specifically, skills practiced during complex operations require more than annual refreshers for firefighters to stay proficient at these skills. There is strong support for this type of hands-on training. Evaluations repeatedly cite the need for this type of training that is focused on skills and techniques used at structure fires or incidents that we respond to daily.

All of us involved in this training strongly believe that it has had a significant impact on the skill level of members of the Seattle Fire Department. Specific training components support this. Over the past seven years, we have increased the basic level of competency of our members. We have gone from having members without the knowledge or skills to complete basic tasks to challenging our members with complex multiple-company operation (MCO) scenarios and then evaluating their decision making.

In closing, to reiterate, under stress, firefighters perform as they train. Thus, the learning environment should recreate the context and content of the environment in which performance is expected. Recent emergency incidents have been evaluated to show the complexity of company assignments and tasks. We can then identify the skills that our members lack when forced to make critical and complex decisions. Those critical decisions directly affect the safety of those personnel. We cannot and should not expect members' skills and decisions to be anything more than the cumulation of their experience and training. This training has a direct impact on the safety of firefighters at emergency incidents. We provide our members with the knowledge, skills, and abilities to perform their assigned duties in a safe and competent manner. This is why OSET is so important and should be our number-one priority for training.

—Rick Verlinda (Safety Chief, Seattle Fire Department)

THE 75% SOLUTION

After Bret Tarver's death in 2001, the Phoenix Fire Department and IAFF Local 493 formed a joint labor/management recovery team to take a long, hard look at how the Phoenix Fire Department was preparing its operations firefighters to fight fires and rescue firefighters who got into trouble inside structures (RIT capability). Fire Chief Alan Brunacini, recovery team co-chairs Steve Kreis and Brian Tobin, Todd Harms, and many other officers and firefighters on the recovery team uncovered Phoenix's shortcomings and came up with innovative solutions.[4] One of the most important and progressive changes suggested was the *75% theory*, or *75% solution*.[5]

The 75% solution states that 75% of a fire department's training budget, resources, and energy should be focused on preventing firefighter deaths. Specifically, these programs should concentrate on firefighter health and fitness, safe driving, air management, basic firefighting skills, and good fireground decision making. The remaining 25% of a fire department's training budget, resources, and energy should go toward training programs that concentrate on firefighter rescue. These intervention programs must train firefighters in rapid intervention and firefighter self-rescue techniques. Why? Because no matter how good prevention is, bad things do happen.

Fire prevention in modern buildings provides an excellent example of why training firefighters in rapid intervention and firefighter self-rescue cannot be overlooked. Fires still occur in buildings with the most modern sprinkler systems, fire alarm systems, and detection systems. While they don't happen as often as they did when there were no fire prevention systems, they still happen occasionally. This is why fire departments must use 25% of their training resources and time on intervention and self-rescue. Firefighters must know what to do when things go wrong, because things will go wrong.

The 75% solution was a completely new idea to the U.S. fire service and has already influenced progressive fire departments around the globe. The idea is simple and makes perfect sense: Spend 75% of your training time teaching firefighters how to stay out of trouble; spend the remaining time teaching firefighters what to do if they (or one of their own) get into trouble. Stress prevention over intervention, but make sure that firefighters know how to save themselves and other firefighters should something bad happen in a fire building (fig. 13–17).

We believe that fire departments should adopt the 75% solution if they have not done so already. The evidence is there for everyone to see, and it is tough to refute.

Fig. 13–17. Instruction in self-rescue. Firefighters learn how to use a personal emergency escape rope should they have to bail out of a structure in an emergency. Self-rescue training should comprise 25% of a fire department's training budget, with the other 75% of the training budget focused on preventing firefighter injury and deaths.

Fire departments must address firefighter health and fitness. Firefighters need yearly medical physicals that are thorough and complete. Firefighters need to be screened for heart problems, cancers, joint injuries, and other ailments and illnesses that are a part of our very stressful job. Firefighters need proven and varied physical fitness programs to help them keep their weight down. They need programs to teach them how to eat right and make healthy food choices. Comprehensive health and fitness programs help prevent heart attacks, the number-one killer of firefighters, and should be implemented by every fire department.

DRIVER TRAINING

Departments must also provide safe-driving programs for their members. Firefighters need to learn that racing to a fire or other emergency without regard for speed or intersection safety is no longer acceptable. Right now, across the United States, fire departments are being sued by citizens and the families of firefighters who have died or been seriously injured in accidents involving fire department vehicles. The time has passed when chauffeurs of fire apparatus could fly down the road with red lights and sirens blaring, blowing traffic signals and taking oncoming traffic lanes without a thought to speed or to the safety of private citizens and the firefighters riding on the apparatus.

Firefighters must put on their seatbelts every time they get into their rig. Furthermore, company officers must make sure that their firefighters are wearing their seatbelts at all times. Too many firefighters are being ejected from their rigs and getting seriously injured or killed in vehicle accidents because they are not wearing their seatbelts.

Chauffeurs need to learn how to park the fire apparatus on roadways so that it can protect the firefighters at the emergency scene from getting hit by reckless or inattentive drivers. Almost every day, a firefighter gets injured or killed while responding to or returning from an alarm or while working at an emergency scene on a roadway. Because many of these injuries and deaths could be prevented through safe-driving programs, fire departments should provide them to their firefighters.

AIR MANAGEMENT: A KEY COMPONENT
OF THE 75% SOLUTION

Air management is a key component of the 75% solution. Training on and following the ROAM prevents firefighters from running out of air in structure fires. Remember that NFPA 1404 mandates that fire departments have an air management program and that firefighters must be out of the structure before the low-air alarm activates. Fire departments that are not training firefighters how to manage their air are betting that nothing will go wrong to their firefighters in a structure fire. This is a losing bet to make with firefighters' lives. By not training their firefighters in air management, these fire departments are placing all their faith in intervention resources and preparedness. As we have shown throughout this book, rapid intervention is often tragically unsuccessful.

Giving firefighters meaningful, realistic, ongoing, hands-on training in firefighting basic skills is another essential part of the 75% solution. Firefighters must practice like they are expected to play; they must train the way they are expected to fight. In other words, they need to know how to perform their jobs. Firefighters need to constantly practice how to safely and effectively put out the structure fires that they will encounter during the course of their careers.

Basic engine and ladder company skills must be a major element of any training program. Firefighters need to pull and advance hand lines (fig. 13–18), flow water, properly place the aerial ladder, pull ceilings, cut ventilation holes on pitched and flat roofs, use air management, spot and use portable ladders, conduct rescues from inside and outside the structure, search a room, search multiple rooms, check void spaces for fire extension, communicate over the radio, use the thermal imaging camera, and understand fire behavior, as well as countless other basic fireground skills.

*Fig. 13–18. Firefighters advancing a hose up to the second
floor of a multifamily occupancy. Firefighters need to
continually practice the basics so that they can perform
safely and effectively at fires and other emergencies.
(Photograph by John Lewis, Passaic [NJ] Fire Department,
and Chief Rob Moran, Englewood [NJ] Fire Department)*

Finally, fire departments must provide training and tools to their officers
and firefighters so that they can make good decisions on the fireground.
As discussed earlier in this chapter, the Seattle Fire Department teaches
decision-making skills by letting firefighters talk about their decisions
during debriefing sessions at our yearly hands-on training (OSET). Much
more can be done, however. We would like to see training that uses
computer simulations of real-time structure fires to help train officers and
firefighters in how to make good fireground decisions. Experiential training
is necessary to give fire officers and firefighters the experiences on which
they can draw when they go to a fire and are called on to make split-second
decisions—decisions that could be life or death. Realistic, hands-on,
experiential training is the only way to give firefighters the knowledge and
skills needed so that they can be successful and safe at fires.

Training firefighters to be successful is the foundation of the 75% solution. If we train firefighters to stay out of trouble, keeping them from becoming part of the problem, then we help them be part of the solution. Prevent problems from occurring in the first place. This proactive, positive approach to training works and makes sense. Progressive fire department leaders and administrators would be well advised to seriously consider using the 75% solution as a training model for their departments.

Effective, real-world, hands-on training in the basics must be the foundation of every fire department. Firefighters must be training in basic firefighting skills every day. If firefighters are providing EMS to their communities, they must train on these skills every day as well.

ROAM: THE NEW BASIC SKILL

Air management is the new, basic firefighting skill. Fire departments should add the ROAM to the list of basic skills that their firefighters are taught in recruit school and are expected to train on throughout their careers. If firefighters train on how to manage the air in their SCBA, then they will be safer and healthier. The ROAM needs to be practiced over and over again so that firefighters become proficient at it. As we have seen, running out of air in a structure fire can be deadly, so we must train our firefighters for success and continually train them on this new basic skill.

Fire departments must also provide line officers with the training, tools, equipment, and support to allow them to train their crews. Fire departments need to provide department-wide, real-world, scenario-based, hands-on training in basic firefighting skills, so that every firefighter in the department receives the same training and hears the same message. Fire departments and fire administrators must never fall into the trap of providing WMD and special operations training at the expense of basic fire operations training. Just because government grant money exists for WMD training doesn't mean a fire department should replace all basic firefighting skills training. To do so is both irresponsible and foolish.

Prevention is a far better option than intervention. Fire departments should teach firefighters how to keep from getting into trouble on the fireground. Fire department administrators must make a commitment to prevention and work to make prevention a reality. Adopting the 75% solution as a training model is a great first step for fire departments to take.

Realistic training in basic firefighting skills is essential. Firefighters must train in the basics every day. Firefighters should train like their lives depend on it—because they do.

MAKING IT HAPPEN

The following training exercises should be undertaken in an acquired structure. Every so often, structures will become available in your city, town, district, or mutual-aid districts. Keep in touch with the department of planning and development for demolition permits that have been issued. Contact the owners/developers of these vacant structures and arrange a training time before demolition. Large commercial structures present great opportunities for multiple-company drills. It is also important to drill in single-family dwellings and apartment houses, because most of the fires that we fight occur in these structures.

1. Black out the structure and have the crews follow a pre-laid, charged hoseline inside the structure, taking a charged backup line with them. This is a great air management drill. For more free air management drills, go to Manageyourair.com and click on the "SMART Drills" tab.

2. Have crews ladder the structure for access, rescue, and ventilation operations. Have them cut ventilation holes in the roof. (See John Mittendorf's seminal book, *Truck Company Operations,* for an excellent and thorough description of how to cut ventilation holes in both pitched and flat roofs).

3. Place a firefighter in a remote location without any remaining SCBA air, with his or her facepiece removed, and an activated PASS device. Black out the structure or cover the facepieces of your crew with tint or wax paper. Have your crew go in as an RIT, transfill the downed firefighter, and get out of the structure.

4. Use a stopwatch to find out how long it takes for a 2-person RIT to locate and rescue a downed firefighter (who is 100 ft from the entry point and whose PASS device is activated) in a blacked-out environment with realistic obstacles in the way. Do the same with a 4-person RIT, a 10-person RIT, and a 14-person RIT.

5. Have crews use forcible-entry tools to open all exterior doors, roll-up doors, and windows of the structure.

6. In an acquired two-story single-family dwelling, have companies ladder the upper-story bedrooms and rescue several dummies from windows. Go over the different options for bringing conscious and unconscious victims down ground ladders.

7. Black out the structure and have companies deploy an attack line to the first floor, a backup line, and a line above the fire to the second floor of the structure. You can make a fake fire by encasing a red light in Plexiglas and securing this structure in place, to serve as a target for the hose teams. Practice replacing a burst section on the attack line. Discuss pump pressures with your crews. Have the backup driver do the pump operations and get the supply.

8. Smoke up the acquired structure with a smoke machine and practice PPV, ventilating one room at a time and one floor at a time. Discuss how PPV works, what its positives and negatives are, and when it would and would not be appropriate to use.

STUDY QUESTIONS

1. What should be the foundation of any fire department?

2. Basic firefighting skills must be taught in what type of environment?

3. Name several basic firefighting skills that need to be continually trained on.

4. According to Mark Cotter, how is training different from learning opportunities?

5. What is the new basic firefighting skill?

6. Today, can fire departments count on their firefighters receiving all of their fire operations training as on-the-job training? Explain.

7. How did the Seattle Fire Department decide to train its firefighters in basic fireground operations?

8. Why is training firefighters in the ROAM considered practicing prevention?

9. Explain how prevention and intervention are incorporated into the 75% solution.

NOTES

[1] Cotter, Mark J. From The Jumpseat: Teaching vs. Training/Walk the Walk. *Fire Engineering*. Web article, www.fireengineering.com.

[2] Ibid.

[3] NIOSH Fire Fighter Fatality Investigation 99-F47, p. 5.

[4] Phoenix Fire Department. 2002. Final report, Southwest Supermarket fire.

[5] Ibid, p. 44.

READY CHECKS

INTRODUCTION

Beginning correctly is the hallmark of any good fireground operation. Nothing beats a crew that is mentally, physically, emotionally, and professionally ready to tackle the emergency scene. A great beginning requires that the participants be fully prepared. Preparation meets opportunity, and the results are the stuff of which fireground lore is made. Preparation means that we do not enter the hazard area until we've accomplished goals that enhance our ability to safely mitigate the dangers of the work that we do.

One key area of preparation is checking the basics such as radios and assignments. They pop up time and again in reports of line-of-duty injuries and deaths. The general term used to encompass these items is a *buddy check*. This usually entails a quick check of the team's PPE and readiness for entry. Common problems with preentry preparation were identified that are often overlooked by firefighting teams. The READY check provides a quick, easy, and effective preentry check to remember and resolve many of the factors that cause pain and death on the fireground.

ARE YOU READY TO RUMBLE?

Most company officers spend a lot of time preparing themselves and their crew to fight fire. The job of the firefighter and the fire officer has expanded dramatically in the past 10 years to include many new disciplines and an expanding array of emergency types. Each of these new disciplines requires the knowledge, skills, and abilities to respond effectively. It is essential that firefighters be personally prepared prior to any emergency. This includes showing up fit for duty, doing appropriate equipment checks, and running effective drills.

What flies in the face of preparation is lacking the skills necessary to be ready when it is time to make entry. A team of firefighters that neglects to do a preentry buddy check jeopardizes not only themselves but other teams as well. It takes only a few seconds to confirm the readiness of the team. These few seconds can make all the difference in the outcome of the emergency.

Fireground reports reveal a recurring trend in firefighter injuries and deaths: The same mistakes are being made over and over. Many of these errors can be easily corrected prior to entry. However, the same mistakes may be difficult or impossible to correct once the team is committed to the hazard area. The following method of buddy checks has proven to be effective, reliable, and easy to use. The READY check asks you to verify the following factors:

- *Radio.* The radio is turned on and tuned to the correct channel, and you know who you report to.

- *Equipment.* The team has the proper equipment and is trained to use it.

- *Air.* The air level has been checked and is adequate for the assignment.

- *Duties.* The assignment is known, your role within that assignment is clear, and the team is trained to do it.

- *Yes!* If all of the preceding have been verified, then the crew is ready to enter. If any of the READY check items are not in place, fix the problem and then proceed.

R IS FOR RADIO

If you need any convincing that a huge problem on many emergency scenes is lack of appropriate radio discipline, here's a suggestion. Run a search-and-rescue drill with your department and have each member take responsibility for radio reports while engaged in their various tasks. These can be progress reports, requests from the IC, or Maydays. You will find that some firefighters are not on the right channel, do not know who they report to, or do not even have their radio turned on. The radio is often overlooked in training, and this is a dangerous habit. The *R* in READY check prompts the team leader to pause and verify that radios are on and tuned to the correct channel for the assignment. The team leader also ensures that each member of the team knows who to report to, so that critical messages get to the appropriate place (e.g., operations or Division Charlie) (fig. 14–1).

Fig. 14–1. A team leader ensuring radio status before proceeding

Often we get away with sloppiness in this area because the radio transmissions are not of a critical nature. A progress report that is transmitted on the wrong channel can be repeated once the mistake is recognized and fixed. Lack of practice in speaking clearly while wearing an SCBA is overcome by repeating garbled messages in a different manner. Even radios that are turned off can be turned on when the messages are not critical and the individual is informed of his or her mistake.

Communication problems arise when things are not going well or when a set of variables converges in a deadly way. The most obvious of these is the need to transmit a Mayday, Urgent or other firefighter emergency. Clearly, if an emergency message of this nature is being transmitted, something has gone dramatically wrong. The last thing you can afford to be thinking about, as you are looking up from the fiery basement into which you've just fallen, is how to change the channel on your radio to summon help. The worst thing you can do for a partner who has just been buried in a collapse of a dropped ceiling is send out a Mayday that goes nowhere because your radio is off or dead. The haunting messages from the Hackensack Ford fire are a testimony to how tragic this can be. In his excellent paper on fireground radio problems, Chief Curt Varone shared the following:[1]

> *Among the most well-documented cases of a communications failure contributing to firefighter fatalities, was the July 1, 1988, fire at Hackensack Ford in Hackensack, New Jersey. In 1988, Klem wrote the NFPA investigative report on the Hackensack fire, detailing the circumstances that led to the deaths of five firefighters when a bow-string truss roof collapsed at a fire in an auto dealership.*
>
> *Approximately one minute before the roof collapsed, the IC ordered over the radio for companies operating on the interior to "back your lines out." This message was not acknowledged by any of the companies operating on the interior of the building, nor was it acknowledged*

and/or repeated by the dispatch center. When the collapse occurred, three firefighters in the building were pinned by falling debris. Two other firefighters were able to escape into an adjacent tool room.

Approximately three minutes after the roof collapsed, radio calls for help were made by the two trapped firefighters who escaped into the tool room. These calls initially went unanswered by either the IC or the fire alarm dispatcher. However, the calls were heard clearly by civilians with scanners who were monitoring the incident and were recorded on the dispatch office's tape recorder. Some listeners even called the dispatch center on the telephone to inform the dispatcher of the trapped firefighters. By the time the IC became aware of the calls for help, an effective rescue effort could not be mounted to save the trapped members.

He also listed two other incidences which clearly make the point:[2]

A case in point was the Brackenridge, Pennsylvania, fire, where the deaths of four firefighters were attributed in part to the fact that they were operating on a separate radio channel and did not hear progress reports on the main channel that warned of worsening fire conditions.

Another case was the Pittsburgh fire, where the use of different radio channels by fire and emergency medical personnel contributed to confusion over who was missing and who was rescued. The confusion led to the erroneous conclusion that all firefighters had been accounted for, when in fact three firefighters were missing in the building. As a result, no effort was made to initiate a search for downed firefighters.

Likewise, the best progress reports in the world are worthless if the firefighter sending them does not know how to use the radio in a clear and effective manner (fig. 14–2).

Fig. 14–2. A partner helping a firefighter in trouble. A partner is required to get a good radio transmission to the IC. (Photograph by Lt. Tim Dungan, Seattle Fire Department)

The final aspect of the radio portion of the READY check is knowing who we are reporting to, such as "Division One" or "James Command". As with most things that happen at emergency scenes, the actions of each one of us affects everyone else to some degree. A common occurrence on today's fireground is the erroneous transmission of necessary information to the wrong place. This may cause a cascade effect of the transmission's needing to be repeated or redirected—or the radio message may not be received at all. Our channels are normally swamped with information, especially early in an emergency, and the unnecessary repetition of messages, resulting from not knowing your assignment, is undisciplined and preventable.

Team leaders must ensure that their team members have their radios turned on and tuned to the correct channel and that members know how to use their radios. The first part of this merely requires practice and discipline. The second part will be realized only if you drill your crews in the proper usage of their radios and appropriate ICS. Radios must be used under conditions that require physical exertion. They must be used in areas that are noisy and full of distractions. Also, with the advent of technological advances, such as voice amplifiers, we need to be certain that our members can get their messages out with the new gadgets, as well as without them (when they break, as they surely will).

E IS FOR EQUIPMENT

The best intentions of an aggressive firefighting team can be undone by neglecting to bring the correct tools for the assigned task. The second part of the READY check is to ensure that the members of your team have the equipment necessary to accomplish their task without time-consuming interruptions or dangerous improvisation. This works regardless whether your department leans toward predetermined assignments or prefers a more spontaneous approach to the fireground. If you are tasked with pulling ceilings for a ripping attic fire and bring a set of irons and a thermal imaging camera, a highly irritated hand line crew will chase you back out to get pike poles. Conversely, pike poles in the hands of search teams will be much less effective than a halligan or an ax. The biggest mistake is not bringing any equipment with you at all. Firefighters have all probably heard a variation of the adage "A firefighter without the appropriate tools is merely a well-informed civilian." (fig. 14–3).

Fig. 14–3. Firefighters verifying that they have the appropriate equipment for their task (Photograph by Lt. Tim Dungan, Seattle Fire Department)

As a consequence of the lack of fireground discipline, valuable time is lost as crews retreat to get the equipment that they should have brought in the first place. This can be embarrassing when other crews move into the gap and take care of business for the unprepared crew. Having to send a member out to get a forgotten tool can be detrimental to team integrity and accountability. This comes at the expense of the task and scene safety. Finally, not having what you need before you enter can be deadly. Many are the tales of trapped or otherwise endangered firefighters who used an ax or a halligan to cut their way out of potentially deadly situations. That's going to be extremely difficult if your search team enters with a battle lantern and a buck knife.

The second phase of the READY check is to have your team leaders determine and verify that the equipment necessary for the assignment is in hand prior to entry. Drilling on this can be as simple as holding chalk talks, giving your crews varied assignments, and determining what will best help the team to be successful.

A IS FOR AIR

As the cornerstone of an effective air management program, the ROAM dictates that you have to know what you've got before you go in. We are long past the days when team leaders should feel comfortable winging it in regards to their air supply. If you want to end up as a close call, a last call, or an example of what not to do, nothing will get you there quicker than haphazard air management practices.

Knowing your air prior to making entry accomplishes two goals and can mean the difference between life and death for your crew. The first of these is recognizing whether your equipment is functioning properly. Good daily and weekly checks are usually sufficient to ensure that masks are ready to go. However, many of us have had malfunctions occur in SCBA even though they had already been thoroughly checked that day. No matter if it's a slow leak, damage during response, or a spontaneous malfunction, we rely much too heavily on our masks to not be sure of them before we enter the IDLH environment (fig. 14–4).

Fig. 14–4. A preentry air check by the team leader

Additionally, the team must have enough air to accomplish their assignment, to calmly and safely exit the IDLH environment, and to leave their emergency reserve intact. Each member of the team should have a full bottle prior to entry, and with a quick check, the company officer can verify that the team has the required air. Knowing the status of each team member's air can assist the team leader to best utilize the crew, getting the most work done while operating in a safe manner. Having full cylinders starts the operation off right and will prompt the crew to perform incremental checks of air levels as they complete the task.

A recent article in the *St. Louis Post-Dispatch* looked at some of the lessons learned from the tragic deaths of two St. Louis firefighters. One of the key findings emphasizes the need to know your air on entry:[3]

> *Air tanks weren't full in some companies' breathing equipment. Morrison [Robert Morrison, who perished in the fire] was in the building a short time before his low-air alarm sounded.*

Air checks become critical as we seek to get away from the dangerous practice of waiting for our low-air alarm to activate before exiting the IDLH atmosphere. A critical component of the ROAM is that the last 25% of the bottle be reserved for true emergencies. Starting off with an exact knowledge of what you've got will enable the team to operate safely while aggressively pursuing their assignment(s) toward a positive conclusion.

You can best train for this element of the READY check by making an air-level check mandatory on every run where you and your crew are wearing your SCBAs. A great time to engrain the habit of a quick air-level check is on routine false alarms. Prior to entry on a confirmed false alarm or a single automatic fire alarm, all members of the crew should verify their air prior to entry. Every drill should include a discussion of how air usage will affect completion of the assignment and team safety.

D IS FOR DUTIES

Getting everyone on the same page is often difficult to do in the fast-paced world of firefighting. The READY check allows for this by reminding the team leader to ensure that members know the team assignment and their individual duties within that assignment. Once inside the IDLH environment, communication becomes more difficult, and task-level detail is impossible to impart efficiently. By establishing duties at the outset, you give your team a great start toward a positive outcome.

In addition, assign all members of the team duties that they are trained to perform. While this can be handled in numerous ways, many officers have found that basing crew assignments on position and/or seniority is helpful in the confusion of initial operations.

A team without clearly defined duties usually ends up with assignments going uncompleted, unnecessary turmoil in situations that are already confusing enough, and added risk for the team. When the conditions that have brought us to the particular incident are not getting better and necessitate immediate intervention, the more time that is spent trying to recover from an inadequate or faulty start, the more time the factors working against us will get to play their role (fig. 14–5).

Fig. 14–5. A team leader confirming assignments and ensuring that each member knows his or her role

With all that is going on prior to entry, it is essential that team leaders take the time to ensure that all members of the team have information regarding what they are expected to do.

Y IS FOR YES!

OK. You are about to make entry on a well-involved house fire, and the IC has assigned you to primary search on the second floor. Your partner has just finished masking up, and you both begin a quick READY check:

"Do we have our radios turned on, on the correct channel, and know to whom we report?"

"Radios on, turned to channel 1, and we are reporting to Emerson Command."

"Do we have the correct equipment, including appropriate PPE for the assignment?"

"Full PPE in place, we have a thermal imaging camera and halligan bar."

"Do we know our air status and is it sufficient to make entry?"

"All members with full bottles."

"Do we know our assignment and how to do it?"

"Primary search on second floor with all positions assigned."

If the answer to each of these questions is *yes*, then you have completed the READY check and are prepared to tackle your assignment.

The *Y* in READY check is an affirmation that yes, it is OK to proceed. If any of the preceding questions cannot be answered in the affirmative, then you must stop and address them before entering the hazard area.

WHY WE CHECK

We are told to check our apparatus, firefighting tools, and PPE at the start of each tour in the firehouse. Why do we do that?

While it may be true that the previous crew checked their gear at the start of their tour and found no problems with it, you still need to check it again at the start of your shift. The previous crew might have had a fire or emergency on their shift. Can you be sure that they cleaned, refueled, and recharged the tools and equipment used and put everything back in the right place? Did your SCBA tank develop a leak at the last fire? Was it reported and fixed? Is it fully charged as you start your tour? Do you want to enter the fire with less than a full tank of air? Has dirt or debris fallen into your facepiece so that when you put it on, the debris will be blown into your eyes? If you don't test the fit of your facepiece, you might end up with air leakage at your next fire or emergency. Are the backpack straps set up for the smallest firefighter in your department while you are the largest? How much of a delay will that cause when you discover they need to be adjusted at the fire scene? Does your PASS alarm work, or is it defective? Are the air hoses abraded or cracked and ready to fail when you need them most? It just makes good sense to check your assigned SCBA at the start of the tour before you are required to use it.

How about the saw: Is it fueled? Will it start on the first try? How about your Jaws of Life: Will it start, and is it fueled? Are the hoses and attachments where they belong? How about your generator: Are the cables where they belong? Did the pump operator check to see if his hose fittings and nozzles are where they are supposed to be? Are they even on the apparatus or did another company inadvertently pick them up at the scene of last night's fire? Is the apparatus fuel tank full? Are all of the warning devices working? Are your gloves in your turnout pocket, where they belong, or did someone borrow them and forget to put them back? It would be a shame to suffer the pain of burned hands because you did not take the time to check that you had your gloves. What else do you keep in your pocket? Is it still there, or did it magically disappear? Are the forcible-entry tools where they should be, or will you have to go looking for them when you arrive at the scene of a fire? How about the water extinguisher, is it full of water and charged with air?

I could go on and on with what should be checked at the start of each tour, but I think you get the idea by now. This constant checking is tedious but necessary. If you make it a habit, you will not be unpleasantly surprised if a tool is not working properly or is missing or damaged. We must work as a team, relying on each other, but there are nevertheless some things we should do for ourselves.

—*Frank Montagna (Chief, FDNY)*
http://www.chiefmontagna.com

In some instances, a decision may be made to proceed in spite of the specific deficiency. If this is done—and it should be a rare exception—at least you are doing so with full knowledge of the problem. This is in stark contrast to blindly going forward and discovering the problem in an atmosphere or situation in which recovery would be difficult or impossible.

MAKING IT HAPPEN

1. Conduct basic fireground drills and evaluate the success of radio transmissions and equipment issues without any form of buddy check. Determine how many instances of inadequate radio transmissions (including no transmissions owing to dead or nonexistent radios) or inadequate equipment arise.

2. Perform drills that include READY checks. Time how long it takes your firefighters to complete this procedure over the course of various drills. A READY check should take only 5–10 seconds—and eliminates a lot of wasted time on the interior.

STUDY QUESTIONS

1. Who is jeopardized when a team of firefighters neglects to do a preentry buddy check?

2. What does the "R" in READY checks stand for?

3. Why is it critical to know who you are reporting to?

4. What does the "E" in READY checks stand for?

5. True or false: Utilizing READY checks works only if your department utilizes predetermined assigments.

6. What does the "A" in READY checks stand for?

7. When is a great time to engrain the habit of a quick air-level check?

8. What does the "D" in READY checks stand for?

9. What are the potential consequences when a team lacks clearly defined duties?

10. What does the "Y" in READY checks mean?

NOTES

[1] Varone, J. Curtis. 1996. Firefighter Radio Communications and Firefighter Safety, p. 80. http://www.usfa.dhs.gov/downloads/pdf/tr_96cv.pdf.

[2] Ibid., p. 98.

[3] Kohler, Jeremy. 2007. Official story of fire deaths is withheld. *St. Louis Post-Dispatch*, May 4.

INCIDENT COMMAND
AND AIR MANAGEMENT

INTRODUCTION

The role of the Incident Commander (IC) is a difficult one. The IC must be able to make decisions based on limited information in a short time frame. The IC must also maximize fireground effectiveness, rotate crews efficiently, monitor and manage radio communications, prepare for any possible firefighter emergency, and conduct an ongoing risk-benefit analysis.

The fire service often looks at the Incident Command System (ICS) as boxed positions filled by individuals with a specific job. The reality of ICS is that functions at the strategic, tactical, and task levels are intertwined and inseparable. This chapter explains how a good air management program can greatly assist any IC in supervising firefighters on the incident scene, managing the risks inherent in fireground operations, and forecasting and providing for the needs of the ongoing incident. The appropriate integration of incident command and air management eliminates many of the breakdowns that can lead to firefighter injuries and deaths on the fireground.

INCIDENT COMMAND— A HISTORICAL SNAPSHOT

During the week of November 6, 1961, the city of Los Angeles experienced the most disastrous brush fire in the history of Southern California. The Bel-Air fire began on the north slope of the Santa Monica Mountains and raced through the tinder-dry chaparral. The fire was driven by 50-mile-per-hour winds and soon covered much of the coastal mountains in a firestorm. Thermal air currents and winds swirled countless thousands of burning brands into the air, to later deposit them far from the main fire front. When the fire was finally contained, over 6,000 acres of watershed had been consumed, along with over 480 homes and 21 other structures. Thankfully, no one was killed or injured in this fire.

In 1971, as a result of the Bel-Air fire and other costly and destructive California wildfires, Congress mandated that the U.S. Forest Service design a system that would increase the capabilities of Southern California wildland fire protection agencies. This system came to be known as Firefighting Resources of Southern California Organized for Potential Emergencies (FIRESCOPE). This ICS grew out of these wildland firefighting development efforts and was the primary on-scene command and control system for managing day-to-day response operations.

Also in the early 1970s, the Phoenix Fire Department developed the Fire Ground Command (FGC) system. This ICS was similar to FIRESCOPE but used different terminology and organizational structure. FGC was specifically designed for structural firefighting and proved to be a very easy system to use on the fireground.

By 1981, FIRESCOPE was widely used throughout Southern California by the major fire agencies and was being applied to nonfire incidents as well. Soon FIRESCOPE was adopted by the National Wildfire Coordinating Group and became known as the National Interagency Incident Management System (NIIMS). This system was marketed as an all-risk response system. By the 1990s, the National Fire Service Incident Management System Consortium began to develop operational protocols within the ICS, so that fire and rescue personnel would be able to apply the ICS as a single common system. In 1993, the consortium completed its first document, *Model Procedures Guide for Structural Firefighting*. This document allowed the nation's fire and rescue personnel to effectively apply the ICS to any emergency in any part of the country. The National Fire Academy (NFA), having already adopted FIRESCOPE in 1980, incorporated this material into its training curriculum as well.

In March 2004, after close collaboration with state and local government officials and representatives, the Department of Homeland Security (DHS) issued the National Incident Management System (NIMS). NIMS incorporates many existing best practices into a comprehensive national approach to domestic incident management that can be used at all jurisdictional levels and across all functional disciplines. NIMS has five major functions:

- Command
- Operations
- Planning
- Logistics
- Finance and administration

A potential sixth functional area is intelligence, to gather and share incident-related information (fig. 15–1).

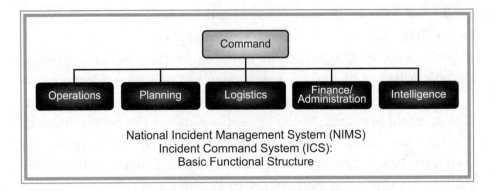

Fig. 15–1. Basic functional structure of the NIMS ICS, showing what the general staff looks like. A potential sixth functional area covering intelligence could be established for gathering and sharing incident-related information. (Source: Federal Emergency Management Agency [FEMA])

In today's fire service, in concept and practice, the ICS is recognized as a necessary organizational tool. The ICS provides for centralized leadership, span of control, increased measures for responder safety, communications, and a methodology for multiagency interoperability.

LEVELS OF INCIDENT COMMAND

The basic configuration of a command structure includes three levels:

- Strategic. Overall command of the incident.
- Tactical. The use of intermediate-level leaders to supervise individual areas of the incident.
- Task. Specific job assignments related to the incident.

At the strategic level, the IC's responsibilities involve the overall command of the incident. This includes establishing major objectives, setting priorities, allocating resources, predicting outcomes, determining the appropriate mode of operation (offensive or defensive), and assigning specific objectives to tactical-level units (fig. 15–2).

Fig. 15–2. The IC operating at the strategic level (Photograph courtesy Seattle Fire Fighters Union, IAFF Local 27)

At the tactical level, intermediate-level officers direct activities toward specific objectives. Tactical-level officers include division or group supervisors, who are in charge of grouped resources and operate in assigned areas or supervise special functions at the scene of an incident (fig. 15–3).

Fig. 15–3. A district chief operating at the tactical level. This district chief is assigned as a division supervisor. (Photograph by Dennis Leger, Springfield [MA] Fire Department)

The task level refers to those activities normally accomplished by individual groups or specific personnel. Company officers or team leaders routinely supervise task-level activities. In combination, task-level activities should accomplish tactical objectives. For structural fire fighting, examples of these tasks include ventilation, search and rescue, extinguishment, and overhaul.

INCIDENT COMMAND AND RISK MANAGEMENT

While the IC has a host of assigned duties, they all depend on the ability of the IC to recognize and manage risk. The safety and welfare of the firefighters and other first responders at the emergency takes precedence over all other aspects of effective incident scene management (fig. 15–4).

Impact	Risk Management Actions		
Significant	considerable management required	must manage and monitor risks	extensive management essential
Moderate	risks may be worth accepting with monitoring	management effort worthwhile	manage effort required
Minor	accept risks	accept, but monitor risks	manage and monitor risks
	Low	**Medium**	**High**
		Likelihood	

Fig. 15–4. Risk management model for incident scene management. ICs must recognize that risk management incorporates both impact and likelihood factors. This basic model directly folds into ongoing size-up of all emergency events.

Risk management is defined in NFPA 1561 (Standard on Emergency Services Incident Management). NFPA 1561 also outlines the factors an IC must consider to minimize the risks facing firefighters. For example, the IC must provide a sound strategic plan based on SOGs, ensure that members have been effectively trained and are provided with and utilize proper PPE, establish an effective communication process, and have a safety plan in place—to name only a few risk management considerations.

BE A VISIONARY

As an IC, many things cross my mind when I run a fire. To be sure, the safety of my crews is paramount among them.

Here's my—or any IC's—dilemma. I send a crew inside a building. I immediately lose visual contact with my firefighters the moment they pass through the door. (Please refrain from sending firefighters through anything but a door. Always ask yourself, why can't they go in through the door? If the answer is that fire conditions demand it and no life safety is involved, then rethink your actions; something is wrong with that picture.) I can only assume what they are doing and where they are at any given time. That requires a lot of trust on my part. If I don't train with them often, then it could be anybody's guess what they're doing or where they might be. To calm my nerves, I have to assume that they are doing what I would be doing under the same circumstances. Again, I want them to be as safe as possible.

To lessen my concerns, anything that can be done to ensure that my crews go in and come out OK is a godsend to me. Air management is one of those things. With it, we are finally beginning to think about what it is that allows us to be in there to begin with—namely, air!

For me, when I crawled around inside burning buildings, air management meant waiting until my low-air alarm began to ring and then reaching behind me and putting my hand on the bell to muffle the sound as I continued doing what I was doing. Today, that would never happen. There are departments that consider it a safety violation when firefighters exit a building with their low-air alarm sounding, unless there is a dire emergency.

It's your air, it's your lungs, and it's your life. Every measure possible should be taken to ensure that you leave this job the same as you came on. We have the packs on our back—so why not use them in the safest manner possible? Besides, knowing that you are looking out for yourself and your crew members makes my job as IC easier. Anytime we get into trouble, it's too late! We should strive to always be visionaries and avoid being reactionaries.

—*Skip Coleman (Assistant Chief, Toledo [OH] Fire Department)*

INCIDENT COMMAND AND AIR MANAGEMENT

The textbook version of ICS identifies the levels of incident command, assigns each a box, and draws lines between the boxes to demonstrate the relationships between strategic, tactical, and task levels (fig. 15–5).

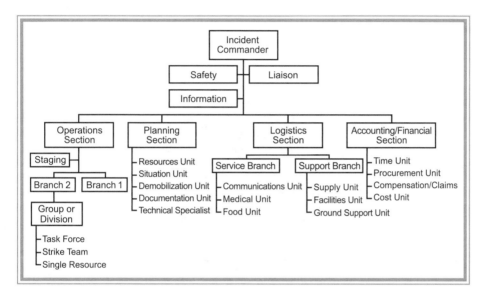

Fig. 15–5. A textbook version of ICS, assigning a box as needed on the basis of the scope of the emergency (Source: FEMA)

The real world does not reflect the order of the idealized ICS chart. Consider, for example, the first-arriving company officer at a residential room-and-contents fire. On arrival, the company officer performs at all three levels, identifying the strategy, implementing tactics, and possibly helping at the task level. When a ranking officer arrives and assumes the command function, the initial IC still performs at all three levels, with a reduction in responsibility at the strategic level and an increase at the tactical level. With subsequent changes in the ICS structure, responsibilities continue to be dispersed and redistributed at the strategic and tactical levels.

Rather than considering the ICS as a series of boxes connected by lines, think of the ICS as several overlapping circles. The first-arriving company officer is the initial IC and holds all three circles within close proximity. As additional crews arrive at the location, the circles move apart, with the original IC holding the overlapping center together. Further expansion of the system will allow separation of the more defined roles. Only in rare and large events does the system reach the point where it can truly be boxed out (fig. 15–6).

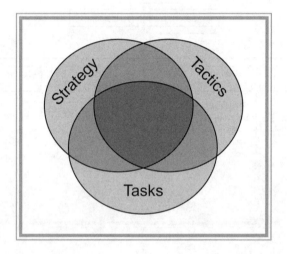

*Fig. 15–6. The ICS modeled as overlapping circles.
The first-arriving company officer, as the initial IC,
holds all three circles of responsibility within close
proximity, with the primary focus on the tactical level.
This is a more accurate representation of the ICS
than the previous model, using boxes (fig. 15–5).*

It is important to understand that the task-level responsibilities do not get reassigned with changes in the ICS structure. The team assigned to ventilate the roof will be expected to complete that task even if a roof division is established. The new IC or division supervisor will assess the scene and provide support to the vent team, but the responsibility to perform the task does not change. The team assigned to search the second floor will be expected to complete that task even if a Division 2 is established. The new division supervisor will support the search team with hoselines, ventilation, or other task-level functional assignments, but the task of searching will remain with the original team.

The process of implementing air management within the ICS can best be correlated to accountability. At its core, incident accountability is a task-level function on the fireground. Firefighters are accountable to and for each other. Company officers are accountable for their team. When the ICS expands above their heads, accountability does not change. In fact, the entire system is dependent on the individual firefighting team to be accountable to and for each other. The strategic and tactical levels can provide support through a personnel accountability report (PAR) request, a roll call, and an accurate status board at the command van, but this can only be accomplished if accountability is maintained at the task level on the fireground. Responsibility to keep track of where the company is assigned can be transferred repeatedly within the ICS while a company is engaged in the fire attack. Never will the responsibility to be accountable to and for each other change hands among the team deployed into the hazard area.

The same is true of air management. Air management is a task-level function. Here, responsibilities and advantages that apply to the classroom version of the ICS chart also apply to air management. The boxes are a fluid representation of the dynamic system that is incident command. The task-level function of air management is the responsibility of each firefighter and company officer. The strategic and tactical levels within the ICS can and will support air management through timed reminders, work cycles, and other methods such as telemetry. The strategic and tactical levels cannot and should not be expected to perform the task of air management for the firefighter or the company in the hazard area. Never will the responsibility for ensuring that a team of firefighters has enough air for entry, work, exit, and an intact emergency reserve, leave the shoulders of those individual firefighters assigned to the IDLH environment.

As noted in chapter 4, the number of firefighters losing their lives in IDLH environments due to running out of air continues to rise, even as the number of structure fires continues to fall. The ICS must have a safety plan that incorporates air management. This safety plan should provide for implementation of air management at the strategic, tactical, and task levels.

STRATEGIC-LEVEL AIR MANAGEMENT

Strategic-level air management begins with the fire chiefs of individual fire departments or jurisdictions. Implementation of an air management policy is a function of the fire chief and the senior staff. The air management policy, like any operational policy, needs to be supported by regularly training operational personnel. The air management policy must receive formal and informal support, positive modeling, and enforcement in a nonpunitive fashion. This process is the foundation for a positive change in firefighter safety. The ROAM (see chap. 10) outlines a very effective air management program. In addition, the fire service is now *required* to have an air management policy in place, as indicated by NFPA 1404 (see chap. 6).

Once the air management policy is in place, the IC can begin to plan the incident around it. While the task- and tactical-level supervisors are working to meet the strategic plan, the IC should forecast outcomes. An offensive strategy can be expected to achieve a predicted result, within a predicted time frame, using a predicted number of resources. The beauty of adding an air management policy to the mix is that the firefighters and company officers functioning at the task level are required to provide information regarding their air supply while operating in the IDLH environment. If that information is included as part of a quick but comprehensive progress report, as in the CARA format (see chap. 16), the tactical level and then strategic level within the ICS structure can compare expected outcomes and crew rotation objectives with actual outcomes and crew rotation objectives. The IC can then adjust the strategic goals around gaps between the expected and actual outcomes.

At an incident, the IC is responsible for ensuring that adequate personnel resources are available to meet the strategic and tactical goals. A forward-thinking IC will anticipate when—and how many—crew rotations can be expected to meet the current goals. Crews report their task-level progress toward the tactical goals assigned to the division/sector supervisor. With an air management policy in place, the crews also report their air use, providing the division/sector officers with a reliable basis for predicting when the crew will need relief in the hazard area. The division/sector supervisor can then make requests from the operations section chief, who can then notify the IC if additional resources are needed. The implementation of an air management policy assists in forecasting resource needs and in recognizing discrepancies between the predicted strategic outcome and reality.

TACTICAL-LEVEL AIR MANAGEMENT

At the tactical level, air management becomes a critical function of team rotation and effective resource management. Depending on the scope of the emergency, the IC and the operations section chief, branch directors, or division or group supervisors must recognize that the work cycle of teams in any IDLH environment hinges on SCBA air time, which has three components:

- Entry time
- Work time
- Exit time

The division/sector supervisor must effectively forecast the need for replacing teams based on these air management elements (fig. 15–7).

Fig. 15–7. A division supervisor proactively managing teams based on task assignment, work time, and SCBA air (Photograph courtesy John Lewis, Passaic [NJ] Fire Department, and Chief Rob Moran, Englewood [NJ] Fire Department)

The division/sector supervisor must ensure that companies operating in the hazard area provide proper progress reports, by reminding them if necessary. When these reports include a measure of the air level remaining, the division/sector supervisor can accurately predict and prepare for crew rotations. In addition, this information provides the division/sector supervisor with enough information to gauge whether the company's assigned task has been achieved. If the task has been completed, then everything is going as planned. If not, the division/sector supervisor should notify the operations section chief or the IC and request additional resources to complete assigned objectives.

Division/sector supervisors provide the bridge between the task level and the strategic level of supervision at the incident. While striving to meet the strategic goals assigned by the IC, they also have direct knowledge of fire conditions and assigned tasks within their area. They can remind crews operating in the hazard area to provide progress reports, including air level, according to the established policy. The division/sector supervisor is close enough to hear if firefighters in the hazard area have extended beyond the acceptable limit of the low-air alarm. Immediate radio communication can be used to quickly identify personnel whose low-air alarms have activated. When radio communication cannot be established with the personnel experiencing a low-air emergency, the division/sector supervisor will adjust the priority to firefighter rescue. With an effective air management policy in place, low-air alarms are a rare occurrence and become a clear, audible warning that something is not right on the incident scene. This notification provides maximum time for the Incident Commander to respond effectively.

TASK-LEVEL AIR MANAGEMENT

Supervisors at the task level are primarily company officers or senior firefighters responsible for one to three other firefighters. Team leaders must fulfill their task assignments while maintaining situational awareness and providing for the safety of the team. One requirement that helps ensure the safety of the team is that no team member should ever breathe smoke. Smoke is toxic and dangerous.

The primary responsibility for air management lies at the task level. Individual firefighters and teams are responsible for managing the air supply they bring with them to the fire fight. While the division/sector supervisor can assist, through time-based reminders and effective training prior to the incident, the firefighter and the company officer in the hazard area must do the task of air management. The firefighter and the company officer should maintain a margin for error in case something goes wrong. The margin for error required by the ROAM, NFPA 1404 (see chap. 6), and common sense is the emergency reserve air. The firefighter and the team leader can use 75% of their air for entry, work, and exit from the hazard area. The remaining 25% of air is reserved for emergencies – it is never to be used unless there is an emergency.

Another aspect of task-level air management is team leaders rotating team members in and out of heavy workload positions based on their air supply and air use. If the entire team uses air at nearly the same rate, the team's ability to perform the assigned objectives will be optimum. This provides maximum work time for the team in the hazard area while maintaining a margin for error.

A team that does not use management cannot be expected to have one individual work very hard for an extended length of time. This hard worker will use air at a much greater rate than the rest of the team. When this hard worker runs low on air, the entire team will need to exit together to maintain accountability (fig. 15–8).

Fig. 15–8. A team practicing air management at the task level. Team leaders must redistribute workload assignments based on rate of SCBA air use. Without redistribution of the workload, overall team work time at assigned tasks will be diminished. (Photograph courtesy John Lewis, Passaic [NJ] Fire Department, and Chief Rob Moran, Englewood [NJ] Fire Department)

An example of effective workload management would be the cycling members of a team of firefighters performing a search of a large residence, for example. The team leader must ensure that the workload is effectively shared on the basis of SCBA air levels. To accomplish this, the team members must communicate air levels at natural breaks to the team leader. The team leader will reassign members in heavy-workload situations, based on air usage, to lighter workload positions.

Balancing the workload not only maximizes the team's ability to effectively accomplish assigned tasks, but also enhances situational awareness by having all members check their SCBA air. With this short air check and the subsequent communication, several important safeguards are completed:

- All members are accounted for
- Members can monitor the smoke and heat conditions
- Members perform a physical check of each other
- A pause allows for reorientation (if needed)
- The team leader has time to quickly recognize whether more resources are needed

Teams that manage their air supplies proficiently will reap these benefits. Overall, achieving a level of situational awareness at the task level—by doing nothing more than checking the air in each member's SCBA—works wonders. It increases firefighter safety and improves incident scene management.

INCIDENT COMMAND FAILURE

NIOSH provides a historical perspective of incidents in which the ICS did not effectively manage an emergency. The reports in this section document firefighter fatality incidents in which a number of factors contributed to the outcome. In these examples, mismanagement of personnel and the inability to factor air management into the strategic, tactical, and task levels of the emergency played a critical role.

CASE STUDY 1

A 39-year-old career captain died after he ran out of air, became disoriented, and then collapsed at a residential structure fire. The medical examiner listed the cause of death as smoke and soot inhalation and noted a carboxyhemoglobin level of 22.7%.[1]

RECOMMENDATION 1. Fire departments should enforce SOPs for structural firefighting, including the use of SCBA, ventilation, and radio communications.

RECOMMENDATION 2. Fire departments should ensure that the IC completes a size-up of the incident and continuously evaluates the risks versus the benefits when determining whether the operation will be offensive or defensive (fig. 15–9).

As with all firefighter fatalities, a number of factors contributed to the tragedy. The most significant factor was that this team of firefighters did not practice air management. The team became disoriented owing to heat and smoke conditions, and one firefighter ran out of air and ultimately died. An air management system that requires firefighters to be out of the IDLH environment before their low-air alarm activates would have given this firefighter more time for survival—either time for rescuers to find him, or time for him to seek shelter and/or find his way out of the single-family residence.

Fig. 15–9. The IC conducting a size-up of the
incident as part of a continual risk-benefit
evaluation to determine whether the operation
will be offensive or defensive (Photograph by
Dennis Leger, Springfield [MA] Fire Department)

CASE STUDY 2

On arrival, firefighters began fighting the fire and searching for two children who were reported to be in the building. Shortly after arrival, a firefighter went alone to the third floor to search for victims. He became lost in one of the apartments and subsequently ran out of air and died before rescue crews were able to locate him.

An autopsy showed that the firefighter suffered burns to his body and inhaled soot and carbon monoxide. Blood tests revealed a carboxyhemoglobin level of 65%. The official cause of death was listed as asphyxia due to smoke and carbon monoxide inhalation.[2]

RECOMMENDATION. Freelancing at the highest levels cannot be permitted. Command of an incident must be undertaken by one individual; if more than one agency has jurisdiction or if the lines of authority in an organization are not clear, then unified command must be established. It is vital that operations be coordinated through the command structure to avoid situations in which opposing strategies and tactics are employed to the detriment of the overall operation (fig. 15–10).

Firefighters can develop tunnel vision, especially in critical, high-stress situations, as was the case with this fatality. A sound air management policy that heightens situational awareness and team integrity is a step in the right direction toward eliminating such tragedies.

Fig. 15–10. A company officer managing his team at the tactical level based on the strategy that the IC has established (Photograph by Dennis Leger, Springfield [MA] Fire Department)

CASE STUDY 3

On November 6, 1998, two volunteer firefighters died while performing an interior fire attack in an automobile salvage storage building. As the crews began to exit, an intense blast of heat and thick black smoke covered the area, forcing firefighters to the floor. Two firefighters were knocked off the hoseline, and their SCBA low-air alarms began to sound as they radioed for help and began to search for an exit.

The two firefighters departed in different directions, and one of them eventually ran out of air. He was found immediately after collapsing and was assisted from the burning building. As firefighters pulled the unconscious firefighter to safety, another firefighter reentered the structure to search for this victim. During his search, this firefighter ran out of air, became disoriented, and failed to exit, becoming another victim. Additional rescue attempts were made but proved to be unsuccessful. According to the medical examiner, the cause of death for the first victim was carbon monoxide poisoning and smoke inhalation. The cause of death for the second victim was given as carbon monoxide poisoning.[3]

RECOMMENDATION. Fire departments should ensure that fire command always maintains close accountability for all personnel at the fire scene.

DISCUSSION. One of the most important aids for accountability at a fire scene is an incident management system. Many firefighters who die from smoke inhalation or flashover, or are caught or trapped by fire, actually become disoriented first.

As emphasized previously, in the event that firefighters become disoriented, are caught in a collapse, are exposed to high temperatures, or encounter other unforeseen events, air is their main limiting factor. The more air firefighters have at the point when they get into trouble, the greater their survivability profile is. Air management is critical to structural firefighting.

Case study 4

On March 14, 2001, a 40-year-old career firefighter-paramedic died from carbon monoxide poisoning and thermal burns after running out of air and becoming disoriented while fighting a supermarket fire. The medical examiner listed the victim's cause of death as thermal burns and smoke inhalation. The victim's carboxyhemoglobin level was 61% at the time of death.[4]

RECOMMENDATION 1. Fire departments should ensure that SOPs are followed and that refresher training is continually provided.

RECOMMENDATION 2. Fire departments should ensure that firefighters manage their air supplies as warranted by the size of the structure involved.

DISCUSSION. Air consumption will vary with each individual's physical condition and level of training, the task performed, and the environment. Depending on the individual's air consumption and the amount of time required in order to exit a hostile environment, the low-air alarm may not provide adequate time to exit. Working in large structures (e.g., high-rise buildings, warehouses, and supermarkets) requires that firefighters be cognizant of the distance traveled and the time required in order to reach the point of suppression activity from the point of entry. When conditions deteriorate and visibility becomes limited, firefighters may find that it takes additional time to exit as compared to the time it took to enter the structure (fig. 15–11).

The common thread that runs through all of these cases is that all the firefighters perished from asphyxiation when they ran out of air. Unfortunately, these fatalities represent only a small sample of the real-world examples that demonstrate the importance of incorporating air management at all levels of the ICS. We can only speculate as to more-favorable outcomes had the firefighters in these cases managed their SCBA air more effectively.

*Fig. 15–11. Fires in such large structures,
such as this school, require an
awareness of SCBA air.*

Air management must be incorporated into all levels of the command structure. Failure to do so will lead to increasing firefighter injuries and fatalities. Remember, structural firefighters are dying at an increasing rate because they are running out of air. The U.S. fire service must begin managing its air.

At the task level, firefighters and company officers must provide for their own safety by managing their air in the hazard area. Task-level air management uses 75% of the SCBA air for entry, work, and exit from the hazard area. The remaining 25% of air in the SCBA is to be used only in an emergency.

At the tactical level, division/sector supervisors must coordinate the activity of teams at the task level while maintaining a safety posture and ensuring adequate and timely relief. Division/sector supervisors need to recognize when expected outcomes and personnel needs are not matching actual outcomes and personnel needs. Division/sector supervisor must also treat any low-air alarm as a significant event within their assigned area and deal with it immediately.

At the strategic level, at every emergency where firefighters are using SCBA, air management plays a critical role. Air management needs to be supported by the strategic-level officer before the incident, during the incident, and after the incident. The strategic-level officer is also responsible for addressing any low-air situation during the post-incident analysis, to identify how it happened, why it happened, and how to prevent it from happening again. To reiterate, air management is a safety process required by the latest NFPA respiratory standard; moreover, it is unquestionably the right approach to take.

MAKING IT HAPPEN

1. As a company, research how the ICS was developed.

2. As a company- or battalion-level exercise, determine what constitutes a strategic-, tactical-, task-, or control-level component of recent emergency events within your department.

3. Have an open discussion regarding risk management within your organization and identify your department's policy regarding firefighter safety as it relates to risk.

4. Ensure that all members are well versed in NFPA 1561—Standard on Emergency Services Incident Management and NFPA 1500—Fire Department Safety and Health Program.

5. At the individual, company, or battalion level, have members research events in which firefighters were injured or died due to complications with an inadequate ICS and the firefighters ran out of SCBA air.

STUDY QUESTIONS

1. In 1971, as a result of the Bel-Air fire and other costly and destructive California wildfires, who mandated that the U.S. Forest Service design a system that would increase the capabilities of Southern California wildland fire protection agencies?

2. In the early 1970s, what fire department developed the Fire Ground Command (FGC) system?

3. In the NIMS ICS, what are the six major functions?

4. The basic configuration of a command structure includes what three levels?

5. What NFPA Standard outlines the definition of risk management and the factors an IC must consider in order to minimize the risks that firefighters are faced with?

6. True or false: Another aspect of task-level air management is for team leaders to be aware to rotate team members in and out of heavy-workload positions based on their air supply and air use.

7. Why must air management be incorporated into all levels of the command structure?

NOTES

[1] NIOSH Fire Fighter Fatality Investigation F2005-05.

[2] New Jersey Division of Fire Safety. Report issued March 12, 2003.

[3] NIOSH Fire Fighter Fatality Investigation F98-32.

[4] NIOSH Fire Fighter Fatality Investigation F2001-13.

CHAPTER 16

COMMUNICATION AND AIR MANAGEMENT

INTRODUCTION

This chapter describes the impact that inadequate or nonexistent radio reports have on emergency operations. History has shown that communication is a critical function of any emergency event, especially when things go bad. SCBA air readings, as a key component of good progress reports, need to be included in the ICS. The CARA format provides an easy-to-remember model that covers each of the key components of a good progress report:

- Conditions
- Actions
- Resources
- Air

METHODS OF COMMUNICATION

All creatures on this planet communicate in some form, verbally or nonverbally. Communication is broadly defined as the effective exchange or expression of ideas, messages, or information using a common system, such as speech, signals, or writing. In the animal kingdom, this is achieved mainly through physical gestures. Human beings, for the most part, use verbal interaction as our primary method of communication.

But, how do humans verbally communicate? The process can be broken down into a step-by-step method. Each step is important in this model, because the communication process is a cycle that requires continual completion: The sender encodes and transmits the message; the receiver decodes and feeds back to the sender. A failure at any point of this process will cause a communication breakdown. Communication does not truly happen unless the message that is sent is also received, understood, and responded to in a way that lets the sender know that the received message is the same as the sent message(fig. 16–1).[1]

Simply giving an order or relaying information does not guarantee that the message was received or understood—which is the ultimate goal of communication. Safe and effective fireground communication is bidirectional. Case studies, including reports of firefighter fatalities, demonstrate that it is critical for firefighters to understand the importance of two-way communication for ensuring safety and sound decision-making on the fireground. Unfortunately, a variety of circumstances can result in a communication failure. For example, a message may not be received by the intended recipient; a message may be received but not understood; a message may be received but misunderstood; a message may be received and understood but not acted on; and a message may be received and deliberately, or selectively, ignored. For a variety of reasons, all of these failures do occur on emergency scenes, and a single failure can prove deadly to firefighters and rescue personnel.

Fig. 16–1. A training lieutenant (left)
providing a classroom session on SCBA. The
message has been sent and received, and the
students are completing the communication
cycle by verbally repeating the lesson so that
the officer knows they have understood the
message. (Photograph courtesy Seattle Fire
Fighters Union, IAFF Local 27)

Fire departments can take several steps to help prevent negative consequences arising from a failure to complete the communications loop. First, formal acknowledgments should be required for every message. While people are unaccustomed to using formal acknowledgments during ordinary conversation, their use is vital to communicate critical messages when nonverbal cues cannot be used to avoid ambiguity.

The Wildland Firefighter Safety Awareness Study recommended the use of levels of formal acknowledgment.[2] With slight modification, these same levels can be utilized by structural firefighters (fig. 16–2).

Fig. 16–2. The company officer on the first due engine at a working house fire giving direction to other responding units and the dispatcher with a portable radio. Level 1 should be used for routine messages, and level 2 or 3 should be used for critical messages. Senders and receivers must request clarification if messages are not clearly understood. (Photograph courtesy Seattle Fire Fighters Union, IAFF Local 27)

LEVELS OF COMMUNICATION

Level 1 is simple acknowledgment of routine information. For example,

Sender: "Engine 1 responding."

Receiver: "OK, Engine 1."

Level 2 is acknowledgment and feedback of key information. For example,

Sender: "Engine 1, report to the 10th floor."

Receiver: "Engine 1 copies, reporting to the 10th floor."

Or,

Sender: "Engine 1, advance a 1¾" line into the fire apartment."

Receiver: "Engine 1 copies, advancing a 1¾" line into the fire."

Or,

Sender: "Engine 2, pick up my supply line at Maple and Sycamore."

Receiver: "Engine 2 copies, picking up the line at Maple and Sycamore."

Level 3 is used to acknowledge more complex instructions. This level of acknowledgment also provides an opportunity to conduct dialogue on interpreting the instructions or may be used to request clarification. The receiver may repeat the order on where to go and what to do and clarify what is expected if unclear. For example,

Sender: "Engine 1, advance a 2½" line into exposure 4 and attack the cockloft fire. If unsuccessful within five minutes, pull out."

Receiver: "Engine 1 copies, advance a 2½"into exposure 4 and hit the cockloft. Pull out if we can't knock it in five minutes. Does this building have a truss roof?"

Or

Sender: "Engine 1, we now have reports of victims on the seventh floor. Do you want a second alarm?"

Receiver: "Engine 1 copies, victims on the seventh floor. Send a second alarm and two additional truck companies."

The sender (including dispatchers, command officers, and company officers) has the obligation to actively monitor the radio until acknowledgment is received and the sender believes the message has been fully understood by the recipient. If there is no response initially, the sender can probe for understanding by asking questions like, "Did you receive?," or by giving a prompt such as, "Engine 1, acknowledge." If the sender believes that the receiver is unclear about a vital piece of information, the sender must continue to probe until satisfied the message is completely understood.

Receivers have the responsibility to acknowledge messages and to request clarification if they do not completely understand. Failure to acknowledge by the receiver should be considered cause for concern, and the sender should follow up to determine why confirmation was not given. Dispatchers are a critical component of the communication loop. They must have an in-depth understanding of the fireground environment to triage messages according to importance and rebroadcast vital messages to all of those en route to or at the scene.

Without positive communication, emergency events cannot be coordinated in a safe and effective manner, and this is one of the most common criticisms on the fireground. Despite technological advances in two-way radio communication, important information is not always adequately conveyed on the emergency scene. Inadequate fireground communication is repeatedly cited as a contributing factor in incidents reported by the USFA, the NFPA, and NIOSH. Additionally, the number of near-miss incidents in which fireground communication was ineffective may be higher than generally realized.

NFPA STANDARDS PERTAINING TO COMMUNICATION

Nationally recognized standards serve as the cornerstone for developing SOPs pertaining to communication. The following NFPA standards—and specific sections—address fire service radio communication:

- NFPA 1201—*Standard for Developing Fire Protection Services for the Public*
- NFPA 1221—*Standard for the Maintenance and Use of Public Fire Service Communication Systems*
- NFPA 1500—*Standard on Fire Department Occupational Health and Safety Programs*
- NFPA 1561—*Standard on Emergency Services Incident Management System*
  - 6.1.4: "An Emergency Service Organization (ESO) shall provide one radio channel for dispatch and a separate tactical channel to be used initially at the incident."
  - 6.2.1: "The incident management system shall include SOPs for radio communications that provide for the use of standard protocols and terminology at all types of incidents."
  - 6.2.2: "Clear text shall be used for radio communications."
  - A.6.2.2: "The intent of the use of clear text for radio communications is to reduce confusion at incidents, particularly where different agencies work together."
  - 6.3.2: "To ensure that clear text is used for an emergency condition at an incident, the ESO shall have an SOP that uses the term *emergency traffic* as a designation to clear radio traffic."

– A.6.2.3: "A change in strategic mode of operation would include, as an example for structural fire fighting, the switch from offensive strategy (interior fire fighting with hand lines) to defensive strategy (exterior operations with master streams and hand lines) or establishing a perimeter around an active crime scene. In such an instance, it is essential to notify all affected responders of the change in strategic modes, to ensure that all responders withdraw from the area, and to account for all responders."

– 6.3.4: "When a member has declared an emergency traffic message, that person shall use clear text to identify the type of emergency, change in conditions, or tactical operation."

– A.6.3.2: "Examples of emergency conditions could be 'responder missing,' 'responder down,' 'officer needs assistance,' 'evacuate the building/area,' 'wind shift from north to south,' 'change from offensive to defensive operations,' or 'firefighter trapped on the first floor.' 'Mayday' is another term that could be used; however, an organization that routinely responds on wildland or maritime incidents should avoid using this signal in that it could cause confusion at these types of incidents. In addition to the emergency traffic message, the ESO can use additional signals such as an air horn signal for members to evacuate as part of their SOPs."

– 6.3.5: "When an emergency has been abated or all affected members have been made aware of the hazardous condition or emergency, the incident commander shall permit radio traffic to resume."

– A.6.4.3: "Some emergency services organizations might also wish to be provided with reports of elapsed time-from-dispatch. This method could be more appropriate for ESOs with long travel times where significant progress might have occurred prior to the first unit arrival."

- NFPA 1710—*Standard for the Organization and Deployment of Fire Suppression Operations, Emergency Medical Operations, and Special Operations to the Public by Career Fire Departments*

  - 6.4.4: "Standard terminology, in compliance with NFPA 1561, shall be established to transmit information, including strategic modes of operation, situation reports, and emergency notifications of imminent hazards."

The 2006 edition of NFPA 1404 (see chap. 6) requires that firefighters exit the IDLH environment prior to the activation of any low-air alarm. This change in how firefighters manage their SCBA air has a direct correlation on fireground communication at the strategic, tactical, and task levels. Incident Commanders, company officers, and team leaders all will be required to more closely monitor SCBA air levels, and provide timely crew rotation based primarily on progress reports from teams working in the hazard area.

INJURIES AND CASUALTIES RELATED TO INADEQUATE COMMUNICATION

The earliest documented case in which radio communication was implicated in a firefighter casualty was in 1978. Four firefighters in Syracuse, New York, died in a three-story wood-frame apartment building when fire erupted from a void space, trapping them on the third floor.[3] Approximately 16 minutes into the fire, a weak radio transmission, "Help me," was recorded on the master fire control tape at the Syracuse Fire Department dispatch office. There is no indication that anyone on the fireground or in the dispatch office heard the message. Approximately 1 minute later, a second transmission was recorded: "Help, help, help, [static]." This transmission was apparently not heard by any fire personnel on the scene or in the dispatch office. However, an observer with a scanner reported to a fire officer on the scene that he heard a radio transmission, "Help, help, help, third-floor attic." It was not clear what action was taken in response to the information provided by the observer; a second alarm was not called for another 16 minutes into the fire, and the first of the fatalities was not discovered until about 4 minutes after the second alarm was called (fig. 16–3).

*Fig. 16–3. Firefighters being removed from the
second floor after being forced out by intense heat
and low air (Photograph courtesy Dennis Leger,
Springfield [MA] Fire Department)*

THE HACKENSACK FORD FIRE

The July 1, 1988, fire at Hackensack Ford, in Hackensack, New Jersey, is one of the best-documented cases in which communication failure contributed to firefighter fatalities.[4] Before the roof collapsed, the IC radioed companies operating on the interior to back their lines out. This message was not acknowledged by the interior companies or by the dispatch center. In the ensuing collapse, three firefighters were pinned by falling debris, while two other firefighters escaped into an interior toolroom.

Radio calls for help by the firefighters trapped in the toolroom followed the roof collapse. Although these calls went unanswered initially, civilians monitoring the incident on scanners alerted the dispatch center. By the time the IC finally learned of the calls for help, it was too late to mount an effective rescue effort. In 1988, Klem concluded that a major factor contributing to the firefighter deaths at the Hackensack Ford fire was the "lack of effective fireground communications both on the fireground and between fireground commanders and fire headquarters."[5]

THE REGIS TOWER FIRE

In 1995, Chubb and Caldwell reviewed the April 11, 1994, fire at the Regis Tower, in Memphis, where two firefighters died.[6] The fire occurred on the 9th floor of an 11-story fire-resistive high-rise building. The first firefighters to arrive on the fire floor were quickly in peril for a number of reasons, including a decision to take the elevator to the fire floor, a hysterical and violent male victim, and a flashover in the room of origin (fig. 16–4).

*Fig. 16–4. Firefighters attacking a high-rise fire.
Working fires in high-rises are very challenging
and can quickly overcome the resources of any
fire department. (Photograph courtesy Seattle
Fire Fighters Union, IAFF Local 27)*

Companies on the scene of the Regis Tower fire were operating
on an unrepeated fireground channel. At one point, a firefighter who
later died made a series of four urgent radio transmissions attempting

to communicate with his company officer. These transmissions were inadvertently made on the dispatch channel, not the fireground channel. The IC was monitoring the fireground channel on his portable radio, while attempting to monitor the main dispatch channel on the mobile radio in his vehicle, which was serving as the command post. At the time these urgent transmissions were made, the IC was away from his vehicle; thus, he did not hear them. The transmissions were heard by a dispatcher monitoring the dispatch frequency, but no further action was taken by the dispatcher to inform the IC that a member may have been in distress.

THE BLACKSTOCK LUMBER FIRE

In 1990, Isner gave a written report of his investigation of a fire at the Blackstock Lumber facility in Seattle, on September 9, 1989.[7] The fire claimed the life of a Seattle fire lieutenant. The lieutenant had advanced a hand line into an exposure building with another firefighter when conditions rapidly deteriorated. After trying unsuccessfully to find the way out, the officer began calling for help on his portable radio. As the officer got low on air, he passed the radio to the firefighter, who also transmitted repeated requests for help. None of these requests for help were heard by the IC, other personnel on the scene, or dispatch personnel; however, the transmissions were heard by people in the area who were monitoring the incident on scanners. The firefighter was able to make his way toward an exit, where he collapsed and was eventually rescued. At the time the firefighter was rescued, he was incoherent, and no one responding realized that the lieutenant was still in the building. The lieutenant ultimately died of inhalation of the products of combustion.

The firefighter subsequently reported that when he was calling for help over the radio, he could hear the dispatchers providing move-up information to companies that were relocating, so he knew that the radio was working. Isner concluded that the radio was not on the normal fireground channel, since no one at the scene heard the requests for help. He also concluded that the radio was not transmitting through the repeater, without which the portable radio could not have been heard by the dispatch center.

THE 1995 PITTSBURGH FIRE

Finally, Routley investigated the February 14, 1995, fire in Pittsburgh that claimed the lives of three firefighters.[8] During a critical period in the fire, four firefighters ran out of air and became disoriented in the building. One firefighter was located and removed by other personnel. Although only semiconscious, the rescued firefighter reported that other members were still inside. Over the next several minutes, confusion developed as to how many firefighters were still missing and how many had been rescued. The confusion led to the erroneous conclusion that all members were accounted for when in fact three firefighters remained lost in the building (e.g., see fig. 16–5).

Routley cited communication problems as a factor contributing to the failure to realize that three members were still missing. Pittsburgh's fire and EMS were separate municipal departments that routinely responded to fires together. Each department operated on entirely separate radio channels. Consequently, direct radio communication between emergency medical personnel and the fire department IC was not possible. This arrangement contributed to the confusion as emergency medical personnel relayed messages about who was missing and who had been rescued through their dispatcher to the fire dispatcher, and, ultimately, to the IC.

Collectively, these events provide the foundation linking firefighter safety to effective fireground communication. They also show the extent to which the fire service has come to rely on radio communication.

Effective communication has always been an important component of successful fireground operations. However, the modern fire service has come to rely so heavily on radio communication that efficient operations, as well as firefighter safety, now depend on how well our radio communication systems function.

Fig. 16–5. A fire in a single-family residence. A high risk to firefighters exists in such fires because of concern for life safety of the occupants. (Photograph by Dennis Leger, Springfield [MA] Fire Department)

Two categories of messages contribute to radio system overloading. The first category comprises incident-related messages, pertaining directly to the incident at which companies are operating. The second category comprises messages that are not related to the incident. These include dispatching, routine radio traffic, and other incidents taking place simultaneously.

In general, the overloading of a radio channel with non–incident-related messages can be addressed through the use of additional radio channels. However, overloading with incident-related messages is a matter of radio discipline. Without proper radio procedures, fireground channels can become overloaded with incident-related traffic just as easily as can combined dispatch/fireground channels.

Problems with overloading due to incident-related radio traffic must be solved by implementing radio communication and discipline procedures. The Hackensack Ford fire is a good example: The New Jersey Bureau of Fire Safety (1989) cited that 50% of the messages transmitted at the Hackensack fire were never acknowledged.[9] This lead to extreme confusion on the fireground, and was a major contributing factor to these firefighter deaths.

The drawback of relying solely on an IC to monitor a fireground channel is that there are a multitude of factors at the emergency scene that are competing for the IC's attention—command decisions are made, face-to-face and cellular telephone communications take place, reference materials are checked, accountability documentation is prepared, and physical observations are made of conditions and firefighting activities. All of these occur under ambient noise and stress levels that are less-than-ideal for listening to a radio.

TYPES OF COMMUNICATION PROBLEMS

Communication problems commonly encountered by firefighters can be divided into two categories. First, are problems related primarily to mechanical/technical issues, such as; unsuitable equipment, radio malfunction, limited system capacity, or atmospheric interference. The second category of problems is broader, including critical human factors necessary for effective communication (for example, radio discipline and completing the communication loop). While the research literature dealing with fire service communication is sparse, more has been written about the technical aspects than about the human factors. This section addresses both, with particular emphasis on the human factors involved in improving firefighter communication.

TECHNICAL ISSUES

A variety of technical issues can adversely affect communication among firefighters. While the applicability of these challenges to individual departments may vary depending on multiple factors, technological progress holds the promise of solving some of the problems commonly encountered today.

UNSUITABLE EQUIPMENT. The predominant communications-related concern reported by firefighters and company officers is the difficulty in communicating while using the SCBA. The use of SCBA, while critical to firefighter safety, can interfere with effective communication, both face to face and via portable radio. Few firefighters are unfamiliar with this problem, and most have asked, in bafflement, "What did they just say?" after attempting to comprehend a radio transmission sent by an interior firefighting crew. Even face-to-face conversation through SCBA is extremely difficult during a working fire, owing to high levels of background noise and the barrier imposed by the facepiece.

Fireground safety concerns dictate that firefighters perform both functions: use SCBA and communicate effectively. This can create a dilemma when vital messages must be clearly communicated within the fire environment. As a result, some firefighters have found it necessary to momentarily remove their SCBA facepiece to transmit a message over a portable radio or directly to a colleague. The obvious danger of removing the SCBA facepiece, even briefly, is evident, considering the thermal, toxic, and oxygen-deficient hazards posed by a fire and the resulting products of combustion (see chap. 5). A single, unprotected breath of such an atmosphere may be sufficient to cause long-term health problems, incapacitation, or even death. Since firefighters may expose themselves to the potential for serious injury to effectively communicate, there is clearly a need to improve the technology to correct current systems limitations (fig. 16–6).

EQUIPMENT FAILURE. Modern public-safety radio communication systems are complex and highly technical. They may encompass a multitude of fixed antenna sites, CAD terminals, mobile radios, mobile display terminals (MDTs), portable radios, cellular telephones/faxes/modems, and even laptop computers. This equipment must function properly at the emergency scene. However, this is often not the case. For example, many of the 800 MHz radio systems currently in use fail to work properly inside structures, paritcularly in high-rise buildings and basements. Firefighters operating on the fireground should not have to worry about the functionality of the overall communication system.

*Fig. 16–6. A firefighter sending a radio
communication while wearing SCBA. Effective
communication with a portable radio and SCBA both
in use can be very challenging. Firefighters must
practice radio transmissions with these critical pieces
of safety equipment. (Photograph by Dennis Leger,
Springfield [MA] Fire Department)*

INADEQUATE SYSTEM CAPACITY. In several incidents, inadequate capacity of the radio system was deemed to be a factor contributing to negative outcomes. This is most likely to occur during complex, multiple-alarm incidents with many units operating—and attempting to communicate—simultaneously. The sheer volume of radio traffic may overwhelm both dispatchers and Incident Commanders. In numerous instances, vital transmissions went unheard owing to the volume of radio traffic on a system whose capacity was exceeded. Although radio discipline, which will be discussed in detail later, minimizes this problem, radio communication systems with multiple-channel capability are best suited for fire department and emergency operations.

INTERFERENCE. Atmospheric, environmental, and electronic interference may hamper effective communication at the incident scene. This can take many forms, from the "skip" created by solar disturbances and atmospheric fluctuations, to interference caused by topographical features, such as hills or tunnels. Communicating in high-rise buildings, in ships, or below grade is often difficult (e.g., see fig. 16–7). While responding to an incident, radio transmissions may be compromised by the background noise of sirens. Obviously, some of these interfering factors cannot be controlled. Others, however, are preventable.

Human factors

Clearly, human factors are critical for ensuring safe and effective fireground operations. Good human communication skills and procedures will help promote safety even in the face of technical difficulties.

Fig. 16–7. Communication in basements, high-rise buildings or other unusual structures such as ships can pose a significant challenge. (Photograph courtesy Seattle Fire Fighters Union, IAFF Local 27)

RADIO DISCIPLINE. Radio discipline, a vital step in holding down radio traffic, is essential for effective communication among firefighters, dispatchers, and other emergency personnel. As mentioned previously, a lack of radio discipline can overwhelm even the best communication system. Allowing unlimited message transmissions may create a situation where vital messages cannot be heard because of the number of less important transmissions being broadcast. In contrast, restricting radio traffic to only vital messages may prevent important information from being broadcast. The challenge, therefore, is achieving a balance, ensuring that all potentially important information is broadcast, but not at the expense of emergency transmissions from interior crews.

There are several steps firefighters can take to improve radio discipline. An obvious start is not using radios when face-to-face dialogue is possible. Face-to-face communication is generally better than radio communication, since sender and receiver can use nonverbal cues to convey ideas and enhance understanding (e.g., eye contact, physical contact, body language). Distractions are also reduced, and people can ask questions or identify problems more readily during one-on-one dialogue.

Incident management recognizes that transfers of command occur face to face, when possible. This positive step is crucial to ensure that effective communication occur at the command level (fig. 16–8).

Radio communication skills are critical to convey information at the incident scene. One of the most essential of these skills is being a good listener. Although listening to radio traffic while performing fireground tasks is difficult, it is an important skill to develop. By doing so, firefighters can avoid re-broadcasting non-urgent messages that have already been transmitted, and maintain awareness of the overall situation. Listening skills also help firefighters recognize when potentially urgent information has not been broadcast.

Good speaking is another vital radio communication skill. Messages need to be transmitted using a logical format, at the appropriate volume, with good enunciation, and at a moderate pace. Most firefighters are familiar with the frustration of trying to understand someone who is either screaming or whispering into the radio, or who speaks very fast or slow. Prior to transmitting a message, firefighters should collect their thoughts and format the message mentally. Messages should be clearly stated without distracters like "um" or "uh." Messages that are clear and direct minimize unnecessary radio traffic and help prevent urgent messages from being delayed or unintentionally overridden.

Fig. 16–8. Face-to-face communication at the command level (Photograph by Dennis Leger, Springfield [MA] Fire Department)

The best way to develop good listening and speaking skills is through training and continued practice during multiple-company operational drills or simulations. It may also be helpful for command or training officers to review tapes of actual incidents or drills to critique communication procedures and reinforce the importance of radio communication skills.

Another way to improve radio discipline is to create department Standard Operating Procedures (SOPs) that describe standard message formats and distinguish between routine, "Urgent", and "Mayday" messages. In addition, standard terms should be defined for use during radio communication in order to avoid potential confusion and promote brevity during message transmission (fig. 16–9).

Fig. 16–9. Firefighters practicing Mayday and rapid intervention skills at an acquired structure. Effective communication is a necessary part of training. (Photograph courtesy Seattle Fire Fighters Union, IAFF Local 27)

INCIDENT MANAGEMENT. Most fire departments and public-safety agencies regularly utilize a system for managing incident operations and personnel accountability at fire or emergency scenes. Effective communication minimizes potentially negative consequences that are brought on by rapidly changing situations. Fire department incident management systems and operational SOPs often detail assignments for specific units based on their order of arrival at the scene. For example, the first-arriving engine secures a water supply and begins fire attack (fig. 16–10). The second-arriving engine pumps the first engine's supply line. The first-arriving truck company conducts a primary search. And so on. Proper coordination of these functions is vital to protect firefighters and fight the fire. Achieving the required coordination is a function of effective communication and appropriate command and control practices.

Fig. 16–10. A coordinated attack at a working house fire. The first-arriving engine company deploying an initial attack line. Many departments follow predetermined SOPs for initial operations regarding first- and second-arriving engine and truck companies. (Photograph courtesy Walter J. Lewis)

ICs at complicated fire or emergency incidents will need support to ensure the effectiveness of fireground communication. It is not possible for a single individual to manage the scene, ensure accountability, make strategic and tactical decisions, and monitor one or more radio channels. Even at routine fires, the potential for information overload exists. Chief officers should assign aides early in the incident to help them manage communication and other tasks. Multiple aides may be needed to monitor radio traffic if several radio channels are used simultaneously (fig. 16–11).

Fig. 16–11. An aide assisting the ICs with fireground communication. It's easy for an IC to be overwhelmed by emergency radio traffic. (Photograph courtesy John Lewis, Passaic [NJ] Fire Department, and Chief Rob Moran, Englewood [NJ] Fire Department)

The number of significant incidents in which urgent messages have gone unacknowledged underscores the responsibility of everyone on the scene to actively listen to radio communications for key words like "*Mayday,*" "*Urgent,*" and "*Priority*". Firefighters should not hesitate to report these messages directly to the command post if these messages are not acknowledged quickly. This is an especially important duty for members of RITs and firefighter assistance teams.

CHAIN OF COMMAND. Traditionally, fire department communications have been predominantly unidirectional. Stress has been placed on giving orders, following orders, and sending messages. This is due to the need for unity of command and span of control as the primary means of maintaining order on the fireground. Although there is little room for extensive conversation on the emergency scene, maintaining the chain of command creates a potential communication problem, since firefighters may be reluctant to circumvent the chain of command at the risk of being considered insubordinate.

A more common problem, expressed by some command officers, is that firefighters report information to the wrong person because they are unaware of changes that were made to tactical assignments. When such misrouting occurs, it is important that the message recipient first relay the message to the appropriate person and then advise the sender of the proper reporting pattern.

PROBLEM REPORTING. Firefighters have a tendency to not report when they are having difficulty completing their assigned task, whether that be forcing entry, establishing a water supply, or searching the fire floor. Firefighters are sometimes reluctant to report difficulties for fear of being judged as slow, incompetent, or non-aggressive—all of which are contradictory to the established culture of the fire service.

Effective communication is especially important here, since the coordination required to safely accomplish fireground tasks may be compromised. When the IC does not receive a progress report after an appropriate period of time indicating completion of a critical task—for example, an "all clear" on the primary search—the commander should query the assigned company for a report.

Other company officers should listen to radio traffic while performing their assignments to ensure that activities around them will not have adverse impacts. For example, the officer of an engine company assigned to attack a fire should actively listen for any indication that the companies assigned to ventilation duties are experiencing difficulty.

A tiered system of key words should be used to clarify the urgency of critical messages. For example, "*Priority*" messages might be defined as those requiring a swift response without the implication of immediate danger. "*Urgent*" might denote a circumstance where bodily harm is likely to occur in the absence of immediate action.

Severe problems might be reported using "*Mayday*" to indicate that a unit is actively involved in an emergency situation (e.g., a firefighter is lost, trapped, severely injured, or out of air). The transmission of a "*Mayday*" or the activation of a PASS device or low-air alarm should immediately cause the channel to be cleared of all non-urgent radio traffic, so that the IC can determine the location and status of the firefighter in trouble.

Whatever the specific key words, it is vital that all firefighters understand their relative urgency and the actions required for each. To prevent complacency, "*Mayday*" should be reserved for only the direst of circumstances.

PROGRESS REPORTING. Accurate, regular progress reporting is critical for sound decision making and for fireground safety. Command officers need regular progress reports so that they can make sound strategic and tactical decisions. These reports are often given by portable radio–equipped members who, besides their obvious tactical duties (e.g., searching for victims and advancing hoselines), serve as the eyes and ears of the IC. Without such information, the IC might have to make decisions based on incomplete or inaccurate data.

Ultimately, the four key elements of a progress report that paint a clear picture for the IC are: smoke and heat conditions, what actions have been completed or need to be completed, resources needed to complete assignments or potential assignments, and the team's lowest SCBA air reading. A concise way to provide an effective progress report is to utilize the CARA model. CARA stands for:

- Conditions

    - Smoke (visibility, density, color, volume, pressure): Light, medium, or heavy

    - Heat (interior tenability): Low, medium, or high

- Actions

    - Has your team completed its assignment? What other actions need to be completed (e.g., ventilation, primary search)?

- Resources

    - What other resources does your team need to complete its assignment or other forecasted tasks?

- Air

    - Relay the lowest SCBA air reading within your team. The air reading can be expressed in psi or as a percentage. This air check will enhance the safety of the team and improve the overall operation by providing the IC with a snapshot of when to rotate crews in/out based on SCBA air supplies (fig. 16–12).

Fig. 16–12. An interior team, in a training scenario, doing a SCBA air check prior to giving a CARA progress report

A good example of a CARA progress report would be:

> *Engine 1 has heavy smoke with moderate heat on the second floor [conditions], advancing a 1¾" attack line [action], requesting immediate ventilation and assistance with a primary search [resources]; our air is at 75% [air].*

This progress report conveys critical information to the IC in a clear, concise manner. Keep in mind that the CARA format is another tool to add to the previously recognized communication policies of a particular fire department. For example, many departments utilize a "to/from" communication model (e.g., "Second Avenue Command from Engine 20"). In addition, some departments require confirmation of interior team accountability as part of their progress reports. The CARA progress report model is intended to enhance the effectiveness of communication within the command structure.

TIMELY PROGRESS REPORTS

Timely progress reports are critical for the IC. A number of departments utilize a time-on-scene prompt from the dispatcher to obtain updated information from companies involved in the emergency, as well as to remind the IC to provide an ongoing size-up. A progress report from companies working in the IDLH atmosphere gives the IC an opportunity to examine the positives and/or negatives of the strategy, as well as to ascertain the effectiveness of the corresponding tactics. Progress reports also provide the IC with a mental prompt to consider crew rotation, incident support needs, and whether additional resources will be necessary.

Additionally, regular progress reports heighten the situational awareness of the interior teams by making them:

- Recognize whether there have been changes in the smoke and heat conditions
- Inform the IC if the team is succeeding with their assignment
- Consider what other actions need to be accomplished
- Request resources to help them succeed with their assignment or achieve forecasted tasks
- Check their SCBA air levels

Regular progress reports also provide the team leader with the opportunity to check on the physical and mental integrity of the team as a whole and maximize the team's effective work time by rotating the workload on the basis of fatigue and air consumption.[10]

Ideally, all firefighters should have their own portable radio. At the very least, each two-person team entering a fire situation should have a portable radio. More training should be conducted to develop effective firefighter communication skills. These skills, including the CARA format progress reports, should receive a greater training emphasis.

FINAL THOUGHTS

Radio discipline must achieve a balance between limiting nonessential radio traffic and ensuring that potentially important information is regularly broadcast. Policies should define:

- Standard message format
- Important/urgent messages, as opposed to routine messages
- *"Mayday"* procedures
- Procedures for operations conducted on multiple channels
- Roles and responsibilities of those involved in the communication process at every level
- Procedures for regular progress reporting as prescribed by the CARA format

NFPA 1404, with the implementation of the 2006 edition, requires that firefighters exit the IDLH environment prior to the activation of any low-air alarm. This change in how firefighters manage their SCBA air has a direct effect on fireground communication at the strategic, tactical, and task levels. Incident Commanders, company officers, and team leaders will be required to more closely monitor their situational awareness, especially SCBA air levels, and to provide timely crew rotation based primarily on progress reports from teams working in the hazard area.

The effectiveness of any incident management system depends on effective communication between firefighters and the ICS. All personnel assigned to an incident must practice actively listening to radio traffic for information that may affect the performance of their assignments and safety at the incident.

MAKING IT HAPPEN

1. Have firefighters practice portable radio communication with full PPE and SCBA (on air). Find the physical placement for the radio and/or microphone that allows transmission of the clearest possible messages given the equipment used.

2. Develop a CARA communications SOP for your department. Consider, for example, what will be the standard definitions for low, medium, and high smoke and heat conditions.

3. Have firefighters practice the CARA method of giving progress reports, with and without full PPE and SCBA (on air).

4. Develop a department SOP dictating when progress reports will be required (e.g., at timed intervals, on completion of assignments, when conditions change significantly, or when members work into their low-air alarm).

5. Explore further references on fireground communication and how to overcome recognized problems. For example, see Adam Thiel's *Improving Firefighter Communications,* Report 099 of the USFA Major Fires Investigation Project conducted by Varley-Campbell and Associates, Inc./TriData Corporation.

STUDY QUESTIONS

1. Describe the recognized model of communication.

2. Explain the model of communication levels prescribed by the Wildland Firefighter Safety Awareness Study.

3. What are the two categories of messages that contribute to radio system overloading?

4. Overloading the radio channel with incident-related messages is, in large measure, a matter of _____?

5. Name the five human factors that affect radio communication.

6. What four elements of a progress report are denoted by the acronym *CARA*?

NOTES

[1] Advanced Supervisory Practices. John Matzer, ed. 1992. ICMA

[2] *Wildland Firefighter Safety Awareness Implementing Cultural Changes for Safety.* 1998. TriData Corporation.

[3] Demers, D. P. 1978. *Fire in Syracuse; Four Fighters Die.* Quincy, MA: National Fire Protection Association.

[4] Paraphrased from Klem, T. J. 1988. *Five Fire Fighter Fatalities: Hackensack, New Jersey—July 1, 1988.* Quincy, MA: National Fire Protection Association.

[5] Ibid.

[6] Chubb, M., and J. E. Caldwell. 1995. Tragedy in a residential high-rise, Memphis, Tennessee. *Fire Engineering.* 148 (3): 49–66.

[7] Isner, M. 1990. Fire fighter dies in warehouse fire. Fire Command, 57 (8), 30–35.

[8] Routley, J. G. 1995. *Three Firefighters Die in Pittsburgh House Fire, Pittsburgh, Pennsylvania.* Emmitsburg, MD: United States Fire Administration.

[9] New Jersey Bureau of Fire Safety. (1988). *Firefighter Fatalities*: Hackensack Ford, 320 River Street, Hackensack, New Jersey. Trenton, NJ: Author.

[10] NFPA 1404—Standard for Fire Service Respiratory Protection Training, A.5.1.7(5).

WEAPONS OF MASS DESTRUCTION AND HAZARDOUS MATERIALS INCIDENTS

INTRODUCTION

This chapter discusses the importance of air management in both WMD and hazmat events and outlines how your department, district, and crews can use the ROAM to increase their safety at these events. Firefighters must manage the air they breathe when they use specialized equipment, such as Level A hazmat suits and SCBA for nuclear radiation, chemical agents, and WMD.

A CHANGED WORLD

On the morning of September 11, 2001, our world changed forever. That morning, people around the globe watched in horror as a group of radical Islamic fundamentalists from the Middle East used jetliners as missiles, flying them into the North and South Towers of the World Trade Center, in New York City, and into the Pentagon, in Washington, D.C. The passengers and flight crew of a fourth hijacked plane, United Airlines Flight 93, stormed the cockpit in a courageous attempt to retake control of the plane. Sadly, Flight 93 crashed in a field outside the town of Shanksville, in rural Somerset County, Pennsylvania. The terrorists' intended target for Flight 93 is believed to have been either the White House or the U.S. Capitol building, in Washington, D.C.

As the terrorists slammed the hijacked planes into their populated targets, people watched in awe as firefighters from the FDNY, in New York City, and the Fairfax County and Arlington County fire departments, in Washington, D.C., entered the burning buildings to help those in need. That day, New York City's bravest lost 343 firefighters when the twin towers collapsed. Before those heroic members of the FDNY perished, however, they helped with the largest and most successful rescue and evacuation operation the world has ever seen. Their memory and sacrifice will never be forgotten. This chapter is dedicated to those 343 brave FDNY firefighters who were lost that terrible day.

A determined enemy showed its face on 9/11. Fundamentalist Islamic extremists who wish to destroy modern civilization are the new face of evil. These terrorists target innocent civilian populations in the hope of spreading their perverted form of the Islamic faith. To date, Osama bin Laden and the worldwide terrorist organization al Qaeda have attacked and killed innocent civilians in the United States, England, Spain, Indonesia, Afghanistan, Iraq, Saudi Arabia, Yemen, Pakistan, Somalia, Israel, and other countries. Al Qaeda's stated goal is to bring terror and destruction to all societies that do not conform to its fundamentalist Islamic beliefs and/or those who support the state of Israel.

The attacks of 9/11, the Oklahoma City bombing carried out by Timothy McVeigh in 1995, the attack on the World Trade Center by Ramzi Ahmed Yousef in 1993, the sarin and cyanide gas attacks on Tokyo's subway system by Aum Shinrikyo in 1995, and al Qaeda's worldwide campaign of terrorism, have changed the way fire departments must train and operate. We have learned that WMD are now the weapons of choice of terrorists and terrorist organizations. Furthermore, WMD incorporating hazardous materials, such as nuclear, biological, and chemical agents, have the potential to cause countless deaths if terrorists target a large city or population center. Fire departments around the world have formed specialized teams of firefighters to deal with these specific hazards, and governments have earmarked millions of dollars to equip and train fire departments in these specialized areas.

WMD AND AIR MANAGEMENT

Terrorist organizations like al Qaeda have expressed their desire to acquire nuclear, biological, and chemical WMD to terrorize civilian populations of the world's largest cities. Imagine a weekday attack on a major city—such as New York City, London, Berlin, Madrid, Rome, Paris, Tel Aviv, Cairo, or Los Angeles—where a low-level nuclear device is detonated. The injuries and ensuing widespread panic would be unlike anything the world has seen since World War II.

Since fire department personnel are the first on scene at most emergency events in their cities, they alone are responsible to help the wounded and trapped until other governmental organizations, such as the military, can arrive. Moreover, military personnel in numbers sufficient to make a difference usually do not arrive until long after the fire department has begun rescue operations in the hazard zone. We watched this scenario play out in New York City on 9/11. The members of the FDNY were the first responders at the World Trade Center towers and were the ones who sacrificed their lives to rescue the citizens they had sworn to serve. Fire departments must understand that they will most likely be acting alone for an extended period if a WMD event strikes their city—and they alone are responsible for the safety and welfare of their members.

Nuclear WMD

An attack against a city using nuclear weapons is the nightmare scenario that everyone in the world fears the most. A military-style nuclear weapon in the hands of terrorists could bring death and destruction to a city on a scale that we have not seen since the nuclear detonations in Nagasaki and Hiroshima, Japan, in 1945.

The possibility that terrorists could acquire or build a fully functional nuclear weapon is remote. A more realistic scenario is that a terrorist organization could construct a conventional explosive incorporating readily available nuclear material—a radiological dispersal device, or "dirty bomb"—and detonate this device in a population center somewhere in the world. To date, no dirty bomb has ever been successfully detonated. However, two functional dirty bombs attributed to Chechen separatists were discovered—one in 1995, in a park in Moscow, and the other in 1998, outside the Chechen capital of Grozny.

It is plausible that a dirty bomb could be assembled and detonated in a civilian population center. Nuclear material could be stolen from a health care facility or a university or while in transit. This could be incorporated into any type of bomb—pipe bomb, car bomb, suitcase bomb, or suicide bomb—which could then be detonated in the middle of a city center, with a cloud of low-level nuclear fallout spreading for miles.

In any nuclear WMD scenario, firefighters must protect themselves by wearing SCBA. The prognosis for anyone breathing in dust contaminated with alpha, beta, or gamma particle emission sources is not good; death occurs within days or weeks, depending on the amount of exposure. Therefore, SCBA are required at all nuclear WMD incidents or suspected incidents. Furthermore, firefighters wearing SCBA must manage their air. Simply put, firefighters dealing with civilian injuries and decontamination efforts from the detonation of a dirty bomb cannot run out of air in their SCBA (fig. 17–1).

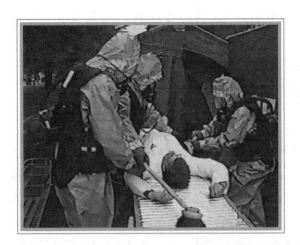

Fig. 17–1. Firefighters decontaminating a patient exposed to a nuclear WMD. Note that the firefighters are wearing rebreathing-style SCBA, which allows them to work much longer than if they were using standard non-rebreathing SCBA with air cylinders. (Photograph by Charles Cordova and The Seattle Fire Department Haz-Mat Unit, Unit 77)

As critical as air management is for structural firefighting, it is even more important for firefighters responding to nuclear WMD. Firefighters who have been exposed to low levels of alpha and beta emissions must be decontaminated, and it is vital that they have enough air remaining in their cylinders to get through the decontamination process without running out. Firefighters absolutely must follow the ROAM during any type of nuclear event. Their health and safety depends on it. Firefighters should be through decontamination, in the warm zone, before their low-air alarms activate. They cannot risk the consequences of inhaling any alpha or beta particles.

A recent example of how deadly alpha-emitting nuclear material can be when it enters the body by either ingestion or inspiration is the 2006 assassination of Russian dissident writer and former KGB agent

Alexander Litvinenko. Litvinenko was poisoned with the nuclear material polonium-210, which he ingested in a drink. This low-level alpha particle source killed him in less than three weeks. No cure for this type of radiation sickness is available.

Biological WMD

Biological WMD are every bit as dangerous to firefighters' health and safety as nuclear WMD. Biological agents are fairly accessible and have the potential to spread rapidly through a population. Anthrax, smallpox, botulism, cholera, tularemia, brucellosis, Q fever, and *Coccidioides mycosis* are all caused by biological agents that could be weaponized as biological WMD; ricin, which is a toxin found in castor beans, and Ebola and Machupo hemorrhagic fever viruses are further agents that could be weaponized and used against the populace. These agents can be spread in several ways: as aerosols, in food and water supplies, by skin contact, or through injection.

The odds are that firefighters arriving at a biological WMD incident will not even know that a biological agent has been released. Biological WMD are much slower acting than nuclear or chemical weapons; symptoms associated with most biological weapons take days or weeks to appear. A good example that illustrates how slow biological weapons act is the anthrax attacks of 2001. Beginning on September 18, letters containing the bacteria causing anthrax (*Bacillus anthracis*) were mailed to several news media offices and two U.S. Senators, killing 5 people and sickening 17 others. The first victim was hospitalized on October 2, with symptoms of disorientation, high fever, and vomiting. This victim died three days later. However, a full two weeks had passed following exposure before this victim developed symptoms severe enough to go to the hospital.

There are four classes of biological agents: *bacteria, viruses, rickettsia,* and *toxins.*[1] Bacteria are single-celled organisms that cause disease in humans, plants, and animals. Anthrax and cholera are examples of bacterial diseases. Viruses live inside cells and parasitize the host cells to multiply. Ebola, Machupo, and smallpox are examples of deadly viruses. Rickettsia are smaller than other bacteria and live inside host cells; Q fever, typhus, and Rocky Mountain spotted fever are examples of diseases caused by these organisms. Toxins are naturally produced by animals, plants, fungi, and microbes; manufactured by these organisms, they can be more toxic than many chemicals synthesized. Ricin, botulinum toxin, and mycotoxins are all examples.

Many of these biological agents can be weaponized and either inhaled, as with anthrax, or ingested in water or food, as with cholera. Firefighters responding to a known or suspected biological weapons attack must wear respiratory protection—namely, their SCBA. This also means that they must also follow the ROAM by exiting the hazard area and passing through decontamination before their low-air alarms activate.

Chemical WMD

Many experts believe that chemical WMD will be the weapons of choice employed by terrorist organizations intent on killing and injuring large numbers of people. This is because these weapons are relatively easy to obtain or manufacture, are comparatively inexpensive, and have already been used by terrorists against civilian populations. On March 20, 1995, the religious group Aum Shinrikyo used sarin gas to attack the crowded Tokyo subway system during the peak of the morning rush hour. This attack killed 12 people and injured more than 5,000. This was the first time chemical weapons were used by terrorists against innocent civilians and illustrates how readily a determined terrorist organization could acquire chemical weapons and terrorize a city.

There are five classes of chemical weapons: *nerve agents, blister agents, blood agents, choking agents,* and *irritating agents.*[2] All of these are relatively fast acting. Victims exposed to these agents manifest symptoms quickly, depending on how acute the exposure is.

NERVE AGENTS. Nerve agents block or disrupt the pathways by which nerves transmit messages. These agents act on neurotransmitters and are similar in nature to organophosphate pesticides, although much more toxic. Sarin, tabun, V agent (VE, VG, VM, and VX), and soman (GD) are examples of nerve agents. Victims who have been exposed to nerve agents will quickly exhibit outward signs of exposure; they will salivate (drool), lachrymate (tear), urinate, defecate, have gastrointestinal pain and gas, and experience emesis (vomit). The acronym "SLUDGE" is used to describe these various symptoms. Nerve agents are liquids and can be easily aerosolized, which is the most efficient way to distribute them. However, they can also be absorbed through the skin.

BLISTER AGENTS. Blister agents, also called *mustard agents* because of their characteristic smell, cause large, painful water blisters on the bodies of those affected. These are corrosive compounds that can easily penetrate clothing and are quickly absorbed by the skin. Sulfur mustard (H and HD), nitrogen mustard (HN1 and HN2), and lewisite (L) are examples of blister agents. Clinical symptoms might not appear for hours or days after exposure. Besides blistering of the skin, symptoms include red, irritated eyes, tearing, runny nose, severe cough, shortness of breath, abdominal pain, nausea, bloody vomit, and bloody diarrhea. These agents are oily liquids and can be aerosolized or vaporized.

BLOOD AGENTS. Blood agents are chemical compounds that contain cyanide, which prevents the cells in the body from using oxygen. Blood agents really don't affect the blood at all; therefore, the term *blood agent* is a misnomer. Another term for blood agent is *cyanogen*. The most common blood agents are cyanogen chloride (CK) and hydrogen cyanide (AC). Perhaps the most infamous blood agent is *Zyklon B* (hydrogen cyanide), which was used by the Nazis during World War II to murder Jews and other undesirables in the gas chambers of the concentration camps.

All blood agents are toxic at high concentrations and cause rapid death. In their purest form, blood agents are gases that are inhaled. Chemical precursors of blood agents are industrial chemicals—cyanide salts and acids—that are commonly available. Blood agents can be liquefied under pressure and aerosolized.

At nonlethal concentrations, blood agents cause victims to exhibit respiratory distress, nausea, sweating, drowsiness, dizziness, and headache. The cyanide in blood agents is an enzyme inhibitor that ultimately interrupts cellular respiration. Inhalation of high concentrations of blood agents causes death in a matter of minutes.

CHOKING AGENTS. Choking agents (also called *pulmonary agents*) suffocate victims by stressing the respiratory tract and making breathing difficult. Common choking agents are chlorine gas, phosgene (CG), diphosgene (DP), and chloropicrin (PS). Exposure to these agents causes choking, coughing, and severe eye irritation. Choking agents are gases and must be stored and transported in bottles or cylinders. Phosgene was used as a chemical weapon during World War I and continues to be stockpiled as a weapon by many countries.

IRRITATING AGENTS. Irritating agents incapacitate their victims. These agents are designed to be nonlethal, though people have died by asphyxiation after having been exposed to some of these agents. The most common irritating agents are the gases pepper spray (OC), tear gas (CS), mace (CN), and dibenzoxazepine (CR). These substances are used as riot control agents by police agencies and military forces. Irritating agents incapacitate their victims by causing immediate tearing, stinging of the eyes, difficulty keeping the eyes open, coughing, choking, and difficulty breathing. High concentrations of these agents can cause nausea and vomiting.

As with nuclear and biological WMD, firefighters responding to known or suspected chemical WMD must use all PPE, which means wearing and using their SCBA. Because most chemical weapons can be inhaled, firefighters need to follow the ROAM at any suspected chemical event. Firefighters cannot afford to run out of air in their SCBA cylinders and breathe in the toxic chemicals of a WMD, no matter how minor the event might first appear. The ROAM dictates that firefighters should be out of the hazard area and through decontamination before their low-air alarms activate. The ROAM provides for firefighters' respiratory safety.

Besides being aware of the primary dangers posed by the chemical itself, firefighters need to be aware that patients exposed to a chemical WMD will most likely be off-gassing, or emitting, the chemical agent that has harmed them, which can cause secondary exposure of firefighters. The dangers of secondary exposure are evident from the Tokyo subway sarin attack in 1995; of 1,364 emergency medical technicians (EMTs) dispatched during the course of this event, 135 experienced secondary exposure to sarin gas and had to be treated at area hospitals.[1] Luckily, no one died from the secondary exposure. This is another reason for firefighters at a chemical event to wear and use their SCBA and to follow the ROAM.

HAZMAT AND AIR MANAGEMENT

Hazmat teams throughout the fire service have been using air management for years. When a team of hazmat technicians gets into their Level A protective suits and makes entry into the hot zone of a suspected hazardous materials event, they know that they must have enough air in their cylinders to get themselves out of the hot zone and through decontamination. These teams are acutely aware of their air supply.

The Seattle Fire Department hazmat team dedicates one member to track the air supplies of the entry team. This is a very important job and should never be taken lightly. The air monitor is responsible for tracking the entry team's air use and makes sure that the entry team is through the decontamination process before their low-air alarms activate (see chap. 10). However, air management is not just another task assigned to a monitor at the hazmat command post. The members of the entry team are trained to manage their air by taking responsibility for their own air supply and checking on their partners' air supply while they are in the hazardous environment. In other words, air management is a personal responsibility.

AIR MANAGEMENT AND HAZMAT RESPONSE

Air management techniques have been practiced by hazmat response personnel for many years. Since the late 1980s, when hazmat response teams began to form across the country, estimating work times in relation to a responder's air supply has been a constant practice. In most situations, especially when wearing SCBA, hazmat responders do not commence work or enter a toxic atmosphere without some idea of how long their mission duration will be. In other words, hazmat responders work within the parameters of their air; they don't keep working until the air runs out. This is a subtle but important distinction. Additionally, it is commonplace to check in with those responders on air to advise them of their remaining work time and/or the need to exit the hot zone—all based on air use.

Hazmat responders know that being zipped inside a Level A protective suit or working in a toxic atmosphere leaves little room for error. To that end, a lot of attention is paid to one's air supply and the belief that the reserve amount of air is just that—reserved for some unforeseen problem. Typically, several hundred pounds of air is maintained as a safety margin, not to be viewed as extra work time. I have never seen a hazmat responder continue working in a toxic atmosphere with a low-air alarm sounding. It's viewed as a miscalculation—or an emergency—to be in the hot zone, actively engaged in mitigating the problem, with a bell or whistle ringing. The bottom line in hazmat is this: you just don't compromise your reserve air.

By contrast, I have seen firefighters working in heavy-smoke conditions with their low-air alarms blaring. The perception in the fire service seems to be that working in smoke is somehow less hazardous than working in an airborne concentration of a dangerous chemical. Keep in mind that the toxic by-products of combustion present as much or more of a hazard than any chemical at a hazmat incident.

Smoke inhalation is responsible for 5,000–10,000 deaths annually and more than 23,000 injuries. In the United States, thousands of firefighters are injured by the ill effects of smoke inhalation each year. To that end, it is the responsibility of individual firefighters to understand the nature of fire smoke and the effects of smoke inhalation; to demonstrate continued proficiency in the use of their SCBA; and most of all, to employ reasonable work practices when it comes to operating in smoke-filled environments.

—Rob Schnepp (Assistant Chief, EMS/Special Operations,
Alameda County [CA] Fire Department)

Hazmat teams understand how vital their air supply is to their health and safety. They know that should they run out of air in either the hot or the warm zone of a hazmat incident, their lives may be at stake (see figs. 17–2 and 17–3). Hazmat team members are careful not to become part of the problem by crossing the point of no return and running out of air (see chap. 11).

If the hazmat technicians are able to practice air management at hazmat incidents, then why shouldn't firefighters battling structure fires do the same? Firefighters should treat today's smoke as a hazardous material. Firefighters cannot afford to take even one breath of the toxins and poisons that compose today's smoke (see chap. 5). If we know that today's smoke is immediately dangerous to life and health—and it is—then why are firefighters not managing theirs supplies in fires the same way specialty teams manage their air at hazmat incidents? Firefighters need to view the smoke from structure fires as the hazardous material that it is and treat it as such. By doing this, firefighters will be more apt to monitor their air supplies.

Fig. 17–2. Hazmat technicians in Level A protective suits rescuing a victim during a hazmat exercise on a Washington State ferry. Hazmat technicians monitor their air supply constantly and have been practicing air management for years. (Photograph by Charles Cordova and Unit 77)

Fig. 17–3. Hazmat technicians rescuing incapacitated victims during a hazmat exercise at a school (Photograph by Charles Cordova and Unit 77)

GRANTS FOR AIR MANAGEMENT

Since 9/11, the federal government has been supporting fire departments and fire districts around the country through grants. These grants were established to give firefighters the training, tools, and resources necessary to better protect the public and fire service personnel from fire and related hazards in this time of terrorist threats. The DHS oversees many of these grants.

One of the largest grant programs is the Fire Act Grant, which is overseen by the DHS. The DHS is responsible for managing three specific grants:

- The Assistance to Firefighters Grant (AFG)
- Fire Prevention and Safety (FP&S)
- Staffing for Adequate Fire and Emergency Response (SAFER)

The DHS also runs the Urban Areas Security Initiative (UASI), which awards grants to high-threat, high-density urban areas that face the highest risks from terrorism. These federal grants exist to equip and train fire departments.

State governments have also recognized the importance of giving money and equipment to their fire departments and fire districts so that they can better protect their citizens. Many states understand that firefighters will be the first emergency personnel at the scene of any major event, such as a terrorist WMD attack. State governments recognize the reality that federal assistance at a WMD attack will not be immediate and that fire departments working at such an event must be prepared to go it alone for quite a while.

Fire departments and fire districts must apply for these federal and state grants to receive them. This means that someone in the organization has to write the grant proposal and follow its progress through the application process. Today's fire department administrators need tounderstand the grant process. They have to be grant-savvy to access the money that is available to fire departments.

Federal and state grant money can be used to train firefighters in air management and equip firefighters with new breathing apparatus. Fire departments and districts can apply for grants that supply their firefighters with the latest chemical, biological, radiological, and nuclear (CBRN) compliant SCBA and for air management training.

The Seattle Fire Department received federal and state grants that allowed our entire department to be trained in the ROAM and to purchase all-new SCBA with 1800L cylinders. The money is out there to get your fire department or district air management training and the latest equipment. If we did it, so can you.

These grants are available to departments and districts that understand the grant process and apply for them. However, these grants are not automatic. Fire departments must exhibit a need and have a plan for how they would use the grant monies. Fire departments and fire districts cannot allow these grants to go to waste or to be hoarded by other departments that are more grant savvy. Administrators owe it to their firefighters—the men and women doing the hard work of answering 911 calls—to apply for these grants so that their people are the best-equipped, best-trained, and best-prepared firefighters they can be.

As stated time and again throughout this book, firefighters are needlessly putting their lives at risk by not managing their air inside structure fires. Inside the hazard area of a structure fire, they can and should manage their air supplies. Firefighters must manage their air so that they are out of the hazard area before their low-air alarms activate.

Notably, special operations teams—the hazmat and confined-space teams—are already practicing air management. Let us spread this positive mentality about the importance of the air supply beyond the specialty operations teams and use it in everyday fire operations, where it is truly needed. Today, firefighters are running out of air in structure fires and are dying or being exposed to toxins and poisons that will affect their health sooner or later. Let us stop these completely preventable deaths and injuries by having every fire department and fire district adopt the ROAM as policy and have it become an SOP.

In the not-too-distant future, air management will be just another basic skill that is taught to every firefighter in every fire department and fire district around the globe. Fire departments and fire districts can either ignore this inevitable change taking place or forgo their resistance and embrace air management today.

MAKING IT HAPPEN

1. Research the Alexander Litvinenko poisoning case, using the Internet or other resources. Review with your crew or department how he ingested a low-level alpha particle source, which eventually killed him. Discuss how important SCBA are to firefighters' health and safety at any nuclear or radiological event. Use this discussion of SCBA to lead into a discussion of how firefighters must manage their air at these events by using the ROAM.

2. Does your fire department or fire district have written policies (SOPs) for dealing with nuclear, biological, and chemical WMD? If so, review these policies with your crew(s). Is there a written air management policy included in the WMD policies? If your department or district does not have written policies for dealing with nuclear, biological, and chemical events, then begin developing these policies for your department or district. Make sure an air management policy is also included.

3. Research the 1995 Tokyo subway sarin attack, using the Internet or other resources. Discuss with your crew(s) how firefighters were exposed to sarin by the off-gassing from patients—that is, secondary exposure. Discuss how these secondary exposures of firefighters could have been avoided had they been wearing their SCBA at this chemical WMD attack.

4. Set up a WMD drill for your crew, station, department, or district. This drill will take planning and resources. The purpose of this drill is to see how your crews or department will perform at such an event. Training must take place beforehand, so that crews have the information, equipment, and resources necessary to handle such an event. Have units respond to a drill location with real people acting as if they have been exposed to a chemical WMD, such as sarin. Record this drill on video and have volunteer recorders track the actions of fire department personnel. Pay particular attention to the air management aspect. Review the results of this drill with your firefighters.

STUDY QUESTIONS

1.   What is meant by *WMD*?

2.   What is a dirty bomb, and why is this type of WMD much more likely to be used by terrorists than a nuclear bomb?

4.   What are the four common types of biological agents? Give examples of all four.

5.   Anthrax is what type of biological agent?

6.   Why do many experts believe that chemical WMD will be the weapons of choice of terrorist organizations?

7.   What are the five classes of chemical weapons?

8.   How do nerve agents work?

9.   Give an example of a blood agent.

10. The gas phosgene is what type of chemical weapon?

11. Hazmat teams have been using air management for years. Why?

NOTES

[1]   *Emergency Response to Terrorism, Self-Study.* 1997. FEMA/USFA/NFA, p. 9.

[2]   Ibid., p. 12.

[3]   Tetsu Okumura, et al. Lessons Learned from the Tokyo Subway Sarin Attack, *The Journal of Prehospital and Disaster Medicine.* 2000, Vol.15 (3): s30, Fifth Asia-Pacific Conference on Disaster Medicine.

RAPID INTERVENTION OPERATIONS AND AIR MANAGEMENT

INTRODUCTION

Rapid intervention was conceived in the early 1990s. This was a positive step towards providing the resources necessary to rescue firefighters in trouble. Rapid intervention provides a team of appropriately equipped firefighters who are staged in a position outside the structure and ready to deploy immediately after receiving a Mayday or other firefighter emergency. Rapid intervention has met with documented success, as well as resistance. This chapter addresses the vital importance of air management and how it relates to firefighter safety and survival during rapid intervention operations.

Rapid Intervention Team
Rescues Tulsa Firefighter in House Fire

11-20-2005 • NICOLE MARSHALL, World Staff Writer • Tulsa World (Oklahoma)

A Tulsa firefighter was injured early Thursday when a weak floor gave out under him while he was battling a house fire and he fell about 12 feet into a basement. Joe Carollo, a fire equipment operator for the fire department, was treated for a torn bicep and injuries to his shoulder and back, said Fire Capt. Larry Bowles. A fire department rapid intervention team rescued Carollo and he received only a small burn on his thigh, Bowles said. Firefighters were dispatched to a house fire at 1150 S. Elgin Ave. about 4 a.m., Bowles said. When they arrived, they saw flames coming from the unoccupied residence, which was being remodeled. Carollo and other firefighters went inside to attack the blaze. A weak floor gave out underneath him and he fell. District Chief Eddie Bell then sent in the rapid intervention team to rescue him. "They have a water supply dedicated solely to such an emergency. They were standing at the front door and responded immediately. In this case, it worked perfectly," Bowles said. Bowles said the firefighters are to be commended for their response to the dangerous emergency. "We practice for these sort of events so when they become a reality, we have the ability to respond quickly, Bowles said. Carollo was examined by the city physician on Thursday and will likely be off from work for some time while he heals.

Fig. 18–1. A successful rapid intervention effort performed by the Tulsa (OK) Fire Department (Courtesy Nicole Marshall, World Staff Writer, Tulsa World [Oklahoma])

THE MANDATE FOR RAPID INTERVENTION

In the mid-1990s, the Occupational Safety and Health Administration (OSHA), in 29 *Code of Federal Regulations* 1910.134, mandated the two-in/two-out rule, which was developed to provide a safety factor for the first-arriving company during the initial stages of an aggressive interior fire attack. Although this rule was well intentioned, the likelihood that a two-person rescue team would be able to find and extract a firefighter in trouble has been proven to be low (fig. 18–1).

Countless nationally recognized departments and instructors have delivered hands-on training and classroom sessions in support of rapid intervention. Thousands of hours and millions of dollars spent on Rapid Intervention Team (RIT) development and implementation have led to a number of documented cases in which RITs have performed successful firefighter rescues.

Today, most fire departments have rapid intervention policies and SOPs. Fire departments routinely drill on rapid intervention, practice rapid intervention, and learn about rapid intervention successes and failures. Yet, the number of firefighter fatalities in structure fires has not been significantly reduced. In the United States, between 1990 and 2000, 29 firefighters died because they became lost inside structures and ran out of air (fig. 18–2).[1]

Fig. 18–2. Firefighters practicing rapid intervention
operations, preparing to enter a zero-visibility training prop
(Photograph by Lt. Tim Dungan, Seattle Fire Department)

THE BRET TARVER INCIDENT

On March 14, 2001, the Phoenix Fire Department's response to the Southwest Supermarket fire resulted in the tragic loss of Firefighter Bret Tarver.[2] This was a shock to the fire service because Phoenix is home to one of the most respected and progressive fire departments in the nation. If it could happen in Phoenix, it could certainly happen anywhere in the United States.

At some point, Tarver and his partner got off the hoseline and moved deeper into the smoke-filled 20,000-plus-square-feet supermarket, away from their exit. Tarver and his partner ended up in the vicinity of the butcher shop, where they eventually became separated (fig. 18–3).

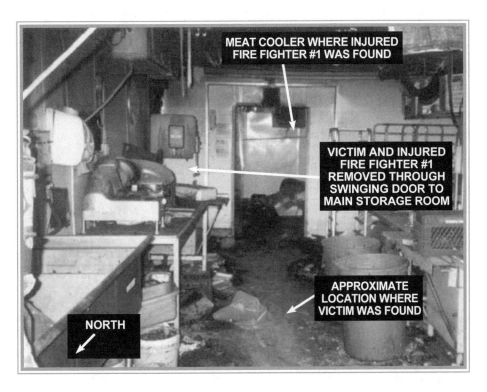

Fig. 18–3. The area where firefighter Bret Tarver
was located and extricated (Source: NIOSH)

Tarver radioed the IC and reported that he was lost in the back of the building. The IC deployed two companies as rapid intervention crews through the front, with no success (table 18–1). Because of the toxic smoke, Tarver became combative with those attempting to rescue him, and the initial RIT efforts failed. Only after Tarver succumbed to products of combustion was he eventually removed, in a state of cardiac arrest. Outside the building, subsequent resuscitation efforts failed. (For a detailed description of this incident, see chap. 7.)

Table 18–1. Time line of Tarver's rescue

Tarver calls for help via radio	
Rapid Intervention deployed	1 minute
Company locates Tarver	9 minutes
Tarver out of the structure	43 minutes
Total Lapsed Time	53 minutes

PHOENIX'S INTERNAL EVALUATION

Over the next year, a total of 269 Phoenix Fire Department companies comprising 1,444 members, as well as companies from 19 neighboring communities, participated in a rescue drill to evaluate the rapid intervention process.[3] On average, it took reporting companies 2.47 minutes to get ready. *"Ready"* was defined as having a full crew with tools assembled outside the building's point of entry. From the ready state, it took an average of 2.55 minutes from the receipt of the Mayday to the rescue team's entry. On average, it took 5.33 minutes to reach the downed firefighter (who was placed 40 ft off the end of a 150 ft attack line). It took 21.8 minutes to locate, package, transfill, and extricate the downed firefighter from a commercial structure in a training environment.

Additionally, one in five firefighters got into trouble during these training exercises. *"Trouble"* was defined as ranging from low-air alarm activation prior to exiting the IDLH atmosphere, to rescuers getting lost.

THE ELEMENT OF TIME

One of the most striking findings in Phoenix's evaluation was the critical role of time. It took a considerable amount of time, both at the actual Southwest Supermarket fire and during training scenarios, to access the structure and locate and extricate the firefighter in trouble. In Bret Tarver's case, it took the first company a total of 10 minutes to locate him after they were activated as a rapid intervention company. During the training scenarios, it took, on average, more than 21 minutes to complete the extrication process (fig. 18–4).[4]

Fig. 18–4. A company officer ensuring that a firefighter properly dons his SCBA in a practice session (Photograph courtesy John Lewis, Passaic [NJ] Fire Department, and Chief Rob Moran, Englewood [NJ] Fire Department)

When working in an IDLH environment, the limiting factor is air. In the Tarver incident and subsequent training scenarios, the time required for the RIT to locate the downed firefighter exceeded the air that remained in the downed firefighter's SCBA cylinder.

The amount of SCBA air consumed by working firefighters is addressed in NFPA 1981—*Standard on Open-Circuit Self-Contained Breathing Apparatus for Emergency Services*. NFPA 1981 (2002 edition and 2007 draft) states that working firefighters will consume between 90 and 105 liters of air per minute. This standard is currently undergoing revision, and cylinder size terminology is being changed to reflect cylinder volume in liter increments. What are now called 30-, 45-, and 60-minute cylinders will be described in terms of cylinder sizes, as 1200L, 1800L, and 2400L cylinders, respectively (see chap. 2).

For the purpose of this discussion, we will assume a standardized use rate of 100 liters of air per minute of SCBA air. This use rate encompasses 98% of firefighters operating in an IDLH environment.[5] To determine how long a lost or trapped firefighter will breathe SCBA air, consider that NFPA 1981 requires that all SCBA low-air alarms activate at 25% of the remaining cylinder volume, regardless of cylinder size (table 18–2). Given the best-case scenario in which a working firefighter calls a Mayday immediately after activation of the low-air alarm, a RIT will have between 3 and 6 minutes to locate and transfill the lost firefighter before they are out of air. The figures in table 18–2 are averages based on several variables. Air consumption will vary by individual, but these figures provide a baseline for estimating the time available to rescue a lost or trapped firefighter.

Table 18–2. Available air and time to rescue a downed firefighter

Standard Cylinder Classification	Amount of Air Available in a Full Cylinder	Amount of Air Available when the Low-Air Warning Alarm Activates at 25%	Time after Low-Air Warning Alarm Activates at 100 Liters/Minute
30 minutes	1,200 liters	300 liters	3 minutes
45 minutes	1,800 liters	450 liters	4.5 minutes
60 minutes	2,400 liters	600 liters	6 minutes

Time is never our ally when firefighters are working in an IDLH environment. Engineered structural components and lightweight construction systems have made working-time in newer buildings even shorter—and far more dangerous. Furnishings and other decorating products in our homes and offices are predominately made from synthetics, causing fires to burn faster and hotter, and exposing firefighters to a higher incidence of flashover and backdraft (fig. 18–5). These same products produce a far more toxic smoke in greater quantities than the natural fibers used in the past. Today, a simple room-and-contents fire will produce upwards of 20,000 parts per million of carbon monoxide (see chap. 5).[6]

Fig. 18–5. A working fire burning modern-day furnishings. Today's fires burn hotter and faster and are far more toxic owing to synthetics. (Photograph courtesy Walter J. Lewis)

NIOSH compiled statistics from investigations of 68 incidents between 1977 and 2000 in which at least one firefighter was killed at a structure fire. The report states that 44 firefighters in 33 incidents died from asphyxiation.[7] Running out of air contributed to all of these fatalities. Additionally, NIOSH recommended the implementation of an air management program after Bret Tarver's demise. This was the first recommended change in behavior for the fire service regarding SCBA use in any NIOSH firefighter fatality report. The recommendation emphasized that air management must take into account the size of the structure involved (fig. 18–6; also see chap. 15).

Fig. 18–6. Firefighters at a large, complex structure fire, where practicing air management is critical (Photograph by Dennis Leger, Springfield [MA] Fire Department)

AIR MANAGEMENT BY A RIT

At a residence fire in your first-due district, the dispatch center advises all responding companies that they have received multiple calls and have upgraded this to a working incident. The dispatch center also advises that a 42-year-old man is known to be in the basement.

The first-arriving apparatus establishes command and reports heavy smoke showing. Fireground assignments are based on order of arrival for engine and ladder companies. You are on an engine company that was dispatched as the RIT. First-arriving companies are met with two victims in the front yard who are handed off to the ambulance crew. Companies begin to stretch lines and force entry. Bystanders on scene confirm that there is a man in the basement as companies are entering the building.

The next few minutes are hectic. The first engine's line stretches straight back through the front door and ends up in a bedroom above the fire. The second line follows them in, making any repositioning of the lines difficult. The first truck company locates the victim in the basement and begins moving him to the stairs. A third attack line enters through a rear door with direct access to the basement stairs and begins advancing down the stairs, toward the fire. As the line is being moved down the stairs, the engine company encounters the first truck company, whose members are attempting to bring the victim up the stairs, and their low-air alarms start sounding.

Things become more hectic. As it turns out, one member of the first truck company is paired up with the officer of the second engine (their line is still on the first floor), and they are the ones moving the victim. Members of the third line, moving down the basement stairs, begin to help with the victim. One member of the third engine company moves off the stairs to free up the congestion. The area is completely charged with smoke, the fire is still burning unchecked, and heat forces members to the floor. A hand line is passed in through a basement window and is operated by another member of the third engine company, whose low-air alarm is sounding.

Unable to make any progress, the truck company member at the base of the stairs is running out of air. The engine company officer of the third line, who stopped to assist with the victim, is also running out of air. Multiple low-air alarms are sounding.

The other member of the first-due truck company who was believed to be in the basement is actually in a low-air situation in an upstairs bedroom. He and his partner got split up and didn't know it. Two additional members—the officer of the second engine company, assisting in the basement with the victim, and another member of the third engine company—are also running out of air.

An RIT is deployed to assist with the victim. The original RIT was the third engine to arrive, so a later-arriving engine has assumed the role of the first RIT. A second RIT had arrived and was performing outside duties prior to RIT deployment.

What's this have to do with air management during RIT operations? In answer, lack of air management often necessitates RIT operations. The preceding account was real, and six firefighters ran out of air inside the structure. The house was less than 2,000 square feet in area. The member operating the hoseline in the basement had to be pulled from the basement window; his mask was off, and his tank was empty. Other members ran out of air on the stairs or as they were exiting. The RIT that was deployed was in place at the top of the stairs, with an RIT air pack, but it wasn't used.

Not one member involved in the preceding incident called for help or reported their out-of-air status until they had exited! Worse, they all continued to operate inside the building while their low-air alarms were sounding. Nobody broke the chain of bad decision making under stressful conditions.

Air management during RIT operations is critical. When the RIT is deployed, every member operating on the fireground should perform an air check. As evidenced by the fireground described here, firefighters will push themselves harder when faced with increasingly difficult fireground operations. While there could (and should) have been six Maydays given on this fireground, there were none. As the conditions worsened, everyone ignored all of the mechanisms put in place to indicate trouble. Tunnel vision does and will take over—as it did here.

During RIT operations, all members of the team must be constantly aware of their air supply and their air consumption. They must know how hard they are pushing themselves and know that adrenaline and emotion will increase their air consumption and decrease their air supply as compared to normal fireground operations. The RIT officer must be concerned not only with his or her air status but also with the air status of all team members and the victim. RIT officers must manage their own air supply and that of their team, so that they remain part of the solution and don't become part of the problem. *Discipline* is essential.

A critical piece of equipment that must be brought with the RIT is an emergency air supply, the RIT air pack. This air supply is intended for the victim. Securing the air supply of the distressed member is a critical part of the overall RIT operation. With poor air management on the part of the RIT, this air supply may have to be used for team members—at the expense of the victim.

The bottom line is to know your air supply limits, as well as those of other team members, when you're operating as part of an RIT. Plan for backup teams early and have them cycled in, so that they arrive at your location before you need them. Make every attempt to have the team out of the building just before their low-air alarms go off.

How long can you work? Perform realistic training and find out; there's no better way. The mental and physical stress created during a RIT operation will cause you to consume air much faster than at any other time on the fireground. Don't forget to factor this in while you're operating on the inside.

When you're operating as part of an RIT, you should always have an RIT air pack that can be used to secure the air supply of the downed firefighter once located. Providing air to the victim can often stabilize the situation and give you—and command—a chance to switch gears from search to removal.

—*Jim McCormack (Fire Department Training Network)*
www.fdtraining.com/

AIR MANAGEMENT FOR RAPID INTERVENTION

Given the fact that firefighters are running out of air and dying in structure fires, why do they continue to work until, or even beyond, activation of their low-air alarms? Is it because of the culture of our job? Is it because firefighters, by nature, are aggressive and task focused? Or more importantly, is it because firefighters were trained to work until their low-air alarm activated? The truth is all of these factors put firefighters at risk of running out of air (fig. 18–7).

Fig. 18–7. A firefighter exposed to toxic smoke
being treated at the scene by EMS personnel

If the fire service continues to push the envelope regarding the safe use of SCBA, and we continue to work in the same manner we always have, then we will continue to lose firefighters just as we always have. The numbers of deaths attributed to asphyxiation and running out of air will continue to climb. The latest figures state that 63% of fireground fatalities in structure fires are a direct result of running out of air (asphyxiation).[9]

Fire departments like Phoenix and Seattle have documented that it takes longer to locate a lost firefighter in a structure than originally believed. In the majority of cases, if firefighters work into their emergency air (i.e., beyond low-air alarm activation) and get into trouble, there will not be sufficient time to reach them before they will run out of air. Remember that firefighters only have approximately 3–6 minutes of air remaining if they work until their low-air alarm activates. We know that it takes about 10–12 minutes to locate a lost firefighter in a training exercise. In an actual firefighter emergency, the time to effect the rescue could be greatly increased owing to thermal conditions, structural instability, and many other unforeseeable circumstances.

Having an air management policy that states that all firefighters working in an IDLH environment with a SCBA must practice the ROAM is a great starting point for increasing firefighter safety on the fireground. The ROAM states , know how much air you have before you enter the IDLH area, and manage that air so that you leave the hazardous environment *before* your SCBA's low-air alarm activates. The ROAM will, without question, increase a firefighter's survivability profile if they get into trouble inside a structure fire.

Rapid intervention is a reactionary process whereby the rescuers are potentially behind the time curve. The firefighter in trouble has only a limited air supply, and, given the statistical evidence, is pushing Murphy's law to its limits.

One of the most alarming realities of rapid intervention is that potential rescuers can and do become victims themselves. During the Worcester, Massachusetts, cold-storage warehouse fire in 1999, four of the six firefighter fatalities were rescuers sent in as two-person rapid intervention teams. During Phoenix's RIT training exercises, one in five firefighters got into trouble.[10] The Seattle Fire Department has taken these two significant events, as well as our own tragic loss of four brothers on January 5, 1995, in the Mary Pang Warehouse fire, as the impetus to develop and implement a progressive rapid intervention function that addresses most if not all of these lessons learned (fig. 18–8).[11]

*Fig. 18–8. Seattle firefighters saluting one of their
fallen comrades whose body is being removed
from a collapsed building (Photograph courtesy
Seattle Fire Fighters Union, IAFF Local 27)*

SEATTLE'S RAPID INTERVENTION GROUP

Important elements of Seattle's Rapid Intervention Group (RIG) program include:

- Adequate personnel and equipment resources
- Established functional groups
- Predetermined expansion of the incident management system
- Separate communication ability for rapid intervention operations
- Specially trained engine and ladder companies, from double houses, that are designated as RIGs
- Ongoing operations-level training for companies designated as RIGs[12]

Air management is critical to rapid intervention for both the rescuers and the rescued. It has been established that there is only a 5–10-minute window of opportunity to get rescue personnel to firefighters in trouble. To that end, Seattle's rapid intervention program follows a six-step model:

1. Preplanning
2. Positioning of resources
3. Initiating the rescue
4. Locating the firefighter(s)
5. Protecting the firefighter(s) in place
6. Extricating the firefighter(s) and rescue teams

At all structure fires, the IC is required to assign a Rapid Intervention Team. The RIT is generally an on-scene engine company with a minimum of four members. In addition, the Seattle Fire Department recently added an additional engine company to all residential structure fire responses to perform RIT functions. The RIT's initial responsibility is to stage equipment (separate water source, adequate 2½" hose to reach the farthest point of the building, 200 ft of 1¾" hose with an appropriate nozzle, thermal imaging camera, rescue air kit, and forcible-entry tools) in a location that will allow rapid access to the structure without interfering with primary fireground activities (fig. 18–9).

At any structure larger than a single-family dwelling, a RIG, consisting of specially trained engine and ladder companies, is added to the initial alarm dispatch to form a firefighter rescue group. This provides a RIG with a minimum of 10 firefighters, including three company officers.

At this point, the RIG will complete steps 1 (preplanning) and 2 (positioning of resources) of the six-step process. Preplanning is critical because the RIG needs to have thorough information regarding the building, fire conditions, and crew locations.

Fig. 18–9. Seattle's RIG assembling equipment and personnel during a training session prior to entering a training prop in an acquired structure (Photograph by Lt. Tim Dungan, Seattle Fire Department)

Armed with the knowledge that time is of the essence to the lost firefighter, a designated entry monitor is established, as in a hazmat operation. The monitor's responsibilities include tracking the entry time, work time, and exit time of the RIG personnel committed to the rescue area.

On initiation of a firefighter rescue, the likelihood that firefighters may experience tunnel vision increases. Team leaders making entry in a firefighter rescue situation are directly responsible for monitoring their team's SCBA air supply. Members are trained to conserve air, whenever possible, through crew rotation and air checks.

Seattle's air management program has been interwoven into the ongoing risk-benefit approach to rapid intervention, and all rapid intervention rescuers must exit the IDLH atmosphere prior to activation of their low-air alarms. The goal is to maximize rescue effort results while keeping the highest level of safety in mind.

RIT's SHOULD BE PART OF THE SOLUTION—NOT PART OF THE PROBLEM

All rescuers must understand that they are there to solve the problem, not become part of it. History has shown that many well-intentioned rescuers failed in their mission by disregarding their own air supplies. No one wants history to be repeated in the form of rescuers perishing while attempting to save one of their own. Seattle's RIG policies are designed to increase the survivability of the lost firefighter and the Rapid Intervention Team sent to rescue them. Adequate replacement RIT resources will be needed to support an ongoing firefighter rescue.

Many departments and fire districts do not have the Rapid Intervetion personnel and resources that larger departments do. However, smaller agencies can pursue mutual-aid agreements with adjacent jurisdictions so that resources can be more readily available for rapid intervention in the event of a firefighter emergency. Additionally, cross-training with these jurisdictions should be done whenever possible, to ensure compatibility of equipment (especially SCBA), communications systems, and command structure.

Air management is a proactive approach to firefighter safety that can be implemented today at the individual, team, company, and department levels—with nothing to lose and everything to gain. Air management can be folded into all elements of emergency operations where firefighters use SCBA. These include rapid intervention operations, where firefighters may become so focused on the rescue attempt that they develop tunnel vision and run out of air themselves. It is imperative that we, the U.S. fire service as a whole, improve on our own safety and welfare.

The Phoenix and Worcester fire departments have our heartfelt thanks for sharing their tragedies and findings. May we all collectively make firefighter fatalities in structure fires a statistical zero!

MAKING IT HAPPEN

1. Have members attend one or more rapid intervention classes provided by qualified instructors.

2. Ensure that your department has a rapid intervention policy that addresses the strategic-, tactical-, and task-level functions of rapid intervention, which includes air management.

3. Have members research the Phoenix event in which Bret Tarver died (NIOSH Fire Fighter Fatality Investigation F2001-13) and determine the critical factors that changed how the Phoenix Fire Department addressed rapid intervention.

4. Have all members become familiar with NFPA 1981—Standard on Open-Circuit Self-Contained Breathing Apparatus for Emergency Services.

5. Have members research the Worcester event (NIOSH Fire Fighter Fatality Investigation F99-47) and determine the critical factors that led to rescuers becoming victims.

STUDY QUESTIONS

1. What organization mandated the two-in/two-out rule?

2. During Phoenix's rapid intervention training, what proportion of firefighters got into trouble?

3. During Phoenix's rapid intervention training scenarios, how long did it take, on average, to finalize the extrication process?

4. SCBA cylinder size terminology is being changed to reflect volume in liter increments. How will 30-, 45-, and 60-minute cylinders be described in terms of liters?

5. NFPA 1981 (2002 edition and 2007 draft), states that working firefighters will consume, on average, how many liters of air per minute?

6. The latest available figures state that what percentages of fireground fatalities are directly attributed to firefighters running out of air (asphyxiation)?

7. During the Worcester, Massachusetts, cold-storage warehouse fire in 1999, how many of the firefighter fatalities were rescuers sent in as rapid intervention resources?

NOTES

[1] US Fire Service Fatalities in Structure Fires 1977–2000, NFPA, July 2002. [2] NIOSH Fire Fighter Fatality Investigation F2001-13.

[3] Phoenix Fire Department. Internal report on the Southwest Supermarket fire, available online at http://www.phoenix.gov/FIRE/report.pdf.

[4] Kreis, Steve, et al. 2003. Rapid intervention isn't rapid. *Fire Engineering*. December. 52–61.

[5] NFPA 1981—Standard on Open-Circuit Self-Contained Breathing Apparatus for Emergency Services. 2002 edition and 2007 draft.

[6] Kruchuk, Scott. Phoenix Fire Department study.

[7] US Fire Service Fatalities in Structure Fires 1977–2000, NFPA, July 2002.

[8] NIOSH Fire Fighter Fatality Investigation F2001-13.

[9] US Fire Service Fatalities in Structure Fires 1977–2000, NFPA, July 2002.

[10] NIOSH Fire Fighter Fatality Investigation F99-47.

[11] Routley, J. Gordon. Four firefighters die in Seattle warehouse fire. USFA report 077.

[12] Seattle Fire Department, Training Guide #10-10 (Oct., 2002).

SURVIVING THE MAYDAY

INTRODUCTION

Despite the best efforts of everyone on the fireground, sometimes things go wrong. While the emphasis of this book has been on prevention, to eliminate many of the elements that are causing tragedy at fires, we can never create a completely accident-free environment. The variables are too diverse, the situations are too dynamic, and the information is too limited for us to make perfect decisions all the time.

No one wants to be involved in a Mayday situation. Usually the Mayday means that something has gone terribly wrong and that firefighters are lost, hurt, or trapped. This is every IC's nightmare and must be anticipated in training if the outcome is to have a chance of being positive.

Fortunately, the stigma of calling a Mayday is becoming a thing of the past, as firefighters are finally recognizing that it is essential to mobilize help as quickly as possible. Once the Mayday has been called, the firefighter who needs help must, for the time being, help themselves. No matter how well trained the RIT is, it will take time to bring the needed resources to the firefighter.

In most situations, the most vital factor for those in need of rescue is air. Without a sustained air supply, no other interventions will matter, because asphyxiation will result in rapid death. This chapter details actions you can take to respond when the unthinkable happens—when the Mayday is *you*.

MAYDAY! MAYDAY! MAYDAY!

The scenarios are many, and the reasons for them are extensive. The reality is that once you need help, calling the Mayday as early as possible is the best way to get it. When you've done that, your comrades will be on the way. Meanwhile, you have to do whatever it takes to give them time to get to your location. In short, you have to survive.

There are many steps you can take, both physically and mentally, to put yourself in a position to be rescued. Throughout, this book has stressed the importance of leaving your emergency reserve intact should you be caught in a Mayday situation. Here that concept takes center stage.

So that's what the emergency reserve is for

Imagine the situation that would cause a firefighter to be incapable of self-extricating from a fire. How extreme would the circumstances need to be for individuals used to tough situations not to be able to help themselves? Serious personal injury, collapse, entrapment, disorientation, and myriad other terrible scenarios come to mind. Nevertheless, in every one of these, survivability is possible if help can get there in time. Often, death is not instantaneous; air is the critical factor and rapid delivery of air to the victim is always top among the priorities of rescue crews.

The emergency reserve, the last 25% of the air bottle, is intended for this situation (fig. 19–1). This emergency air is your air. The rest of the bottle is for the incident and should be fully utilized to get the job done. The emergency reserve is for the Mayday and is the primary tool available so that firefighters can successfully survive.

*Fig. 19–1. The last 25% of the bottle, or emergency
reserve. This air should not be used for regular operations
because it is needed to give RITs time to perform a rescue.
(Photograph by Lt. Kyle White, Seattle Fire Department)*

Depending on cylinder size, anxiety level, and consumption rate,
among other factors, the emergency reserve normally lasts between one
and eight minutes during normal operations. The Mayday environment is
obviously not business as usual, and consumption will vary dramatically
based on what is happening to the firefighter. Falling into a fiery basement
is going to be different from getting lost in a smoky warehouse. Remember
that leaving your emergency reserve intact gives you a good amount of air
time, which will in turn give rescue crews a chance to bring you more. If
you violate the ROAM and use that reserve for routine operations, your
chances of not having air in the event of a Mayday go up dramatically.

A discussion on making your air last longer occurs later in the chapter,
but it is important to mention how long the emergency reserve can be
made to last. In drills across the country, firefighters were put in a Mayday

situation and made to hunker down and attempt to make their emergency reserve last as long as possible. Routinely, the final 25% of air lasted 30–45 minutes, although some firefighters extended that air beyond one hour. How do you like the chances that your RIT will succeed given that kind of time frame?

This is in stark contrast to air management practices of the past, in which the Mayday was called when the firefighter's low air alarm was ringing or the cylinder was already completely out of air. Since the firefighter's anxiety level will be off the charts at this point, any remaining air will be utilized quickly. Decision making is hampered because of urgency and possible exposure to the products of combustion. The decision to use the emergency reserve for regular operation can still be made by individuals unwilling to change; it just can't be rationally defended.

DON'T FREAK OUT

Individuals who are able to control their emotions in extremely stressful situations make better decisions and perform more effectively. It is imperative that firefighters be trained in these types of situations before they actually happen. While training can never anticipate every possible scenario, firefighters can be taught the generalities regarding what to expect and how to react. They can be placed in situations of elevated stress, so that a real-life situation will not be first time they've been tested under pressure.

During the Mayday, it will be critical to purposely slow your breathing and attempt to regulate your actions. The adrenaline burst will significantly increase your heart rate, and you may be compensating for limitations due to injury or entrapment. There is absolutely no positive benefit to losing your cool or blindly performing actions that do not lend themselves to your ultimate rescue. This has to be your focus, and it needs to be purposeful.

MOVE TO A POSITION OF SAFETY AND GET ORIENTED

Getting out of harm's way is a natural response that will probably be triggered automatically. Instinctive responses, like getting free of an entrapment or away from intense heat, will always be priority number one; no one needs to be told this (fig. 19–2). You will also need to ensure that any other members of your team are being helped to the fullest extent possible.

*Fig. 19–2. A firefighter moving through debris and
entanglement hazards to get to a position of safety
(Photograph by Lt. Tim Dungan, Seattle Fire Department)*

Moving to a place that provides refuge from heat, smoke, or additional collapse will increase your chances of survival. It is important to begin figuring out your location within the building. Getting oriented can make all the difference if you can report your location to the IC, who will relay this information to the RIT team that is trying to find you.

Did you fall through the floor? Are you trapped under debris? Did you make it to an exterior wall? These are all invaluable pieces of information that will affect your chances for survival. Again, remain calm, move to the safest position possible given your circumstances, and attempt to get better oriented. It is always preferable to locate an exterior wall, and doing so may provide you with an avenue of egress. In collapse situations or where fire continues to erode the structure, bearing walls are much safer than interior partition walls. Getting into a room where a door can be closed to provide a barrier to heat and smoke may give you the time you need to be rescued. If necessary, you may have to breach a wall to get to a safer location. Be sure that the area into which you are moving is preferable to the one you currently inhabit.

If you are unclear of your current location, look for landmarks that might assist the IC in pinpointing your current position. Relaying features characterizing your location may enable the IC to recognize what floor you are on or where in the building you've moved to. Relaying anything that might narrow down the search is a favorable step. It also keeps you thinking and helps you maintain your calm.

FIREFIGHTER MAYDAY PARAMETERS

Our study of Mayday on the fireground commenced in 2001. The fire service literature at the time did not give clear and specific directives on when and how a firefighter should call a Mayday. Local fire department SOPs related to Mayday typically had more information on what to do after it was called. The directions to firefighters as to when to call it read something like this: "When a firefighter is lost, missing, or trapped and believes his or her life is in danger, he or she should call Mayday." There were no definitions or descriptions of *lost*, *missing*, or *trapped*, and by the time a firefighter believes his or her life is in danger, it may be too late.

Since the fire service did not have an adequate model to guide our study of the firefighter Mayday doctrine, we turned to ejection doctrine for military fighter pilots as a benchmark to guide our research. Fighter pilots are given clear and specific parameters as to when they must eject. The ejection parameters are written as if-then logic statements. For example, "If neither engine can be restarted, then eject"; "if hydraulic pressure does not recover, then eject"; and "if still out of control by 10,000 ft above terrain, then eject." There can be 10 or more ejections parameters for each type of aircraft.

Pilot ejection training consists of classroom and flight simulator time to develop the cognitive and affective skill. Then, the ejection-seat trainer is used to imprint the psychomotor skill. Ejection retraining occurs every 6–12 months throughout a pilot's career, and it must be passed at the 100% proficiency level. Ejection procedures are part of the crew briefing before every flight. The reasons for failure or delay to eject are addressed as part of ejection training:

1. Temporal Distortion (time seems to speed up or slow down)

2. Reluctance to relinquish control of ones situation.

3. Channeled attention (e.g., continuing with the previous selected course of action because other more significant information is not perceived).

4. Loss of situational awareness (e.g., controlled light into terrain).

5. Fear of the unknown (e.g., reluctance to leave the security of the cockpit).

6. Fear of retribution (for losing the aircraft).

7. Lack of procedural knowledge.

8. Attempting to fix the problem.

9. Pride (ego).

10. Denial (e.g., this isn't happening to me).

The *Ejection Seat Training Operations and Maintenance Manual* (Southampton, PA: Environmental Tectonics Corp., 1999) states,

> The first and absolutely most important factor in the ejection process is the decision to eject. … The decision to eject or bailout must be made by the pilot on the ground before flying. … You should establish firmly and clearly in your mind under which circumstances you will abandon the aircraft.

Our next step was to use the if-then logic model to develop Mayday parameters for firefighters. We used a qualitative process to draft a set of Mayday parameters. Our initial work concentrated on the single-family dwelling because this type of occupancy is describable, it is dangerous for firefighters, and most of us routinely respond to this type of alarm. We asked 339 firefighters whether they would call a Mayday under the following conditions in a single-family dwelling fire creating an IDLH environment inside the structure:

Condition	% Who Would Call the Mayday
Tangled, pinned, or stuck; low-air alarm activation	98
Fell through roof	94
Tangled, pinned, or stuck; could not extricate self in 60 seconds	92
Caught in flashover	89
Fell through floor	88
Zero visibility; no contact with hose or lifeline; do not know direction to exit	82
Primary exit blocked by fire or collapse; not at secondary exit in 30 seconds	69
Low-air alarm activation; not at exit (door or window) in 30 seconds	69
Cannot find exit (door or window) in 60 seconds	58

From this work, we concluded that there was no clear consensus as to when a firefighter should call a Mayday. By contrast, the ejection doctrine for fighter pilots teaches that they must eject when an ejection parameter is presented.

Following our early publications on the study of the firefighter Mayday doctrine, several fire departments across the country began testing their firefighters to see if they would call a Mayday when put into a Mayday situation. The result was that the majority did not call a Mayday at all, and those that did called too late, not leaving enough time for successful rapid intervention operations. In addition, 24 line-of-duty deaths were identified in which failing to call, or delaying to call, the Mayday contributed to the fatality.

The Seattle Fire Department studied three near-miss incidents in which firefighters did not call Mayday but survived. This analysis revealed that firefighters can quickly lose their cognitive-thinking and fine-motor skills from carbon monoxide poisoning. This impairs their ability to call a Mayday. The need for rigorous, continual Mayday training for firefighters in the affective, cognitive, and psychomotor domains became evident.

Firefighters and military personnel use a cognitive decision-making process called *recognition-primed decision* when confronted with time pressure, changing conditions, high stakes, unclear goals, and incomplete information. We rely on our past training and experience to address the situation presented to us now. Firefighters will most likely never have to call a Mayday at a real fire—just as fighter pilots will most likely never have to eject for real. Thus, the only way to ensure that firefighters have 100% Mayday-calling competency is to train and drill them over their entire career.

During our study of the Mayday doctrine, we were advised that the fire service had not accepted Mayday as the word to use. NFPA standards state, "The term Mayday should not be used for fireground communications because it could cause confusion with the term used for aeronautical and nautical emergencies." We contacted the National Search and Rescue Committee (NSRC), the federal authority that has jurisdiction over the word *Mayday*. The NSRC reported that it is permissible for the fire service to use *Mayday*, and the fire service's use of *Mayday* will not interfere with aeronautical and maritime frequencies. The NSRC suggested that NFPA and other organizations update their standards.

Instructional materials—to design training and drill programs on calling the Mayday—are needed. With the cooperation of the Anne Arundel County (MD) Fire Department, the Maryland Fire and Rescue Institute, and the Laurel (MD) Volunteer Fire Department, we developed a Mayday calling drill for firefighters. The drill design identified specific Mayday parameters with which a firefighter might be confronted. The four basic mayday parameters are

- Trapped or lost
- Stuck
- Collapse (something falls on you)
- Fall (through floor or roof)

The next step was to develop training props that could be used to simulate each Mayday parameter—without putting the student at risk yet making the experience significant enough to trigger the Mayday-calling response. The props required that the students crawl (in complete PPE, SCBA, and blacked-out facepiece) into a closet or a small bathroom and chock the door closed, to simulate being trapped or lost; a wire lasso was looped over students' SCBA cylinders while they crawled, to simulate being stuck; a piece of chain-link fence was dropped over firefighters while they crawled, to simulate collapse; and the crawling students were dropped into a ball pit, to simulate falling.

Now that we had the *if* part of the logic statement, we needed the *then*. We developed specific radio procedures: Repeat the *Mayday*, then give *LUNAR* (location, unit, name, assignment [and air remaining], and resources needed, respectively). Thus, it might sounds like this:

> Mayday, Mayday, Mayday. Division 2, quadrant A; Engine 24, firefighter Smith; search team; 25% air left; stuck in wires, need cutters.

Depending on the fire department radio system, the student can attempt to push the emergency identification button (EIB) on the radio. If they have an EIB, they can do it with the gloved hand; it aids in their rescue before or after the Mayday is called. In 2004, the Anne Arundel County Fire Department put 1,000 firefighters through this drill.

From this first drill, a complete training package needed to be developed. With the cooperation of Firehouse.com, Fire and Education Training Network, TrainingDivision.com, and the National Fire Academy, a compact disc training package was created that covers the affective, cognitive, and psychomotor domain of the firefighter Mayday doctrine. The package includes 40 minutes of video demonstrating the training, instructor guide, reference articles on the topic, a task analysis for calling a Mayday, cognitive/affective written test, and five "Job Preference Requirement" check-off sheets. Students who pass the course are eligible for National Fire Academy certificates.

Mayday parameters are rules for when a firefighter must call Mayday. Something has gone wrong, the fire is getting bigger, you are running out of air, or carbon monoxide is affecting your thinking and motor skills. It takes time for the RIT to get to you, so Mayday must be called immediately; the window of survivability closes quickly, and other firefighters will risk their lives to get you out. Firefighters need to train and drill on Mayday in rookie school and throughout their careers.

This synopsis of six years of our work does not do justice to the dozens of instructors, hundreds of students, and many organizations that have helped create, test, and present this firefighter Mayday doctrine. We are grateful for all their help. We hope that our work has contributed to the safety of fire service personnel through its content and methodology. This work does not end our study of the firefighter Mayday doctrine, so we look forward to the next chapter.

—Dr. Burton A. Clark, Raul Angulo, Steven Auch, David Berry, Brant Batla

CALL THE MAYDAY

Calling the Mayday should be your first priority if you hope to survive. You will obviously need to get out of immediate harm's way and get to a safe place. No one is going to think about a Mayday if they are burning or running from falling debris. However, the sooner you can get the message out—who you are, where you are, and what your basic situation is—the more likely help will arrive in time.

When calling the Mayday, firefighters need to address the same issues fighter pilots are trained to consider when deciding whether to make an emergency ejection:

- Temporal distortion (i.e., time seems to speed up or slow down)
- Reluctance to relinquish control of the situation
- Channeled attention (i.e., continuing with a previously selected course of action because other, more significant information is not perceived)
- Loss of situational awareness (i.e., controlled flight into terrain)
- Fear of the unknown (i.e., reluctance to leave the security of the cockpit)
- Fear of retribution (for losing the aircraft)
- Lack of procedural knowledge
- Attempting to fix the problem
- Pride (ego)
- Denial (i.e., this isn't happening to me)

It is easy to imagine any of these factors also becoming a problem for the firefighter who is deciding whether to call the Mayday. Training should directly address these factors, and the culture within the department needs to make calling the Mayday early a priority.

The Mayday message itself is critical, as it contains most of the information that the IC will use to direct the rescue effort. Hopefully, there will be good accountability at the command level, and RIT crews will have been assigned and will be ready for action. Still, it is the message that you send that will determine how the operation proceeds. There are four elements that must be in every Mayday message:

- Who you are (Engine 25, Ladder 12, Firefighter Johnson, etc.).
- Where you are (lost on the second floor, fell through the floor into the basement, etc.).
- What your situation is (out of air in a closet, trapped under debris with injuries, etc.).
- Acknowledgment of the message. An unacknowledged Mayday is no different than one that was never sent.

While there are more details that can certainly be added to Mayday messages, the preceding four are essential.

POSITION YOURSELF TO BE RESCUED

Once you've moved to a position of safety and have received acknowledgment of your Mayday, it is time to enhance your position in the building. While the instinctive movement away from heat, smoke, falling debris, and other hazards will put you in a safer position, areas of the structure that provide better locations for rescue should be your next goal.

Paths of egress are ideal because they provide an obvious path to searchers coming from the outside. There is also the chance that the path may not be so adversely affected as to prevent you from self-extricating. Any area that provides good landmarks and access for rescuers is preferred.

Exterior walls are often cited as good locations, because of strength of construction, possible means of egress (window and doors), and ease of approach for rescuers. Getting to a window may allow the RIT to gain access via ladders, which is particularly important if interior collapse has occurred. Being able to radio out that you're on the third floor on an exterior wall eliminates a large area for the search teams to cover.

Finally, it is beneficial if you can get to a room with a door that can be closed if necessary. Barriers to fire are important, and a closed door can buy critical time for crews to respond. You must do your best to relay any information about your location to the IC, particularly if something changes.

After finding a good position to hunker down, concentrate on conserving your air and taking actions that will assist the searchers in locating your exact position. A good example would be ensuring that your PASS device is activated. However, this will need to be weighed against the need to communicate with the IC. It is difficult to hear transmissions with the PASS alarm sounding, so turning it on and off may be a necessity. Furthermore, using your ears could make all the difference in hearing an oncoming crew. You might hear the noise made by rescuers long before they could hear you. That information should be immediately relayed to the IC, so they can notify the search crews that they are close to your position. Thus, while turning on the PASS alarm may be beneficial, letting in run constantly may not be the answer (fig. 19–3)

The various tools that you've brought into the structure fire can be of help. Tapping with a tool against a wall or floor could provide the sound needed for crews to hone in on your location. A light shined up against a wall or ceiling may be seen by the searchers. Positioning yourself in the line of sight of rescuers, without unnecessary exposure to hazards, is another useful tactic. This holds true even if you lose consciousness; you do not want to be lying on top of your PASS device or blocking the egress door should this happen. If you sense that you may be close to passing out, it is best to position yourself next to a wall, a hallway, or a door opening. This will maximize the effectiveness of the PASS device signal. As with every situation, this will depend on the hazards in your surroundings.

*Fig. 19–3. A firefighter beginning air
conservation after breaching a wall to
get to a safe location (Photograph by
Lt. Tim Dungan, Seattle Fire Department)*

CONSERVING YOUR AIR

Several tactics can be employed to make your remaining air last longer than it normally would. It cannot be emphasized enough that keeping the emergency reserve intact for emergencies needs to be mandatory. Having 25% of the air in your cylinder remaining would have a dramatic impact on RIT operations (fig. 19–4).

Fig. 19–4. An RIT preparing to make entry. Leaving the emergency reserve intact gives a well-trained RIT a fighting chance of making the rescue. (Photograph by Lt. Tim Dungan, Seattle Fire Department)

Once you consciously decide to control your breathing, your survivability profile expands dramatically. It is important to switch from breathing through your mouth (necessary during periods of high exertion and effort) to breathing through your nose. Exhalation can be through your mouth, but should be controlled and unforced. Make every effort to relax and recognize you are getting good air despite any perception based on your situation that would cause you to panic. Panic will deplete your air supply and diminish your chances of survival.

SKIP-BREATHING

Two techniques of *skip-breathing* are routinely used across the country. The first, from the diving community, involves taking an extra breath during the skip. The method can be summarized as follows:

- Inhale fully
- Hold breath for normal exhalation time
- Take an extra breath before exhaling
- Exhale slowly
- Repeat
- Remain mentally and physically calm
- Visualize good air exchange occurring

The second form of skip breathing is simpler and proceeds as follows:

- Inhale fully
- Hold breath for normal exhalation time
- Exhale slowly
- Repeat
- Remain mentally and physically calm
- Visualize good air exchange occurring

Another simple technique is the *counting method*, which allows flexibility and provides the firefighter with something to concentrate on while conserving air. This can be invaluable in helping to calm down and steady our emotions. The technique is performed as follows:

- Inhale—slowly and fully—for five seconds, using either skip-breathing method
- Hold for five seconds
- Exhale for five seconds
- Hold for five seconds
- Repeat cycle
- Remain mentally and physically calm
- Visualize good air exchange occurring

The duration that each breath is held can be varied depending on your exertion level, conditioning, and stress level.

Finally, there is a technique being researched by Frank Ricci and Kevin Reilly called the Reilly-Emergency Breathing Technique (R-EBT) that is performing well in medical studies. We recommend reading the author's article "Rethinking Emergency Air Management: The Reilly-Emergency Breathing Technique" (*Fire Engineering*, December 2007) and visiting their Web site at firefighersafety.net. This technique, which is called "Hum Breathing" by some teachers, is performed as follows:

- Reilly- Emergency Breathing Technique

1. Technique
   A. Inhale as you normally would.
   B. While exhaling, "hum" your breath out in a slow, consistent manner.
   C. The hum is low, and usually cannot be heard over the low air alarm. In situations when a firefighter needs to disentangle his/her SCBA or rapidly move around obstacles, it may be difficult to continuously hum after each breath. Breathe as you normally would and intermittently use the R-EBT. The more you use the R-EBT the more it will increase your survival time.

As mentioned earlier, these and other techniques have been proven to make the emergency reserve last from 30 minutes to well over an hour. Thus, they can provide ample time for the RIT to make entry and provide rescue. The extent to which you violate the ROAM will determine how much air you have when all hell breaks loose.

I'M OUTTA AIR!

That was the desperate plea of Captain Eric Abbt of the Houston Fire Department as he awaited the arrival of the RIT. He was out of air on a fifth-floor stairwell and could not exit the building. The recorded radio transmission is haunting; you hear the fear in his voice, amid the noise and chaos that is all around him, and realize that he does not have long to live if help does not arrive. His story encapsulates the messages of this textbook.

Captain Abbt was in the middle of a rescue and risked a lot to save a lot. Whatever reasons bring you to an out-of-air situation, the brutal fact remains that you must improvise to survive. No matter how bad it has become, you can make it. Do everything in your power to search for that last breath and hold on for the RIT.

It is important to prepare yourself for the emotional and psychological shock that is inevitable when you run out of air. Your only chance of survival is to approach this terrible situation with a deliberate belief that you will survive no matter what. There are measures that have proven to be effective in prolonging survival in no-air situations.

First, you should stay low, as that's where you have the best chance of getting decent air and, more importantly, not inhaling superheated air and gases. Pillows, your hood, or any other filtering device can be used to keep from choking on smoke. It will not keep you from inhaling carbon monoxide or hydrogen cyanide, but does provide some filtration of the denser components of smoke.

Getting additional breaths of good air should be your next focus. These will depend entirely on what type of structure you are in and the contents of that structure. The following is a list of sources of air that may buy you the extra minutes that you'll need while waiting for help to arrive:

- Punch a hole in the wall (e.g., through the wallboard) with your forcible-entry tool (fig. 19–5). Smoke may not have penetrated the wall, and fresh air may be inside.
- Punch a hole through an exterior wall and seal the opening with your mouth to allow airflow without smoke.
- The insides of cabinets, drawers, or closets may have pockets of clean air.
- Faucets, pipes, and drains may yield multiple breaths.
- Balls can be punctured and give a controllable source of air.
- Tanks, including the toilet tank, may have clean air.
- Sealed bags may contain good air.
- Any item with a closed door, such as an appliance, may provide air.
- The firefighters' fog nozzle has been used as a filtration device. There may be breathable air at the point where the fog comes out of the opening.
- Closed boxes of any kind may have pockets of breathable air.
- Oxygen cylinders in hospitals, residential, and nursing homes are sources of air.

*Fig. 19–5. A firefighter punching a small hole to release
trapped air in the void space. This action could buy
additional time for RIT crews to arrive. (Photograph
by Lt. Tim Dungan, Seattle Fire Department)*

The preceding list is certainly not complete, but it demonstrates the
very real possibility that you can find other sources of air and thereby
prolong your survival and improve your chances of rescue. Remember,
if you have activated the RIT, there are firefighters coming to get you.
They are coming with everything they've got, so do everything possible to
survive until they arrive.

TRANSFILLING, BUDDY-BREATHING, AND SHARING AIR

The Seattle Fire Department uses a transfill system on all of its Mine Safety Appliances air packs that allows any member to deliver up to 50% of his or her current capacity to another member (fig. 19–6). This is the primary method used to share air when someone runs out in the IDLH atmosphere. Current protocol requires that the donor have a reading of 2,000 or more on his or her gauge before transfilling. This keeps both members from operating at 1,000 psi or less and is the manufacturer's recommendation. The strengths of this system are that it can be done quickly, with practice, and that it allows the members to be disconnected from each other once the air has been shared. Conversely, it is not always wise to give up half your bottle, and the chance exists for something to go wrong with the connection, resulting in depletion of the entire air supply.

Fig. 19–6. A firefighter transfilling a downed member

Many systems of buddy-breathing exist, all of which have strengths and weaknesses. Currently, attachments are being added to SCBA that allow another member to hook a regulator line directly into the air supply. Using this arrangement, air can be shared as needed by both members. This system's benefit is that it allows for an incremental sharing of air (as opposed to a flat 50%) and can be done quickly. The downside is that the members must remain attached, and that makes it difficult to move or do anything else to improve the situation.

Some departments train in a buddy-breathing system in which members share a regulator back and forth. They attach and reattach to each other, sharing whatever air is left in the donor's bottle. This would certainly be a last-resort measure, to be used only when there is no other alternative.

Finally, it is now required that all new SCBA have an RIT attachment that allows a donor to give air to a fellow firefighter. These attachments vary depending on the type of SCBA and should be an area of emphasis for SCBA training programs. Whatever method is utilized by your department, it is imperative that you train realistically and often. The place to learn how your transfill or buddy-breathing system works is not in zero visibility, with your partner screaming for air.

PREPARE FOR THE WORST

As the fire service experiences more Maydays, our knowledge base grows as to how best to survive them. The preceding thoughts provide a common-sense approach to bettering your chances of getting out alive. We highly recommend relevant training such as John Salka's *Get Out Alive* class, which will greatly enhance your survivability in a Mayday event.

It is imperative that firefighters train in advance for the emotional and physical challenges that a Mayday situation will present. It is equally important that we aggressively train to never put ourselves in such a position.

GET OUT ALIVE

Many aspects of firefighter survival interact with each other and collectively have an effect on an interior structural firefighter's chance of survival. Along with air management, there are several other skill sets that increase a firefighter's chance of surviving a dangerous change in conditions. Although these skills are taught individually in many programs and texts, collectively they have become well known as the *Get Out Alive* program:

- *The ladder bail.* This tactic allows a firefighter to exit through a window onto a portable ladder under extremely hostile conditions. Firefighters are taught that if the heat and fire conditions at the exit window are so severe as to prevent climbing out and onto the ladder in the conventional manner, they are to crawl over the windowsill and down the ladder, headfirst. They are further instructed to maintain a tight grip on the rungs to prevent a free fall until they are fully out of the window and on the ladder. At this point, a quick turnaround maneuver is used by the escaping firefighter to assist any other escaping firefighters or to climb down the remaining rungs to the ground.

- *The wall breach.* This tactic allows a trapped or disoriented firefighter to escape a room or area by going through a wall. Ordinary tools such as an ax, halligan, or six-foot hook can be used to break through the wall, which can be sheetrock, lath and plaster, or paneling. After breaking through the wall, the firefighter is taught how to physically move through the wall studs. This process is taught so that the escaping firefighter can rapidly get out of an area without performing any SCBA emergency procedures, such as reduced profile or emergency escape.

- *The rope slide.* This allows a trapped firefighter to make an emergency slide from a window using a rope and a basic hand tool. The bail-out rope is removed from the escaping firefighter's pocket and quickly attached to whatever ordinary hand tool is being carried at the time; performing a basic body wrap with the rope, the firefighter rolls out the window. No special equipment, belt, or other hardware is required, and the firefighter descends to safety in a matter of seconds.

All of these tactics are to be used as last-resort efforts when no other options are available. These skills, along fireground orientation, search rope procedures, and air management, are the backbone of the firefighter survival curriculum. Firefighters are encouraged to stay aware of their surroundings and avoid getting into a position where these tactics are needed; however, should the situation arise, these specific tactics have been used successfully by firefighters to get out alive.

—John Salka (Battalion Chief, FDNY)
http://www.firecommandtraining.com

MAKING IT HAPPEN

1. Review the manufacturer's technical manual of your SCBA, to become familiar with how sharing air is done using the equipment used by your department. Conduct drills on the appropriate procedures as described by the manufacturer.

2. Conduct a drill in which you run your SCBA down to the emergency reserve level (25%), and then switch to conservation mode. Using skip-breathing or another technique of conserving air, time how long you can make the bottle last.

3. Conduct Mayday drills that include noise (activated PASS, radio traffic, etc.), restricted spaces, and zero visibility. Ensure that the scope of the drills includes associated functions, such as situational awareness, radio practice (quality and content of messages), and updated radio communications.

STUDY QUESTIONS

1. How long have members routinely been able to make the emergency reserve last under drill conditions?

2. Individuals who are able to control their emotions in extremely stressful situations do what?

3. During a Mayday situation, moving to a place that provides refuge from heat, smoke, or additional collapse will increase your chances of what?

4. If you are unclear of your position or location in a structure, what can you look for that may assist the IC in helping to locate you?

5. An unacknowledged Mayday is no different than what?

6. What two features do exterior walls provide that interior partition walls do not?

7. Name two techniques that can be utilized to conserve air in a Mayday situation.

8. True or false: A filtration method, such as breathing through a pillow or hood, will eliminate the problem of carbon monoxide poisoning.

9. How much air is transferred using the transfill method of sharing air?

10. What negatives are associated with hooking a buddy-breathing attachment to another member's SCBA?

NOTES

[1] Quinn. 2007. KTRK news report, March 29.
    http://abclocal.go.com/ktrk/story?section=local&id=5165021.

IMPLEMENTING AIR MANAGEMENT IN FIRE DEPARTMENTS LARGE AND SMALL

INTRODUCTION

Fire departments and firefighters come in all shapes and sizes. From the largest fully paid departments to the smallest all-volunteer organizations, firefighters are willing to put it on the line, perhaps even make the ultimate sacrifice, to protect and serve the citizenry. Challenges exist in every department, no matter it's size or makeup. These include both challenges that are unique to your organization and challenges that are the same as those of other departments across the country.

Personnel/staffing challenges, funding challenges, apparatus challenges, and political challenges are familiar refrains from coast to coast. Departments also share many positive attributes: Excellent people providing excellent service, a trusting relationship with the citizens, a large family atmosphere, and the incredible bond that the fire service provides, bring us together in a manner unique among public service organizations.

Air management is yet another way that firefighters can demonstrate their commitment to each other and the communities they serve. Air management increases the level of firefighter safety and the level of service to the community; air management decreases the number of firefighter injuries and illnesses and decreases firefighter medical costs. Implementing air management in your organization demonstrates a commitment to the future of your firefighters and the community. This is independent of the size of the department, the size of the community that it serves, and the demographics of the firefighters, whether paid, volunteer, or a combination of both (fig. 20–1).

Fig. 20–1. Firefighters from paid, volunteer, and combination departments meet to receive training in the ROAM. Training together promotes safety on the fireground and camaraderie within the service.

Many departments, both large and small, have successfully implemented air management. This chapter provides tools to assist you in moving your department toward a future that includes air management. The truth is that air management can work in any fire department, regardless of its size.

THE SEATTLE STORY

Seattle's (population 570,000) municipal fire department has more than 100 years of proud service to the citizenry. The department formed as a paid service after the catastrophic Seattle Fire of 1889 destroyed many downtown structures. Great fires occurred in many larger cities during this era, including Troy, New York, in 1862; Chicago, in 1871; Boston, in 1872; and Seattle, in 1889. In the wake of the Great Seattle Fire, a fully paid department was formed, as well as one of the charter locals (fig. 20–2) of the International Association of Fire Fighters (IAFF)—namely, Local 27. Seattle has a proud tradition of being an aggressive, interior-based firefighting force. The reason for this stroll down memory lane is to demonstrate that even old dogs can learn new tricks. Seattle adopted air management as practice and as policy, thereby demonstrating that a proud tradition does not prevent positive change from occurring.

Fig. 20–2. Firefighters from the Seattle Fire Department conducting interior operations at a structural fire in the era of intermittent use of SCBA (Photograph courtesy Seattle Fire Fighters Union, IAFF Local 27)

How did Seattle change?

Seattle adopted air management in reaction to a cascade of events. Those events include several near-miss incidents, regulatory oversight, changes to NFPA standards, and equipment mandates.

Let's look at one of the firefighter fatality "near miss" incidents. Chris Yob, a lieutenant on Ladder 8, in the Ballard area of Seattle, responded to a fire in the Sunset Hotel a three-story, ordinary construction hotel. The Sunset Hotel was only a few blocks from Station 18, and when Ladder 8 arrived on scene, heavy smoke was showing and multiple victims at windows needing immediate rescue. In addition, there were reports of people trapped on the fire floor. Lieutenant Yob directed Ladder 8 team B, which included the driver, to use the aerial to rescue victims trapped at windows. Lt. Yob and his partner went inside to perform a primary search of the third floor.

When they arrived at the third floor, Lieutenant Yob and his partner found heavy smoke in the hallway. They conducted an oriented search of the floor, apartment by apartment. While they were searching one of the apartments, Lieutenant Yob's low-air alarm activated. Standard procedure at that time indicated that the low-air alarm meant that it was time for his team to leave the building. Lieutenant Yob turned back the way he came, heading toward the access stairwell. Unfortunately, due to the mazelike construction of this turn-of-the-century hotel, he and his partner became disoriented. Even when his partner's low-air alarm activated, Lieutenant Yob was still sure that they could find the access stairwell and their way out.

Other companies from the Seattle Fire Department were operating in the hallway with Lieutenant Yob and his partner. These firefighters heard the low-air alarms of Yob and his partner, but they took no action to help. These companies took no action because the low-air alarm was a common and expected sound on our fireground at that time. Eventually, both Lieutenant Yob and his partner ran out of air and were forced to breathe heavy smoke. Lieutenant Yob has described the effects of the carbon monoxide as a surreal and detached experience that made action and decision making increasingly difficult.

Lieutenant Yob and his partner sought refuge inside an apartment on the third floor. Eventually, Yob and his partner were located by Engine 9. Since they were both out of air, Yob directed that his partner, a probationary firefighter, should receive a transfill from Engine 9 personnel. Engine 9 began an emergency transfill and called a Mayday. Yob recounted, "Once my partner had received a transfill and I knew he would be OK, it felt like a ton of weight had been lifted off my shoulders." Engine 9 then turned their attention to providing Yob with an emergency transfill.

At the time of this incident, the Seattle Fire Department used Mine Safety Appliances SCBA with a belt-mounted regulator and a low-pressure tube connecting the facepiece to the regulator. When Lieutenant Yob was out of air, he disconnected the low-pressure tube from the regulator, as he was trained, to filter-breathe (see chap. 5). Yob saw a window and thought he would be able to get a good breath of air prior to performing a transfill with Engine 9. Yob did not succeed in getting a good breath of air but did see an adjacent roof about six feet below the window. Yob brought his head back into the room, and Engine 9 personnel began the transfill; however, once the transfill hose was connected, air began to escape from Yob's regulator. Yob had not shut down the main line valve when he disconnected the low-pressure tube to filter-breathe. The member from Engine 9 immediately disconnected the transfill hose to prevent all of his air from escaping.

With the added stress of an unsuccessful transfill, breathing thick smoke, and the effects of the carbon monoxide, Lieutenant Yob remembered the roof that he saw when he stuck his head out of the window. He decided to hang from the window and drop onto the roof. He told Engine 9 to get his partner out of the building, then went out the window. What Yob did not realize was that the roof of the building next door was not connected to the fire building. There was a three-foot separation between the buildings, and Yob fell three stories to the ground. Engine 9 personnel called a Mayday and quickly exited to bring Yob's partner to safety and report the incident to the IC face to face.

When Lieutenant Yob landed from the three-story fall, he suffered devastating injuries to one of his arms and his back. Yob was now trapped within a blind alley, alone and injured, with the fire building looming three stories above him. Yob located a locked door of the adjacent structure and broke out the reinforced glass with his good arm. Yob made his way through this building, out the front, and around the block to his assigned apparatus and crew. The crew could see the state Yob was in and arranged fire department paramedic treatment and transportation to the hospital. Yob was able to rehabilitate and return to shift work at Ladder 8 after more than a year.

While this was a traumatic experience for Lieutenant Yob, it also served as a reminder to the Seattle Fire Department that things go wrong. In this case, a series of events led to the near fatalities of a well-respected company officer and a probationary firefighter. Quick thinking by Engine 9 prevented Yob's partner from the same fate Yob experienced. The real learning moment for the Seattle Fire Department—and the impetus for establishing the ROAM—was the lack of action taken by crews within touch, sound, and sight of Yob and his partner when their low-air alarms were activated. These crews were performing commensurate with their training and experience. Seattle Fire crews had a false-alarm mentality about the low-air alarm. The false-alarm mentality nearly cost Yob his life. The Seattle Fire Department learned this lesson and implemented air management and the ROAM. Low-air alarms are now considered a firefighter emergency and trigger RIT activation unless they are dealt with immediately through radio communications.

REGULATORY INTERVENTION

Regulatory oversight came to the Seattle Fire Department in the form of investigations by the Washington State Department of Labor and Industries. Seattle had several instances in which firefighters were injured as a result of dehydration, overheating, or some combination of factors related to rehabilitation during fireground or drill activities (fig. 20–3). The issue boiled down to recognizing and implementing a system to ensure that firefighters adhered to a rotation that balanced work and rest cycles.

Fig. 20–3. Fire buffs setting up a rehabilitation area at a fire scene to support members of the Seattle Fire Department. Multiple agencies can be seen in the rehabilitation area while the second-arriving truck company prepares to go in service. (Photograph courtesy Seattle Fire Fighters Union, IAFF Local 27)

As a result of the Washington State Department of Labor and Industries investigations, fines were levied against the City of Seattle and the Seattle Fire Department (fig. 20–4). These fines prompted the department to approach the problem of injured firefighters with an open mind. The fines also provided an economic incentive to use the money, designated to pay the fines, to solve the problem instead. Air management was part of that solution.

L&I News • April 21, 2004

Seattle Fire Department
Cited for Three Workplace-Safety Violations

TUMWATER — The Washington Department of Labor and Industries (L&I) today issued citations to the Seattle Fire Department for three workplace-safety violations. The citations carry a combined penalty of $63,000.

One of the citations, stemming from an L&I investigation of a training drill at Station 25

1. A repeat "serious" violation with an associated penalty of $10,000 for failing to develop a formal accident prevention program tailored to the needs of operations personnel in warm weather working conditions. This is a repeat violation because the fire department was cited in March 2001 following a similar incident. ("Serious" violations exist where a violation of a safety standard is likely to result in serious injury or death.)

2. A "willful" violation with an associated penalty of $50,000 for failing to implement accident prevention program procedures for warm weather working conditions for operations personnel. This violation was determined to be willful in nature because the fire department failed to implement procedures for safeguarding its employees, despite being aware that the weather conditions required it to do so. ("Willful" violations exist if an employer intentionally violated safety or health rules, was plainly indifferent to the rules, or knew that a violation was occurring and took no steps to correct it. Penalties for willful violations are first calculated as serious and then multiplied by ten.)

A separate L&I investigation was prompted by a complaint in October 2003. The complaint alleged that the fire department's safety officer was directed to stop his investigation into whether there was a relationship between workplace conditions and a number of cancer cases among people who worked at Station 31.

This investigation led L&I to issue a citation to the fire department alleging a "serious" violation for inhibiting the work of the safety investigator, and assessing a $3,000 penalty.

"Fire department safety officers must be free to pursue safety investigations without intervention from superiors," said Michael Silverstein, L&I assistant director for workplace safety. "This independence is essential to a successful workplace injury and illness prevention program."

The Seattle Fire Department has 15 business days to appeal.

http://www.lni.wa.gov/news/2004/pr040421a.asp

Fig. 20–4. Public notification of a citation issued to the Seattle Fire Department by the Washington State Department of Labor and Industries. This citation led to changes in the air management and rehabilitation guidelines.

Air management provides a framework to assist firefighters, company officers, and command staff in identifying the length and number of work cycles a firefighter has performed. Using this information, they can ensure that appropriate rest cycles balance the work cycles to provide for firefighter safety through effective company-level or formal rehabilitation procedures.

There can be no doubt that working hard in full PPE is a physically and mentally challenging activity. This activity results in increased body core temperatures, dehydration, and muscle fatigue. Scientific studies have demonstrated a decrease in effectiveness of the human body when subjected to the physical effects of working in full PPE. Still, firefighters love to fight fire, and the opportunity to participate in an active firefight occurs less often, feeding the natural tendency for a firefighter to want to "get some" when an opportunity presents itself (fig. 20–5). This desire to do the job must be balanced against the natural laws of thermodynamics and the physical limitations of the human body. Work intervals must be balanced with periods of rest, rehydration, and an opportunity to decrease body core temperatures.

*Fig. 20–5. Sequence showing the impact of an aggressive
interior fire attack. Heavy fire showing from the D side of the
building quickly turns to steam as firefighters apply water
from the uninvolved side. (Photographs by Mark Pedeferri,
courtesy Seattle Fire Fighters Union, IAFF Local 27)*

Seattle Fire Department took advantage of air management to implement an effective work/rest interval and appropriate rehabilitation policy. The current practice is outlined in figure 20–6.

Guidelines for rehabilitation of firefighters adapted from the SMOKE INSERT: When and how unit's are assigned to rehab should be dictated by formal department SOP's. Minimum standards should include:

- Identified work-to-rest intervals before company level rehab are listed below and should require a 10-minute company rehab including rest, hydration, and an evaluation of the company's readiness for re-assignment at the completion of the 10-minute rehab.
  - 1 - 1200 liter cylinder without air management
  - 1 – 1800 liter cylinder following the ROAM
  - 20 minutes of intense work.

- Identified work-to-rest intervals before assignment to the rehabilitation area.
  - 2 – 1200 Liter cylinders without following the ROAM including a 10-minute rest and hydration period between cylinders.
  - 2 – 1800 Liter cylinders following the ROAM including a 10-minute rest and hydration period between cylinders.
  - 1 – 1800 Liter or 2400 Liter cylinder work cycle without following the ROAM.
  - 1 – 1200 Liter cylinder without following the ROAM or 1 – 1800 Liter cylinder following the ROAM after having rotated through rehab previously. (This requirement recognizes the cumulative impact of repeated work-rest intervals over the course of an incident and promotes coordinated company rotations and incident accountability.)

- In addition to the work-rest interval considerations, any SOP should include the following for assignment to rehab.
  - The company officer recognizes the company needs to move to the rehab area at any time.
  - The incident commander assigns the company to rehab.

Fig. 20–6. Cutout box from "SMOKE: Perceptions, Myths, and Misunderstanding" (Reprinted courtesy Cyanide Poisoning Treatment Coalition [http://www. cyanidepoisoning.org/pages/documents/SMOKE_ Supplement_forweb.pdf])

IMPLEMENTATION OF THE ROAM

On December 18, 1999, I was in my apparatus returning to my district when frantic radio traffic caught my attention. The radio traffic was from Kansas City, Missouri, battalion chief John Tvedten, transmitting as he became lost in a warehouse fire and ran out of air. My coworker and friend Sam Persell and I will never forget the last transmissions from Chief Tvedten and the heroic rescue attempts his brothers and sisters made to locate him. At that moment, I promised myself that if I ever was fortunate enough to be in a position to prevent a similar situation from occurring, I would.

In September 1946, while speaking about the surrender of Japan at the end of World War II, President Harry Truman said, "It is our responsibility—ours, the living—to see to it that this victory shall be a monument worthy of the dead who died to win it."

President Truman's words live on today in the military and are equally applicable in several aspects in the fire service. Truman's quote has always been especially important to me and resonates in many of my decisions as a chief officer. It is a critical function of fire service leadership to do everything possible to guarantee the safety of our members. When operating in a potential IDLH, the ROAM is the industry's best practice and standard.

When I first met the air management advocates from Seattle five years ago, I was still searching for solutions to implement in order to prevent air management–related catastrophes and to guarantee that the loss of John Tvedten was not in vain. The outcomes of my relationship with the Seattle group were learning, information sharing, and the realization after several experiences that these fire officers have already made one of the greatest contributions to the fire service in the 21st century.

I have been very fortunate in my career to have the opportunity to implement the ROAM both in a smaller, 85-person all-career fire department (Yakima, WA) and in a larger, 350-person metropolitan organization (Spokane, WA). In both organizations, the personnel were (and are) professional, aggressive, competent, and dedicated firefighters. However, as anyone who has more than a few weeks in the fire service realizes, change is tough—especially when it involves the fireground. It has been my experience as a change agent that only about half of administration-forced programs become successful, owing to the difficulties posed by cultural integration. Nevertheless, I am convinced that the same fate should not befall the ROAM.

Cultural integration is the largest impediment to organizational acceptance of the ROAM program. However, three important factors increase the chances for successful implementation: integrity, involvement, and information. After the establishment of an organizational environment rich with involvement and information, I recommend the five-phase process outlined here.

Phase 1. Break with the past and focus on the future

Provide legitimate, externally validated reasons for change—for example, NIOSH reports, this book, and hands-on training programs, as well as other sources that note the weaknesses of the fire service's traditional mode of operating. It is normal to expect ambiguity; nobody can know all of the answers all of the time. Establish a nonpunitive learning environment where people can ask questions and find answers.

Phase 2. Mobilize for change

Begin by clearly signaling that change is coming. This begins at the top. For career organizations, this may be a written agreement between labor and management; in volunteer organizations, it may be a vision statement by the leadership or a vote by the membership. Change can take many other forms, including bringing in outside experts in the ROAM program and sending members from a training or safety committee to a hands-on training session. There are many ways to set the vision and prepare the organization for change while making sure that people are informed, involved, and ready before proceeding.

Phase 3. Select the leaders/salespeople

Look to your star frontline supervisors (lieutenants and captains) and middle managers (battalion/division chiefs) to become the experts in ROAM philosophy. The role of crafting SOPs and SOGs should fall chiefly to the people performing the job on the street. True leaders rise to the occasion and become your best salesperson(s) to the organizations. In this phase, you are looking for a few good trees from which to develop a forest.

Phase 4. Training: Dedicate the resources

Expect time, sweat, toil, and a significant amount of financial resources to do it right. I cannot emphasize this enough. For the bean counters who can be your worst enemy (or best advocate), compare the cost of losing a firefighter. For the 5% who are internal dissenters, tie in subtle (or sometimes not-so-subtle) reminders that their children and family need them and that following this program is the best guarantee that they will go home at the end of their shift—every time.

Phase 5. Live it

After the initial training and policy implementation, live it daily. Make sure that everyone from the front line up to the fire chief sends consistent messages about the new ways of doing business and that they reinforce their words with performance and training activities. Inconsistency breeds uncertainty, and uncertainty will destroy any policy or training. Discipline, training, and commitment are the keys to institutionalizing the ROAM philosophy.

Often, change and transition consume a great deal of personal energy and time. In fact, overcoming resistance to change in the fire service may actually be the toughest lesson to learn for chiefs or union leaders of any age or experience. For those intending to implement the ROAM in their organizations, I encourage them to embrace the power of we. The sooner you substitute we for us versus them in your philosophy, the sooner you will bring success to everyone. This process should not be a bargainable item, resulting in lawyers' charging a higher fee. This is about preventing the loss of firefighters' lives and realizing a safer environment for everyone. It really is that simple.

Because of my experience with ROAM, I can respectfully say that the contribution that the authors of this book and my friends have made to the fire service is a monument worthy of the dead who have fallen before us.

—*Brian Schaeffer (Assistant Chief, Spokane [WA] Fire Department)*

NFPA STANDARDS
AND EQUIPMENT MANDATES

THE PROBLEM

The NFPA produces consensus standards for the fire service. These standards outline the accepted best practices for fire service organizations and have been repeatedly upheld, when challenged, in court. The applicable standard in this case is NFPA 1404 (see chap. 6). This standard assisted the Seattle Fire Department in implementing air management over the long term because Seattle trains its own recruits. Training-division personnel are aware of the NFPA standards and strive to follow those standards to provide safe and effective training to new Seattle firefighters. Failure to follow NFPA standards in recruit training has resulted in serious injury to recruit firefighters within the history of the Seattle Fire Department and in injuries and fatalities of recruits in other fire service organizations. Suffice it to say that NFPA 1404 requires air management training according to the ROAM, and failure to do so exposes a fire department to litigation.

During this same time frame, Seattle had several equipment issues. First among these was the imminent expiration of the service time for nearly all of the department's SCBA cylinders. Seattle was using 1200L (30-minute) cylinders that carried a service life of 15 years. Over 80% of the cylinders were due to expire within 12 months, putting the department in the position of having to purchase more than 800 cylinders immediately and an additional 1,000 cylinders over the next year. This situation influenced staff decisions to replace the 1200L (30-minute) cylinders that were expiring with new 1800L (45-minute) cylinders and implement air management and the ROAM.

The solution

The Seattle Fire Department was already implementing some of the mandates of NFPA 1404 and the ROAM in operational training and recruit training exercises. Firefighters at the company level had been introduced to the concept of the ROAM and had demonstrated the ability to manage their air supply as part of interior operations, during hundreds of drill scenarios in realistic simulated IDLH environments. Regulatory authorities (Dept. of Labor and Industries) demanded a system of work/rest intervals to manage firefighter rehabilitation. Purchases of new cylinders were planned owing to expiration of the service life of the existing cylinders. This combination of circumstances made air management and the ROAM a natural fit for the department.

While the department staff was now on board for a change to the 1800L cylinder, the Joint Safety Committee—and the union representatives in particular—raised concerns about increasing the work-cycle times of firefighters without increasing the rest-cycle times. IAFF Local 27 had good reason to question the implementation of the ROAM and the department's commitment to safe work cycles. Remember that several members of the department had been injured or nearly killed because of inattention to firefighter rehabilitation. This resistance was easily overcome by educating the union representatives about the ROAM. This is especially true when it could be demonstrated that firefighters who follow the ROAM with an 1800L cylinder experience a very short or no increase in the work cycle.

Air management also provides an exponential increase in the margin of error for companies operating in an IDLH environment. The Joint Safety Committee examined near-miss situations in which firefighters' low-air alarms were ignored by other firefighters. They recognized that removing the low-air alarm from the incident scene would prevent this problem. As a result, the Joint Safety Committee agreed to the purchase of the 1800L cylinders in combination with implementation of the ROAM.

Initial implementation of the ROAM and improvement of rehabilitation guidelines were achieved through changes to the policy and operating guidebook and training guides following regular revision cycles. The incident scene was more of a challenge, since changing decades of ingrained behavior takes time. The first incidents after the policy change were rife with violations of the ROAM. Consistent education, as well as increased training opportunities, started to have a positive impact in a short time period. Firefighters and company officers began to see the positive effect that the ROAM was having on crew rotations, implementation of rehabilitation, scene management, and firefighter safety.

Note that the hard chargers who continued to disregard the ROAM and work until their low-air alarms activated were making regular trips to rehabilitation for extended rest cycles. Firefighters who chose to have longer work cycles, by extending their work into the low-air alarm, were required to take long rest cycles per department policy. This action was all the discipline necessary to instigate the change of behavior within the department.

The activation of a low-air alarm is a rare occurrence on Seattle Fire Department incident scenes today. The sound of the bell ringing is a clear indication at the command post that a problem exists on the fireground. Low-air alarm activations are immediately addressed by radio report from the affected company. If a radio report does not immediately follow low-air alarm activation, the IC will adopt a firefighter rescue posture while attempting to identify the crew involved and determine their emergency status. While firefighter near-miss prevention is the hallmark of the ROAM, early intervention when the low-air alarm sounds is another strength for an air management department.

IMPLEMENTATION IN SMALL FIRE DEPARTMENTS

Once air management was implemented, the authors expanded the training and shared their story with the fire service through articles published in the leading fire service magazines. The training methods we developed were adapted to produce an eight-hour hands-on training package that provided students with the skills to perform air management effectively at the task level (fig. 20–7). Pilot programs of this training package exposed the idea to other fire departments within the state. Many of these training officers had been paying attention to the near-miss and fatality situations occurring locally and nationally, and, as in Seattle, they had been looking for a solution to the problem of firefighters running out of air in the IDLH environment. Many of these departments began to seriously consider adopting air management and the ROAM as the solution.

Fig. 20–7. A training prop used by the Yakima (WA) Fire Department training center. This prop was built into a storage area. Eight hours of hands-on training allowed the Yakima Fire Department and neighboring departments to experience air management and the ROAM. (Photograph by Assistant Chief Brian Schaeffer, Spokane [WA] Fire Department)

YAKIMA, WASHINGTON

The Yakima Fire Department was excited about the ROAM. Yakima (population 72,000) is a city in eastern Washington with a fully paid department. They also have a good working relationship with local fire districts. This working relationship extends to a collection of fully paid, combination, and volunteer departments.

The Yakima Fire Department recently purchased 1800L cylinders, and because of the very hot summers experienced on the eastern side of Washington, the union representatives to the Joint Safety Committee were concerned that increased work-cycle times, without appropriate rest-cycle times, would create dangerous overheating problems for firefighters. This is the same concern that was expressed by IAFF Local 27 during the implementation of the ROAM in Seattle. With proactive union representatives, training officers, and staff officers, Yakima Fire moved forward by implementing ROAM training. Formal and informal leaders from Yakima and neighboring departments were trained in the ROAM (fig. 20–8). This training sparked the discussion of implementing an air management system within the Yakima Fire Department.

Fig. 20–8. A firefighter checking the SCBA air supply during hands-on training. Trained firefighters can perform an air check and operate according to the ROAM while improving overall scene safety. Good training is the key to top performance. (Photograph by Assistant Chief Brian Schaeffer, Spokane [WA] Fire Department)

Yakima decided to mandate that in all training exercises, crews begin to exit the structure once they reach 50% of the capacity of the 1800L (45-minute) cylinder. This practice provides 900 liters (50%) of air for entry and work in the hazard area; 900 liters is the same amount of air as provided by a 1200L (30-minute) cylinder without following the ROAM (see table 2–1). Yakima also designated 450 liters (25%) of air for exit from the hazard area. The remaining 450 liters (25%) of air is set aside for a firefighter emergency in the hazard area. Through repetitive training, Yakima has been successful in removing the low-air alarm from the incident scene. This demonstrates that even without an official policy or SOP, firefighter behavior on the fireground is directly related to the training that they receive.

TUSCALOOSA, ALABAMA

The Tuscaloosa (AL) Fire Department is a fully paid department that provides fire, rescue, and emergency medical services to a population of 120,000. Tuscaloosa is also an air management department.

The system that Tuscaloosa uses was developed by several training officers on the ride home from the Fire Department Instructors Conference in Indianapolis. Chief Martin and one of the training officers attended a presentation about the Southwest Supermarket fire, which claimed the life of Firefighter Bret Tarver in 2001. In this presentation, Nick Brunacini of the Phoenix Fire Department made the case for Phoenix's air management system by describing how it works. (See chap. 21.)

According to Training Chief Ken Horst, the Tuscaloosa system has three primary components (figs. 20–9 and 20–10):

- At the task level, individual firefighters are expected to monitor their air and communicate their status to the company officer.

- At the tactical level, company officers are expected to maintain situational awareness and accountability for their team. This includes continual knowledge of the air supply.

- At the strategic level, the ICS is expected to maintain adequate on-deck companies and support tactical-level air management in the hazard area through time-based radio communication.

TUSCALOOSA FIRE DEPARTMENT
STANDARD OPERATING GUIDELINE

Section G   10/2006

THREE-DEEP DEPLOYMENT

PURPOSE:

The purpose of Three-Deep Deployment ("On Deck") is to increase fire fighter safety and accountability and reduce stress and fatigue through the implementation of a crew rotation system when operating in IDLH atmospheres.

G-1   DEFINITIONS:

Working Crews: Those crews assigned to tasks inside the hazard area.
On Deck Crew(s): A crew staged in a designated area just outside the hazard area.
Staged Crew(s): Crew(s) kept in Level 1 Staging
Sector: Geographical area of assignment
Sector Officer (SO): An officer assigned to oversee operations in a particular sector.
FIT (Field Incident Technician): Individual assigned as an Accountability/Safety
   Officer for their particular tactical (sector) position.

G-2   RESPONSIBILITIES

A. Working Crews
   1. Working crews shall perform tasks as assigned by the Sector Officer (SO)/
       Incident Commander (IC).
   2. Crew members should monitor their individual air supply. When
       a member's air supply is 50%, the crew should begin to exit
       the IDLH environment.
   3. Working crews exiting the structure will be assigned the On-Deck position
       after the first interior rotation. Following the 2nd rotation into the IDLH
       environment, the working crew shall cycle to rehab.
   4. In the event of a lost, trapped, or downed firefighter, the firefighter or the
       crew of that firefighter shall declare a "Mayday, Mayday" and report all
       available and pertinent information to their Sector Officer.
   5. The Working Crew shall evaluate the situation, perform a rescue as
       needed, and/or if needed, request through the Sector Officer additional
       resources for assistance.

*Fig. 20–9. First page one of the Tuscaloosa Fire
Department SOG for three-deep deployment. Air
management is addressed as the number-two
item for working crews. (Reprinted courtesy
Tuscaloosa [AL] Fire Department)*

Fig. 20–10. Tuscaloosa Fire Department receiving training in air management and the three-deep deployment model. Training included multiple-company operations prior to implementation of the policy. (Photograph courtesy Tuscaloosa [AL] Fire Department)

Tuscaloosa's method of using time checks and its on-deck philosophy are directly linked to the Phoenix system. Having individual- and company-level accountability for air is an application of the ROAM.

The Tuscaloosa system is related to the ROAM in two additional respects. First, teams are expected to be out of the hazard area before their low-air alarms activate. Second, activation of a low-air alarm in the hazard area receives immediate attention from personnel inside and outside the hazard area.

Tuscaloosa firefighters use a *"natural help order"* by which those inside—and therefore closest to the problem—are empowered to solve the problem first. Help would then come from the designated *on-deck* company assigned as RIT by the sector officer (fig. 20-11). Finally, additional companies are sent to the appropriate sector in support of RIT activation of the on-deck team.

Fig. 20–11. The on-deck tarp, used to secure equipment for the RIT. Companies reporting to a sector officer as the on-deck company are responsible to fill the tarp commensurate with the needs of the incident. (Photograph courtesy Tuscaloosa [AL] Fire Department)

The Tuscaloosa system has one additional component that makes it a very strong air management department. When low-air alarms are activated in the hazard area, the situation is specifically addressed in the postincident analysis (PIA). The air management component of the PIA identifies how the low-air alarm activation occurred, what could have prevented the activation in this instance, and areas of additional training that may be provided to the department.

Size doesn't matter

In the end, the size of the department does not affect the size of the cylinder on a firefighter's back; the size of the department does not change how fast a firefighter breathes; the size of the department does not change the products of combustion; the size of the department does not change the risks inherent in fighting fires from the inside; the size of the department does not change the mandates of NFPA 1404; and the size of the department does not change the impact that a firefighter fatality will have on the organization. In the end, the person responsible for managing the air supply of the firefighter is the firefighter; the person responsible for managing the air supply of the company is the company officer; the person responsible for ensuring that everyone goes home safe from the fire is the IC; and the person responsible for ensuring that a department operates safely and in accordance with national standards is the fire chief.

Wherever your department fits on the scale of size, air management can improve your operations. Whatever size cylinder you use in your SCBA, air management can make the operation safer. Whatever kind of SCBA training program you use, air management is a required component. Whatever task you are assigned on the incident scene, you are responsible for managing your air. Whatever tactical assignment you receive, you must ensure that your company gets a round-trip ticket. Whatever strategic decision you make, you are responsible for making sure that everyone goes home.

MAKING IT HAPPEN

1. To identify how your fire department may best implement air management, find fire departments that are a similar size or have similar operational SOPs that have implemented air management. Ask the change makers in that department what their procedure was and model your own plan after theirs.

2. What are the specific regulations in your state regarding compliance to NFPA standards? Is there a legal mandate in your state for compliance with air management?

3. Review your own fire department's recent history and identify incidents where firefighters experienced low-air or out-of-air situations. Correlate these events with the Bret Tarver and Chris Yob incidents to provide the decision makers in your department with the information they need to come on board with air management.

4. Review your rehabilitation policies to make sure that your department has identified the length and number of work-cycles firefighters can perform before rehabilitation is mandatory. Identify how air management can be used to support an effective rehabilitation rotation.

STUDY QUESTIONS

1. When Lieutenant Chris Yob and his partner from Ladder 8 arrived on the third floor of the Sunset Hotel, what were the smoke conditions?

2. At the time of the near-miss incident that involved Lieutenant Yob, what did the Seattle Fire Department's SOP dictate with respect to when a team should leave the fire building?

3. When Lieutenant Yob became lost, his low-air alarm activated. Why didn't other Seattle firefighters in the hazard area provide Lieutenant Yob with assistance?

4. When Lieutenant Yob and his partner ran out of air, what did they breathe?

5. What regulatory agency cited the Seattle Fire Department for failing to correct noted deficiencies related to firefighter rehabilitation?

6. What positive impact can air management have on work cycles?

7. When can firefighters earn a trip to rehabilitation?

8. Why was the Seattle Fire Fighters Union, IAFF Local 27, concerned about changing from the 1200L cylinder to the 1800L cylinder?

9. Why did the Seattle Fire Fighters Union, IAFF Local 27, agree to implementing the larger cylinder size in the operations division?

10. How did the Yakima Fire Department implement air management?

11. After what fire department did Tuscaloosa model their air management program?

12. What will the Tuscaloosa Fire Department do during the PIA if one of their firefighters experiences a low-air situation during a fire?

AIR MANAGEMENT—OTHER OPTIONS

INTRODUCTION

In this chapter we examine several other air management systems and ideas. Specifically, the relative strengths and weaknesses of the Phoenix Fire Department's system, the Swedish and UK systems, technology-driven systems, the time-on-scene system, and the "half-your-air-and-out" system are discussed.

THE ROAM—SIMPLE AND EFFECTIVE

Throughout this book, we have proposed a simple, effective way for firefighters to manage their air—namely, the ROAM. The ROAM states that firefighters need to know how much air they have in the SCBA and manage that air so that they leave the hazardous environment *before* their low-air alarms activate. By leaving the hazardous environment before their low-air alarms activate, firefighters accomplish two things: First, they leave themselves an emergency air reserve should something go wrong on their way out of the IDLH atmosphere. Second, they remove the low-air alarm from the fireground, which allows it to become a true emergency alarm, calling for immediate action.

The ROAM has brought dive industry standards regarding the emergency air reserve and air management to the fire service, which should have been done long ago. The ROAM is nothing more than a scuba-diving air management standard. While the ROAM is the simplest and most effective air management system that exists, other air management systems are also in use in the fire service.

WORK CYCLES AND ON-DECK CREWS— THE PHOENIX SYSTEM

After the death of Bret Tarver in the Southwest Supermarket fire (see chap. 7), Chief Alan Brunacini and the Phoenix Fire Department realized that the way their firefighters managed their air to change. The Phoenix Fire Department, like every other fire department in the United States, did not have an air management system in place before the Southwest Supermarket fire. Unlike the fire service in European countries like Sweden and the United Kingdom, the U.S. fire service has never really understood or appreciated the importance of the air supply. We have taken our air for granted and have been cavalier in how we use it at structure fires. The U.S. fire service mistakenly views the low-air alarm as a signal to leave the fire building. Unfortunately, it took Bret Tarver's death to initiate a discussion about air management in the U.S. fire service.

The Phoenix Fire Department now has an air management system that works for them. Their system begins with the individual firefighters, who are taught to keep track of their air supply when they are in the hazard area. More importantly, the company officer is expected to monitor crew members' air supplies. Air management is also integrated into the command level. Division/sector officers, with the help of a field incident technician (FIT), keep track of their crews' work cycles and have on-deck crews ready to replace those that must leave the hazard area. Another crew is staged and ready to move into the on-deck position once the on-deck crew rotates into the fire building. The purpose of this "three-deep" deployment (i.e., working crew, on-deck crew, and staged crew) is to increase firefighter safety. Work cycles in the hazard area are kept to about 10 minutes, depending on the type of structure involved and the fire conditions. Phoenix firefighters use 1200L (30-minute) air cylinders. After about 10 minutes, the on-deck crew, which has been staged and monitoring interior operations, replaces the interior crew, who leave the hazard area to either get another cylinder or go to rehabilitation.

The work-cycle system has several advantages. Shorter work cycles ensure that firefighters do not get too overheated. Remember that daytime temperatures in the Phoenix area can be well over 100°F for months at a time. Interior crews working at structure fires in these severe temperatures will be heating up and losing fluids quickly. Fireground work cycles also guarantee that firefighters are out of the hazard area before their low-air alarms activate. Having firefighters out of the hazard area before their low-air alarms activate removes the low-air alarm from the fireground and gives crews an emergency air reserve in case they run into trouble. This is exactly what this book advocates. It really does not matter how fire firefighters do it, as long as they are out of the hazard area before their low-air alarm activates. Firefighters are safer if they leave themselves an emergency air reserve. By instituting 10-minute work cycles, the Phoenix Fire Department has again shown its commitment to the health and safety of its firefighters.

Another advantage of the Phoenix system is that the on-deck firefighters are actively monitoring the actions of the interior crew they are slated to replace. The on-deck firefighters know where the crew is inside the structure, what their assignment is, and whether the interior crew will be able to complete its assigned task(s). When the interior crew is ready to come out of the hazard area, the on-deck crew already knows what they have accomplished and what still needs to be done. This on-deck crew can also serve as the initial RIT should the interior crew get into trouble inside the structure. Since the on-deck crew monitors the interior crew's progress inside, they can be launched by the division/sector officer to begin the rapid intervention process if necessary. This on-deck crew would then be supported by a full rapid intervention group as soon as possible.

The Phoenix system works. However, their system cannot be used by many other fire departments or fire districts because it is resource dependent. Whereas Phoenix is a big-city fire department, most smaller fire departments and fire districts do not have the same level of staffing or resources to place a crew on-deck for each interior crew working inside a structure fire. In these days of minimum staffing levels, many fire departments and districts are lucky to have two or three firefighters on each rig.

Another example of how Phoenix's system is resource dependent relates to the 10-minute work cycle. Phoenix sector chiefs have a FIT working alongside them who helps monitor the interior crews' work cycles. But how many other fire departments or fire districts have the staffing for a dedicated firefighter to monitor the work-cycle times of interior crews? While this system works very well for a big-city fire department like Phoenix, its main drawback is its resource dependence.

WORK CYCLES

In 1996, I was assigned to the Phoenix Fire Department Regional Training Academy as a Recruit Training Officer. At that time, our air management training for recruits consisted of two drills. First, we took them out to the grinder to see how fast they could breathe down their SCBA air bottle. Then, the recruits were positioned in a park and requested to conserve their air, to see how long a SCBA bottle would last. At the end of these drills, we stressed that when the low-air alarm rings, it's time to leave the building. By contrast, in the field, when the bell rang, we all knew that meant that you had approximately five more minutes of air before running out completely.

For years, this was the extent of air management training provided by the department. For the most part, we responded, like most fire departments, primarily to residential occupancies. Fortunately, the exits were very close and the buildings were very forgiving.

Since that time, significant advances have been made in preparing and following proper air management protocols for all environments. Today's fireground requires that firefighters, company officers, sector officers, and the IC take responsibility for proper air management. The IC's efforts must be built around providing enough support for the operation to have a safe and successful outcome.

The concept of *fireground work* cycles was introduced within the Phoenix Fire Department about five years ago and since has been continuously refined. At the task level of operations, firefighters need to manage their air in a basic work cycle: air to enter, air to exit, a safety margin, and the remainder of the supply for their interior operations. The members operating on the interior of structures must be aware of the different types of structures within which they are operating and how those structures affect air management.

For the sector officer and the IC, a ready tactical reserve of firefighters is necessary to maintain offensive interior fire attacks. As companies are deployed, resources are moved forward to fill the *on-deck* position. These on-deck firefighters become the tactical reserve for the companies operating on the inside. As the interior companies approach the end of their work cycle, the on-deck firefighters rotate in and begin their own work cycle. For an operation to remain offensive, command must provide a continuous supply of firefighters to rotate into interior positions.

—*Todd Harms, Assistant Chief of Operations, Phoenix Fire Department*

THE ENTRY CONTROL OFFICER— THE UK SYSTEM

The fire service in the United Kingdom is not a national fire service. There are 59 separate fire brigades in England, Wales, and Northern Ireland, each run independently under the command of a chief fire officer. In Scotland, there are 8 fire brigades, each under the command of a firemaster.[1]

Every fire brigade in the United Kingdom is mandated to abide by UK firefighting standards put forth by the Fire and Rescue Service Directorate, which is overseen by the Office of the Deputy Prime Minister. This means that fire brigades across the United Kingdom follow national firefighting standards, and one of these standards has to do with air management. Fire brigades throughout the United Kingdom work under a single national air management system. Every firefighter in every fire brigade follows this system, which has been in place since the late 1950s.[2] Since Australia and Hong Kong both belonged to the United Kingdom for many years, their fire services use variations of the UK air management system (fig. 21–1).

In the UK system, firefighters entering a structure fire are mandated to first hand off their breathing-apparatus tally—another name for SCBA passport—to the *entry control officer*. Entry control officers are staged at each entry point of a fire building. Their job is to keep track, on a status board, of the air in the SCBA of each firefighter who enters the fire building through a particular entry point.[3] The entry control officers calculate the *time of whistle,* or when the low-air alarm will start activating, for each firefighter wearing SCBA. They keep track of times on a clock and advise the team leader to withdraw from the hazard area at a predetermined air pressure reading or time. They also keep track of the location of teams inside the structure fire. Using this system, the British have dedicated one individual at each entry point to maintain firefighter accountability and monitor each firefighter's air supply. The entry control officer's goal is to have all of their firefighters out of the structure before any of their low-air alarms activate.

Fig. 21–1. Australian firefighters preparing to enter a fire building. Note the breathing-apparatus control board located beside the fire apparatus. The officer is a breathing-apparatus sector commander, or entry control officer. (Photograph by Phil Paff, Station 45, Queensland Fire and Rescue)

The entry control officer will send in an emergency team (RIT) and immediately inform the IC if any of the following occurs:

- Any firefighter fails to return to the entry control point by the indicated time of whistle
- A distress signal is given
- A dangerous situation is developing

At larger incidents, the entry control officer is assigned an aid to help with the extra workload and responsibility. This is necessary when the scale of operations is likely to demand greater control and supervision, more than two entry control points are necessary, more than 10 SCBA are committed to the hazard area at one time, or when branch guidelines are used.

The UK fire and rescue services incorporated air management into an incident command function decades ago. For 50 years, they have dedicated the necessary staffing and resources to make the entry control command function work. Moreover, UK firefighters are safer working inside structure fires because of this commitment. By integrating air management into the ICS and by providing training for their command staff and firefighters, the UK fire service has made good on its pledge to firefighter health and safety on the fireground.

THE BA LEADER—THE SWEDISH SYSTEM

Like the United Kingdom, Sweden has been practicing air management for decades. Before anyone can fight fire in Sweden, they must first attend and complete a rigorous firefighting school that is run by the Swedish Rescue Services Agency (SRSA). The SRSA has four colleges in Sweden, which train the firefighters and fire service officers for all Swedish municipal fire and rescue services. These colleges are located in Rosersberg, Revinge, Skövde, and Sando.

In these colleges, firefighters, who are referred to as *smoke divers,* learn the skills of managing their air supplies. Swedish firefighters follow strict BA control procedures. All fire brigades in Sweden must be able to support BA teams by the formation of a smoke-diving team, consisting of two firefighters, and a BA leader, who carries out BA control at the point of entry. The BA leader is equipped with an SCBA, a charged hoseline, and a radio that allows them to maintain contact with the smoke-diving team (figs. 21–2 through 21–5).[4]

Fig. 21–2. Swedish smoke divers battling a house fire. Note the two-person smoke-diving team, the officer, and the BA leader. (Photograph by Stefan Svensson and Peter Lundgren)

*Fig. 21–3. Swedish firefighters vertically
ventilating a steep-pitched roof (Photograph by
Stefan Svensson and Peter Lundgren)*

Fig. 21–4. Swedish firefighters working at a house fire (Photograph by Stefan Svensson and Peter Lundgren)

Fig. 21–5. Swedish firefighters at a training fire
(Photograph by Stefan Svensson and Peter Lundgren)

The primary purpose of the BA leader is to ensure the safety of the smoke-diving team, which includes actively monitoring the air supply of the crews inside the structure. All Swedish firefighters are expected to be out of the IDLH atmosphere before their low air alarms activate.

THE AUSTRIAN SYSTEM

Austria, Germany, and Switzerland have air management systems that are very similar to the UK and Swedish systems. At every fire, Austrian firefighters must give their officers their SCBA passports, which the officer then places inside a computerized status board located at the fire apparatus (figs. 21–6 and 21–7).

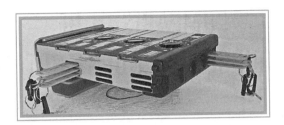

*Fig. 21–6. The Austrian SCBA status board
(Photograph by Hannes Kern, Feuerwehr Vorau)*

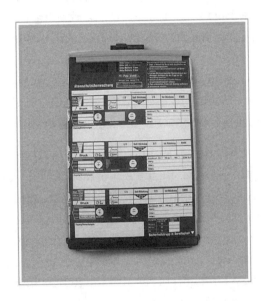

*Fig. 21–7. On-screen view of the Austrian SCBA status
board (Photograph by Hannes Kern, Feuerwehr Vorau)*

The officer's job is to monitor the status board, which tracks each firefighter's time on air and air pressure. Like the Swedish BA leader and the UK entry control officer, the Austrian fire officers never enter the structure. They stay outside to observe fire conditions and keep track of their firefighters' safety. The firefighters inside the fire watch their air supplies, as do their fire officers on the outside.

COMPARISON OF RESOURCE-DEPENDENT OPTIONS

The UK and Swedish air management systems, like the Phoenix system, are resource dependent—that is, it takes dedicated personnel (e.g., the entry control officer or the BA leader) to monitor the interior crews' air supplies. In addition, at larger incidents, each entry control officer in the UK system requires an aid. The UK and Swedish systems work well for their respective fire brigades. Interestingly, per capita, the United Kingdom and Sweden have far fewer firefighter deaths due to running out of air in structure fires than does the United States. However, like the Phoenix system, the UK and Swedish systems will not work unless fire departments and fire districts are willing to dedicate the staffing and resources to the entry control officer position or the BA leader position.

Contrast the Phoenix, UK, and Swedish systems with the costs of implementing the ROAM. Besides the associated training costs, the ROAM does not cost fire departments anything to put into practice. The ROAM is based on personal and crew responsibility—the responsibility of each firefighter and crew in the hazard area to manage their air supply. No commitment to increased staffing on the fireground and no new IC function is necessary. Firefighters can use the ROAM immediately, today—and they will be safer if they do so.

The ROAM is simple, costs nothing to use, and has been proven to be effective by the dive industry for many decades. Most importantly, it will keep firefighters safer on the fireground.

TECHNOLOGY-DRIVEN OPTIONS

Most major SCBA manufacturers have recently developed technology-driven systems intended to help firefighters manage their air in structure fires. Most of this technology relies heavily on telemetry devices, located on the SCBA, that send individual air readings to a central computer that an incident command functionary (the IC, a division supervisor, or a sector officer) must monitor. Some of these systems use Global Positioning System (GPS) technology and can pinpoint the exact location of firefighters inside structures (figs. 21–8 and 21–9).

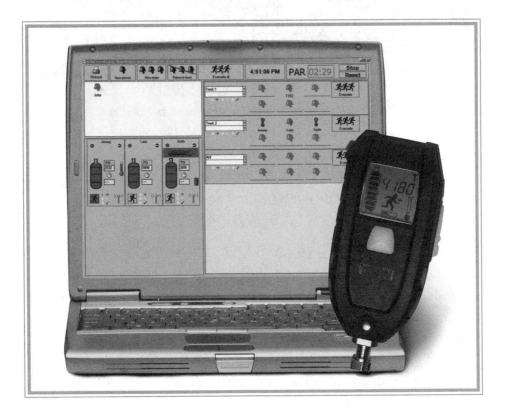

Fig. 21–8. A computer that accounts for each firefighter on the fireground by use of a telemetry unit in the integrated PASS device of the SCBA. This system displays each firefighter's real-time air supply. (Photograph courtesy Mine Safety Appliances)

Fig. 21–9. A technology-driven accountability system at work on the fireground. This system is usually located at the command post and monitored by the IC or another IC functionary. (Photograph courtesy Mine Safety Appliances)

These systems have the potential to revolutionize firefighter accountability and air management. Imagine if the IC or the division supervisor could monitor the exact locations of the firefighters in the structure and keep track of these firefighters' air supplies. If firefighters were to get separated from their crew in heavy-smoke conditions, the IC would immediately know where this firefighter was in the fire building and would be able to monitor how much air remained in their SCBA cylinders. This information would be vital in locating and rescuing the lost firefighters. In fact, the IC might even be able to talk to the lost firefighters over the radio and lead them out to safety.

The key to the success of technology-driven systems is for someone to actually be at the command post actively monitoring the computer screen. The success of these systems is again predicated on having the staffing and resources to support them. A fire officer or firefighter who is a part of the ICS has to be watching the computer screen for this technology-driven system to work.

THE PROBLEM WITH TECHNOLOGY

Manufacturers of these systems have argued that the IC can watch the computer screen and still supervise and run the fire. However, we disagree with this assumption. It may work differently in your department, but in the Seattle Fire Department, the IC is busy enough running the fire and is overworked already. The IC does not have the time to scrutinize the air supply of every individual firefighter inside the fire building. For this technology to work in the real world of structural firefighting, fire departments and fire districts must dedicate the staffing and resources to make it work—and this takes money, time, and training.

Since these accountability/location systems are based on technology, there must be a fallback position, or plan B, when this technology fails (see chap. 8). Technology inevitably does fail, often at the worst possible time. Firefighters need to understand the limitations of technology and must be trained to adapt to and overcome technological failures, particularly when these firefighters are working inside a structure fire. This is why the ROAM is so important. The ROAM does not rely on telemetry or someone outside the structure who is actively watching a computer screen. Instead, the ROAM puts the responsibility of air management on each firefighter inside the fire building.

At its core, air management should be an individual responsibility and a crew responsibility. Although it can and should be incorporated into the incident command system at both the tactical and strategic levels, the

ROAM is not solely dependent on the ICS. This is as it should be. Who would you rather have watching your air supply—you or someone on the outside who may or may not be paying attention?

OTHER IDEAS ON AIR MANAGEMENT

Several other ideas exist about how to solve the air management problem. One such system is to use time-on-scene radio transmissions from dispatch as a cue for firefighters to check how much air remains in their cylinders. Here is how the system works: Every five minutes, dispatch comes over the radio channel and announces the time on scene. This radio report prompts companies inside the fire building, in the hazard area, to check their air levels. Since time-on-scene radio reports come every five minutes, ICs know that everyone will be checking their air regularly, every five minutes.

Like any air management system, training is critical for the time-on-scene system. Time-on-scene radio prompts work as long as the firefighters have been trained to faithfully check their air when dispatch announces the time-on-scene over the fire channel. Firefighters must also be trained to be out of the hazard area before their low-air alarms activate; that is, they need to also follow the ROAM. Firefighters must leave themselves an emergency reserve of air in case something should happen to them on the way out of the hazard area.

One downside to the time-on-scene system is that many smaller fire departments and fire districts do not have dedicated fire dispatch centers. Instead, these fire departments must share dispatch with the police, the sheriff, and the state patrol. Often, these dispatchers are civilians working for the police department or the sheriff, and they have limited knowledge of what firefighters do and what problems firefighters face at a fire scene. The dispatchers in these smaller communities must split their time between police calls and fire calls, and they may not have the ability to give time-on-scene radio reports every five minutes. Again, time-on-scene reports will work, provided that the dispatchers have the time and resources to constantly monitor both the fire traffic and the clock.

The time-on-scene system, coupled with the ROAM, can be an effective and practical air management system. Firefighters check their air every five minutes and know that they must be out of the hazard area before their low-air alarms activate. In addition, everyone knows how much air they have while they are in the IDLH environment. Finally, there are no low-air alarms activating in the fire building unless there is a true low-air emergency.

The "half-your-air-and-out" system

Another idea proposed is the "half-your-air-and-out" system. In this air management system, firefighters work until their SCBA cylinders are half empty and then immediately begin exiting the fire building. Some fire departments and fire districts (e.g., Yakima [WA] Fire Department) have put this air management system into practice with success. These departments and districts rely on the SCBA heads-up display (HUD), located in the facepiece, to signal their firefighters when their cylinders are half empty. Once the HUD light indicates a crew member inside the fire has used half of their air supply, the officer or team leader must radio to command (or an IC functionary, e.g., a division supervisor) that one of their firefighter's cylinders is half empty and the crew is turning around and beginning to exit the structure. Like all air management systems, the crew or team is always limited by the firefighter with the least air in the SCBA cylinder.

Remember that the last half cylinder of air is made up of a quarter cylinder of working air plus a quarter cylinder of emergency air, which is reserved for use only in the event of an emergency. Thus, firefighters using the half-your-air-and-out system only have one quarter cylinder of air to make it out of the hazard area before the low-air alarm activates.

What if firefighters using the half-your-air-and-out system cannot make it out of the hazard area with a quarter cylinder of air? Although this might seem to present a problem, the reality is that most firefighters will be able to exit most structures—except, perhaps, for extremely large commercial structures—using less than a quarter cylinder of air. And if they do not, they have another quarter cylinder of air in the emergency reserve. For fire departments and fire districts that have made the switch to 1800L (45-minute) cylinders, it should not pose a problem to have firefighters exit a structure before breathing into the emergency air reserve.

ROAM VERSUS STAFFING OR TECHNOLOGY

The UK and Swedish fire and rescue services have been practicing air management for decades, and their firefighters are safer as a result. Any fire brigade, fire department, fire district, or fire service that dedicates the resources and staffing to make air management work at the strategic level is truly committed to firefighter health and safety and should be viewed as a model.

In the United States, fire departments and fire districts large and small are beginning to understand how important air management is, and these departments and districts are starting to adopt air management systems that work for them. Fire departments in Seattle (WA), Phoenix (AZ), Bellevue (WA), Portland (OR), Tuscaloosa (AL), Spokane (WA), and many other locations now have air management policies and expect their firefighters and officers to manage their air supplies.

The ROAM is a simple, effective air management system that doesn't cost fire departments or fire districts anything to implement and use. Unlike air management systems that require dedicated staffing and resources in order to work, the ROAM is based on individual firefighter behaviors at the task level. By use of the ROAM, air management becomes an individual and crew responsibility. Fire departments and fire districts that adopt and utilize the ROAM will discover that their firefighters operate more safely on the fireground.

MAKING IT HAPPEN

1. On the Internet, look up departments that practice air management and have air management policies and operating guidelines—Seattle, Tuscaloosa (AL), Spokane (WA), Portland (OR), and Bellevue (WA), among others. Read their air management policies and find out which system might work for your department or district.

2. Have crews in your department or district experiment with several different air management systems. First, black out the apparatus bay or the living quarters of your fire station with black Visqueen (a lightweight, flexible polyethylene film) and duct tape, or black out a few rooms or several floors of a training facility or an acquired structure. Then, have your crews go on air and search the space. Have them practice air management using several of the different systems discussed in this chapter. After these drills, get everyone together and discuss which air management system worked best for them and why.

3. Drill your crew on how to use the ROAM. (For free, time-tested drills, go to Manageyourair.com and click on the "SMART Drills" tab.) After your crew is comfortable with the ROAM, have them use it at every drill in which they are wearing and breathing air from their SCBA.

4. Find out if any of the fire departments or fire districts that have mutual-aid agreements with your fire department are using an air management system. If so, set up a joint training drill to have your firefighters learn about the system that your mutual-aid department or district uses. If your mutual-aid departments or districts are not practicing air management, set up an air management drill for them.

STUDY QUESTIONS

1. How do fire brigades in the United Kingdom practice air management?

2. Explain how work cycles are used as an air management system.

3. How does the time-on-scene air management system work?

4. How does the half-your-air-and-out air management system work?

5. What is the key to success of the technology-driven air management systems that SCBA manufacturers have developed over the past several years?

6. What role does the entry control officer play in the UK air management system?

7. Which air management systems are resource dependent?

NOTES

[1] Fire Service Recruitment. http://www.fireservice.co.uk/recruitment.

[2] The impact on breathing apparatus procedures over the years. FireNet International. http://www.fire.org.uk/FireNet/ba.php.

[3] Breathing-apparatus course notes. Hampshire (England) Fire and Rescue Service, pp. 64–67.

[4] Raffel, Shan. "Compartment Fire Behaviour Training in Queensland, Australia, 1999–2000," pp.3–4. Available online at http://www.firetactics.com/RAFFEL-QFRA.htm.

THE MYTHS OF
AIR MANAGEMENT

INTRODUCTION

When trying to create and implement a new idea in the fire service (or any organization with such a long and storied history), questions arise as to how and why the new idea is necessary. Questions that start with the word "*why*" are the most important type. A "why" question demonstrates that the person asking is interested in the idea, is considering whether it will be accepted, wants to know the reasons behind it , and is ultimately looking to make an educated decision about the new idea. Air management as an idea and as a practice is not new. Rather, through the ROAM, it represents a change in the behavior of firefighters on the fireground.

Since firefighters began using the SCBA, the fire service has been moving toward air management. This text has given many forms and descriptions of the how and why of air management. Readers will learn more as they work within their own organizations to bring about change and increase the safety of their coworkers. This chapter will identify and dispel the misconceptions that have consistently challenged the idea of air management. These are the the myths of air management.

Organizations large and small resist change. Newton's first law of motion states that an object at rest will remain at rest unless acted on by an outside force. (fig. 22–1). The fire service is legendary for the ability to resist change. Change—and resistance to change—is covered in chapter 9, which provides extensive information regarding the reasons why people resist change and the techniques to overcome resistance. Further information can be found by consulting with the change makers within your fire service organization or other groups that count you as a member.

Fig. 22–1. An object waiting to be acted on by an
outside force. Is your department like this taxi?

Fortunately, regulatory authorities and the court system have decreased the ability of fire service organizations to resist change that is mandated by national standards. The modern fire service no longer accepts firefighter fatalities as part of the cost of doing business. While firefighters have been and will continue to be injured or killed in the act of protecting citizens from fire, everyone involved in the fire service should do whatever is within their ability to prevent such incidents. Anything less is unacceptable.

While resistance can be an important and powerful force in ensuring that changes are well designed and implemented, resistance is frequently a reaction to the unknown. Too often, resistance is merely a psychological parallel of Newton's first law of motion. The organization is at rest and wants to remain at rest. Our goal should be to ensure that the outside force that acts on our organization is met with a combination of willingness to learn and a desire to implement new and sound principles within our organizations. This is a much better approach than waiting for some tragic accident followed by statutory regulation. Let's work together to implement positive changes like the ROAM.

MYTH 1: WE DON'T NEED NO STINKING STANDARDS

The fire service has standards of conduct and behavior. Familiar examples include the principle of *abandonment* when providing emergency medical services and the requirement that fit testing be completed prior to SCBA use. (Abandonment is the unilateral termination of care by the emergency medical technician without the patient's consent and without making any provisions for continuing care by a medical professional whose skills are at the same or higher level.[1]) Regulatory standards provide a framework for the health and safety of firefighters. Within the industry, the standards set by the NFPA provide consensus standards that can be readily accessed, adopted, and followed by local jurisdictions. Throughout most of the United States, the NFPA standards represent a baseline of training and performance that is deemed acceptable for firefighters. Basic firefighter training and certification, called Firefighter 1, is taken directly from the NFPA standard; Firefighter 2, Fire Officer 1 and 2, and Driver Operator 1 are further examples of standards for training and performance that are set by NFPA and used throughout the country.

Any Firefighter 1 training program will require the authority having jurisdiction (AHJ) to teach new firefighters how to use the SCBA. Nearly everyone in the fire service received this training in recruit academy. Perhaps nothing is more universally true than that we have all been trained to don, doff, maintain, and properly use the SCBA. Firefighters are trained in the limits of the SCBA and in emergency procedures to employ when the SCBA malfunctions. (Chap. 6 outlines the requirements of the training standard for SCBA.)

The myth in this case is that we don't need to follow or accept the role of the NFPA and other standard-setting organizations in our training or operations. Note that most organizations follow some NFPA standards while clearly violating others. The NFPA standards are consensus standards, and while they are not mandatory (unless your state has formally adopted them), does that really exempt you from their impact?

Whether you or your organization "believe" in NFPA standards will make little or no difference when something goes wrong. You may consider NFPA standards voluntary; however, what the judge and jury decide is what will matter in the end. Most, if not all, states recognize portions of the NFPA standards. Fire departments have a moral and ethical responsibility to consider abiding by NFPA standards as they relate to firefighter safety. There are examples offering numerous lessons: Fire departments have paid out multimillion-dollar awards for training and operational accidents that occurred because the activities that caused them did not abide by the NFPA standard. Training officers have been found personally liable for their actions when they have operated their programs outside accepted NFPA standards.

NFPA 1404 dictates how firefighters use, care for, and train with their SCBA. Chapter 6 provides a thorough review of NFPA 1404. A quick glance through NFPA 1404 will most likely confirm that your department meets most of the requirements related to mandatory training policies. Air management is an integral—and required—component of any SCBA training program.

Training standards such as NFPA 1404 also provide insight into the how and the why of the training exercise. NFPA 1404 gives an interesting example in Appendix 5.1.7:[2]

> *One role of any training program is to generate acceptance of operational evolutions for coordination and skill. The use of proper procedures and the dispelling of false notions concerning the use and application of respiratory protection equipment are equally important. The state of the art in today's fire-fighting environment demands a commitment by each authority having jurisdiction to ensure maximum acceptance in the use of respiratory protection equipment.*

Changing the behavior of the firefighter to facilitate using and following the ROAM will require that organizations embrace the consensus standards of the NFPA and the ideas they represent. The role of the training officer, and the role of the fire service organization that trains firefighters and expects them to respond and operate at emergency incidents, is to require minimum standards of performance. We should not be settling for or accepting minimum performance in any endeavor undertaken by firefighters, but providing for a minimum is absolutely necessary. It is also important to recognize that official accountability to the minimum level of performance will be required following an injury or fatality on the fireground or during training. Beyond the legalities and regulatory language, there is an innate responsibility to each other, to our crew, and to our families to make sure that we are doing everything that we can to operate safely at every emergency.

MYTH 2: IT TAKES TOO LONG
TO CHECK MY AIR

The question of time on the fireground provides fertile soil for discussion. Every aspect of the emergency response takes time. Each step of the operation—from initial dispatch to putting on protective clothing, response, setup, entry, and fire attack—takes time. Increasingly, there has been a focus on the effect that time pressure has on firefighter and company officer reactions during fireground operations. Decision-making models explore how time pressure and stress impact the "how" and "why" of fireground commanders' decisions. Firefighters can easily become overly concerned with the effect of time during the response and the fireground operation. Let's consider some aspects of the fireground emergency response and how time can be an ally, so that we can relate the same process to air management.

MAKING TIME WORK FOR YOU

There are five components to the response time:[3]

- Alarm time
- Dispatch time
- Turnout time
- Travel time
- Setup time

Each of these has an impact on the overall goal of getting water on the fire quickly. Nevertheless, while each of these has an effect on the goal, it is not necessarily true that the company officer or the firefighter can have an impact on them. While you may not be able to affect every aspect of the response time, as an individual, you can and do control certain aspects.

An example of where the individual firefighter and the company officer can have an impact is the turnout time. Turnout time is the time "from when the fire units are notified until apparatus leaves the station."[4] Do the firefighters on your crew start moving immediately to the apparatus when the bell hits, or do they wait to see if the run is for you? Are your firefighters trained to don their protective clothing properly and quickly? Does the company officer have a quick and consistent method of recording the response information? Does the driver know the first-in district, or do they have to look at the map before responding? Is everyone on the crew trained and practiced to the point where the turnout time is as short as possible?

In addition, the turnout time should be used by the crew to begin the size-up and to recognize and counteract the effects of stress. The crew should be able to turn out in a quick, calm, and professional manner.

Travel time is another consideration altogether. Travel time is the time "from when fire units leave the station until they arrive at the scene."[5] While you may be able to trim a few seconds off the travel time, it is imperative that these seconds do not come at the expense of firefighter safety. Fatalities during response are the second leading killer of firefighters.[6] The responsible driver and company officer will recognize that trimming seconds during the response is a consideration only when choosing the most direct response route. Response route selection is a function of pre-incident driver training. The most important consideration with regard to travel time is getting to the location safely. Firefighters cannot have any positive impact on the emergency until they arrive at the location.

Inappropriate response techniques also increase stress and decrease decision making ability, by causing the driver and officer to be too involved during the response. Operate the vehicle in a cool, calm, and professional manner while responding. One sure way to end up beyond the point of no return—becoming part of the problem instead of part of the solution—is to crash the apparatus while responding (fig. 22–2). Wear your seatbelt, stop and control intersections, and drive safely so you arrive alive.

Fig. 22–2. Apparatus that has been in an accident. Accidents have a negative impact on response time and the organization. Getting to the incident in a safe and controlled manner improves performance.

Setup time is the next consideration. Setup time is the time from when fire units arrive at the scene until fire units take effective action (fig. 22–3). This is where calm leadership, good decision-making ability, and regular training unite to make a difference in the amount of time necessary to take effective action.

Fig. 22–3. A high-stress environment, in which setup time is especially important to efficient response and a successful outcome. The amount of time for the company to take effective action after arrival is dependent on training and leadership. (Photograph courtesy Steve Victory)

Time efficiency is enhanced by

- Properly positioning the apparatus the first time
- Making the correct tactical decision the first time
- Deploying the right line, to the right place, the first time
- Having and communicating an effective plan that includes access, ventilation, hoseline placement, firefighter safety (RIT), and search assignments, the first time
- Making a good risk-benefit analysis the first time

Each of these significant time elements can be improved through training and experience.

Once entry is made into the IDLH environment, time should take a back seat to operational effectiveness. Being in a hurry is one step toward the point of no return. Remember, the incident scene is not an emergency for the fire department; the emergency scene is what we do. It does not matter whether you are paid or a volunteer, having a professional attitude toward the work is what matters here. Professionals work in a calm and coordinated manner to get the job done safely, effectively, and efficiently. Well-trained firefighters, operating with good leadership and a coordinated plan, will have a positive impact on any emergency scene in short order. The same factors will significantly increase the time required in order to achieve that impact when a fire department is poorly trained, is poorly led, and takes a haphazard approach to the fireground.

Air management is a required component of SCBA training. How long it takes to check your air supply is a function of training (fig. 22–4). Properly trained firefighters can check their air and communicate the information to the company officer in two to three seconds. Considering all the time elements that go into an efficient operation, taking two to three seconds to verify that every member of the team has enough air to operate safely seems reasonable; taking two to three seconds to improve situational awareness seems reasonable; taking two to three seconds to ensure accountability seems reasonable; and taking two to three seconds in order to go home alive seems reasonable. Failing to perform a skill that enhances your safety because you have not practiced enough to achieve proficiency is unacceptable. This applies to many of the tasks that we do on the fireground—and it certainly applies to air management.

Fig. 22–4. A firefighter taking a quick look at the air gauge during a training exercise. This gives the firefighter the information needed to understand how much time and air he has left to operate in the IDLH environment.

Officers must take the time to train their firefighters, the modern-day smoke divers, to manage their air just as scuba divers do. Dive instructors have described how beginning dive students receive instruction from day one to check their air supply on a regular basis. Like smoke divers, scuba divers have to consider a variety of factors related to air supply. They must consider time underwater, depth, time at depth, and partner's air supply, and they must leave an acceptable margin for safety. The margin for safety is dictated by the dive master and is usually 500 psi.

The fire service should also set a standard margin for safety. Firefighters should follow the ROAM and exit the hazard area before their low-air alarms sound. We will then be operating at the fire the way our confined-space rescuers and hazmat technicians in Level A protective suits already do.

MYTH 3: WE ARE TOO BUSY TO CHECK OUR AIR

Whenever a firefighter claims to be "too busy" to operate safely, the correct response is to be nervous about making them part of your team. The professional approach to any situation is to begin with a mind-set geared toward operating safely, operating effectively, and completing the task. No matter whether you are a volunteer, part-time, or full-time firefighter, you must take the time necessary to perform the basic functions of the job that allow you to operate safely on the fireground (fig. 22–5). Would you accept firefighters who claim "We don't have time to ventilate" or "We don't have time to take the hydrant"? If the answer to this question is *no*, as it should be, then why would you accept that they are too busy to perform other basic tasks that improve safety on the fireground? Air management is one of these basic tasks that must be done every time you are in harm's way.

Fig. 22–5. Well-trained firefighters performing vertical ventilation during a house fire. Having the knowledge, skills, and ability to perform basic firefighter tasks during a fire is a function of training. (Photograph courtesy Dennis Leger, Springfield [MA] Fire Department)

As with the preceding myth ("It takes too long to check my air"), dispelling this myth is a function of training. Thousands of firefighters have learned in training exercises to mange their air supplies successfully. These same firefighters, in departments across the country, have taken this training and made it a part of their emergency operations. When you enter the hazard area, you and your team will operate consistent with the training that you have received. Make air management part of this training, and it will be part of your operation.

You can also learn a lot by watching the training exercises of others. Take the opportunity to observe a training session in which firefighters are wearing their SCBA and practicing basic firefighting techniques for use in a hazardous atmosphere. Keep a close eye on one or two members of the team. Note times when they are not actively working for a two- to three-second block of time. It will be more often than you think. These are perfect opportunities for the members of the team to check their air without having any adverse impact on the team's effectiveness.

What is true for training is true for real operations. For example, consider a two-person primary search team operating in low visibility. The team leader is maintaining orientation and, while waiting at the door, is sending a crew member into each room to conduct the search. The team leader will maintain orientation with the door, search in the immediate vicinity, monitor radio traffic, and keep an eye on conditions in the search area. Team leaders can easily take the two to three seconds to check their own air while crew members search the rooms. As members return, each reports the status of the search to the team leader. If they have been trained, they can quickly tell the team leader how much air they have left. This information allows the team leader to make informed decisions about the primary search assignment: Can the team continue on with this search? Does the team have enough air to finish the search? Will they need help to finish?

The company officer should continually evaluate the progress of the team in meeting their objectives. While the individual firefighter focuses on completing the task assigned, the company officer must also function at the tactical level. The company officer should be asking, on a continuous basis, questions such as:

- What are the heat and smoke conditions?
- Are they changing?
- How much of our assignment have we completed?
- Can we finish with the air that we have left?

- Are there other tasks that need doing?

- Where are we in the structure?

- Where is our exit?

- How much air will it take to get there?

- Is there an alternative exit?

- How much air will it take to get there?

- Who is lowest on air?

- How hard are the firefighters working?

When the company officer notices a change to the environment, significant changes in the crew's location (e.g., changing floors), or completion of an assigned task necessitating that they must contact the IC, the crew should take the time to check their air and communicate with the company officer. In addition, this prompts the company officer to check on the crew with a brief "How you doin'?" before moving on to the next objective or work area. In many organizations, this is referred to as getting a personnel accountability report (PAR) or a roll call.

MYTH 4: I'LL DO IT WHEN THE SITUATION CALLS FOR IT

Perhaps no one is so misguided as firefighters who believe they will operate in a haphazard or lazy fashion during training exercises, and then expect to operate as trained, competent firefighters at the "Big One." The Big One is a fire big enough to wow even the veterans of your department. Fires like these are only seen a few times in a career. Stress decreases your ability to make decisions. The Big One is a stressful event precisely because it is the Big One. Believing that you will operate at the Big One any differently than your training and experience dictate is the most insidious myth of air management and is frequently echoed by students and others in the fire service who argue against air management. If you take nothing else from this book, remember the following advice: Your performance in an emergency will be consistent with the way you train and the way you perform on a daily basis (fig. 22–6).

Fig. 22–6. The Big One. Expecting firefighters to perform differently at the Big One than they do at a bread-and-butter fire is a recipe for disaster. Make air management part of your everyday operation and it will serve you well at the Big One. (Photograph courtesy Dennis Leger, Springfield [MA] Fire Department)

The notion that you can disregard air management at house fires, dumpster fires, car fires, and other bread-and-butter fires and then perform air management at the Big One is ludicrous. There are plenty of adages that refute this myth; "Practice like you play" or "Train as if your life depended on it—because it does" both refute the notion that you can employ this or any technique on a selective basis. Either you check your air when you use your SCBA or you do not. Either you make an informed decision, based on knowledge and skills, of when to leave the hazard area or you leave when your bell rings.

"Practice like you play" should apply to every action that you take in your fire service career. When the situation turns bad, will you be calm, prepared, rational, and leave a margin for error, or will your air run down to empty as the point of no return looms? There is no middle ground. Firefighters must practice air management at every fire where they are breathing air from their SCBA. Firefighters must practice air management at every opportunity. If they do not, they might well become a statistic—namely, the next line-of-duty death. Practice incident command at every incident and training exercise; practice accountability at every incident and training exercise; practice decision making at every incident and training exercise; and practice risk assessment at every incident and training exercise. Review your ICS, your accountability, your decision making, and your risk assessment after each incident and training exercise. Make your operation better by learning the lessons of the past. Diligence in every aspect of the operation ensures that lessons are paid for only once and not repeated. The same holds true for air management as it does for any other fireground task: Practice, play, and learn—every day, at every incident, every time.

MYTH 5: NOBODY HAS MANDATED THE ROAM

The ROAM states, know how much air you have, and manage that air, so that you leave the hazard area before your low-air alarm activates. While it is true that no one has specifically mandated the ROAM, the question remains: Do you want be proactive with your crew's safety, or do you want to be reactive? In other words, how do you want to go about your business? Is it your contention that anything that is not mandated is unnecessary?

Chapter 6 demonstrates how NFPA 1404 mandates training in air management techniques. What are you waiting for—an engraved invitation or just another fireground fatality? Narrow escapes have been and continue to be made on a daily basis. If you doubt this, visit Gordon Graham's and Chief Billy Goldfeder's Web site Firefighterclosecalls.com, on which a whole section deals with low-air and out-of-air situations.

As you read this sentence, somewhere in the United States, a firefighter is running out of air in the hazard area. This firefighter will probably make it out OK and keep the near miss quiet. Or perhaps, get a ribbing from the crew. The only difference between a close call and having the bagpipes played at a funeral is the content of the smoke one is forced to breathe. Chapter 5 gives you the information you need to understand just what is in the smoke and what it can do to you. Unless your physical makeup is different from every other human on the planet, you cannot breathe smoke safely. You may survive an out-of-air situation, but you will still pay for the mistake. No one in the fire service can afford to make mistakes on a regular basis. The piper always comes to call eventually. Perhaps running out of air in the hazard area is one mistake you and your crew can avoid. Take the leadership role. Be the change maker. Be the expert. Be the leader.

MYTH 6: SOMEONE OUTSIDE THE HAZARD AREA CAN MANAGE AIR FOR FIREFIGHTERS

This is the easiest myth of air management to dispel. Picture the last time you approached the command post after exiting a fire building. Did it look like the IC or the aide would be able to monitor every firefighter's air supply from the command post?

ICs have a tough job outside the structure just as firefighters do inside the structure. ICs are good people who work hard and want to perform well. They want everyone to go home, and they want the fire to go out just like the rest of us. In addition to managing the ICS structure, managing the fire building, performing the risk-benefit analysis, monitoring the accountability board, talking to the dispatcher, and working with the Public Information Officer (PIO), the safety officer and division supervisors, should the IC also take the time and expend the mental energy necessary to monitor the air consumption levels of a dozen or more individual firefighters operating in multiple teams in the hazard area?

Even if your department is large enough to have the staffing available to provide a Field Incident Technician (FIT) for the IC, the FIT already has plenty of work to do. Having a dedicated air officer is a system employed by many fire departments in other countries (see chap. 21). There are excellent systems that can assist in implementing a strategic component of a comprehensive air management program. Telemetry systems are available to assist an air control officer to actively monitor personnel in the hazard area (fig. 22–7). However, unless you can afford to have a person specifically assigned to do nothing but monitor air supplies, you can dispel this myth of air management. For situational awareness to be maintained, air management must be an individual responsibility.

Fig. 22–7. Officers training on the Draeger Merlin air
management and accountability system. Strategic
air management systems are a great tool to assist
officers in the command structure and support
firefighter air management. Individuals in the
hazard area must still manage their own air and the
air of their team even when additional support is
available. (Photograph courtesy Draeger Safety)

MYTH 7: IF WE HAVE A LONGER BELL TIME, WE CAN SOLVE THE PROBLEM

Within the fire service, there is a move to change the cylinder bell time to 50% of the rated capacity. Changing the bell time is another myth of air management. This myth is based on the idea that having the low-air alarms on SCBAs set to ring at 50% of the rated cylinder capacity will be the magic bullet that prevents firefighters from running out of air in the hazard area. In reality, the time-to-exit decision is part of an active mental process, not a passive reactionary act.

Setting the bell to ring at 50% will not solve the problem of firefighters running out of air in the hazard area. While some departments may experience a large degree of success and reduce the number of low-air and out of air situations, in the long run, there is little doubt that the technology of the low-air alarm will eventually fail. When this happens to firefighters who are not actively managing the air brought with them into the IDLH environment, they will experience a low-air or out-of-air situation. Technology will not solve this problem. If we could rely on technology to solve this problem, the myths of air management would already be debunked.

Take a quick look at the history of the PASS device. Original versions of the PASS device were not integrated into the SCBA. The first generations relied on the self-discipline of the firefighter. Firefighters had to turn on or activate the PASS device prior to entering the hazard area. Fire departments added checking the status of the PASS device to the preentry buddy check.

The PASS device is not automatic, and at times, firefighters did not turn them on. Perhaps the firefighters considered themselves to be too busy; perhaps they did not have the time. Regardless of the reason, the

PASS device did not get turned on, and firefighters were found dead with unarmed PASS devices. NIOSH and the NFPA began to mandate integrated PASS devices.

The fire service, as a group, lacked the self-discipline to employ the PASS device properly. But, how does this relate to the low-air alarm? The answer to this question can be found by examining how your department operates at fires today. If you went to a building fire right now, would the firefighters in your department leave the hazard area immediately when their SCBA low-air alarm activates? Firefighters often rationalize, "I can work another minute before I leave." Is it reasonable to believe that this situation will improve if firefighters have 50% of their air left when the low-air alarm activates?

THE FALSE ALARM MENTALITY IS KILLING US

Is it possible to train firefighters to distinguish between the low-air alarm of a firefighter exiting the hazard area and the low-air alarm of a firefighter in trouble? The Seattle Fire Department has documented near-fatality situations that occurred when activated low-air alarms of firefighters in trouble were ignored by other firefighters in the hazard area. This happens because the low-air alarm is a regular occurrence and is therefore treated as a false alarm on most firegrounds in the American fire service.

Having low-air warning alarms sounding in the hazard area encourages firefighters to adopt a false-alarm mentality. When the low-air alarm is a regular part of every incident, firefighters accept the noise as a standard—not an emergency—part of the operation. The fire service sees the impact of a false-alarm mentality, by analogy, when responding to automatic fire alarms (AFAs) at apartment buildings. When the fire department arrives at an apartment with regular AFAs, where are the residents? Are they quickly exiting the fire building in response to the alarm? No, because they too have adopted the false-alarm mentality. They calmly roll over and pull the pillow over their head, curse another false alarm, and wait for the noise to go away. Removing all low-air alarms from the fireground will eliminate the false-alarm mentality.

When firefighters exit the hazard area before their low-air alarms activate, the alarm is removed from the fireground. Once the low-air alarm is removed from the fireground, it is no longer normal to hear activation of an alarm. Without regular activations of the low-air alarm, we can eradicate the false-alarm mentality and instead provide immediate intervention for firefighters who are experiencing a low-air situation.

Without the low-air alarm and the false-alarm mentality, the operational fireground will see additional benefits. In particular, the low-air warning alarm becomes a true emergency alarm. When fire departments accept the mandate of the NFPA 1404 by adopting the ROAM, low-air warning alarm activations during an incident will be regarded as a signal that a firefighter may be in trouble. Effective training in air management techniques, implementation of air management practices on the fireground, and appropriate SOGs will determine how each fire department will deal with low-air alarm activations. In the Seattle Fire Department, a firefighter whose low-air alarm activates must immediately notify the IC of the situation, including where and when an exit will be made. Low-air alarms not addressed by a radio communication are presumed to be firefighter emergencies, and teams in the immediate area are expected to provide initial rapid intervention and assistance.

The decision-making ability of a well-trained firefighter, supported by appropriate technology, should prevail over the use of technology alone. Implementing air management training is a zero-cost, zero-technology initiative. Let's dispel this myth of air management and eliminate the false-alarm mentality from the U.S. fire service.

MYTH 8: MY FIRE DEPARTMENT DOES NOT HAVE AN AIR MANAGEMENT PROBLEM

If you have ever had any of your firefighters run out of air in the hazard area, then you have an air management problem in your fire department. Many fire departments are rolling the dice with regard to safety every time they use SCBA in an IDLH atmosphere. Structure fires in modern buildings or buildings with modern contents produce carbon monoxide levels in excess of 30,000 parts per million. Just one or two breaths of this air can render you incapacitated, and even the best of paramedics will be unable to resuscitate you (table 22–1).[7] Modern fire smoke, derived from burning plastics, also contains hydrogen cyanide and other toxic chemicals. Table 22–1 outlines the effects of carbon monoxide gas on human beings. (The contents and effects of smoke are detailed in chapter 5.)

Table 22–1. Toxic effects of carbon monoxide
(Source: NIOSH FACE report 2004-05)

Carbon Monoxide (CO)(ppm)	Carbon Monoxide in air (percent)	Symptoms
100	0.01	no symptoms—no damage
200	0.02	mild headache; few other symptoms
400	0.04	headache after 1 to 2 hours
800	0.08	headache after 45 minutes; nausea, collapse, and unconsciousness after 2 hours
1,000	0.1	dangerous; unconscious after 1 hour
1,600	0.16	headache, dizziness, nausea after 20 minutes
3,200	0.32	headache, dizziness, nausea after 5 to 10 minutes
6,400	0.64	headache, dizziness, nausea after 1 to 2 minutes; unconsciousness after 10 to 15 minutes

Fire departments must critically evaluate air management and its role in day-to-day operation. Approach smoke exposure from a risk assessment profile. This will quickly demonstrate that breathing smoke is an unacceptable risk when it can so easily be avoided by using the SCBA and following the ROAM.

Each near-miss a fire department experiences increases the tragic probability of a fatality. Figure 22–8 illustrates a standard model used by risk management professionals.[8] The base and the second level of the pyramid address what property owners can do to prevent incidents. The property owner calls 911 at the third level of the pyramid. As you move up the pyramid, you are playing a numbers game; an incident on one level of the pyramid increases the likelihood of reaching the next level.

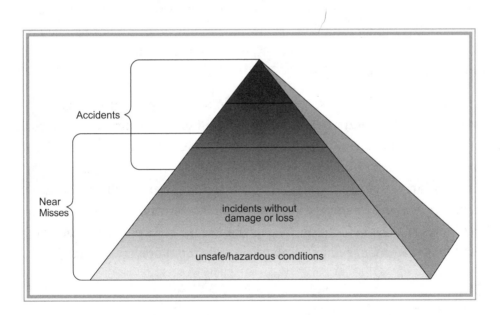

Fig. 22–8. Risk management pyramid

Each industry experiences a different relationship between the levels of the pyramid (fig. 22–8). For example, in a pyramid representing the petrochemical industry, 600 incidents without damage or loss result in 60 incidents with property damage, which results in 10 minor injuries, which result in 1 major injury. The width of the base of the pyramid and the steepness of the sides are influenced by the risk management decisions made by individual organizations prior to incidents occurring and by their response to incidents that do occur.

An exercise can be conducted as a company or battalion drill that will easily demonstrate how close a department comes to the point of no return on a regular basis. Assemble a couple of crews, and ask them to raise their hands if they have ever run out of air or been in a low-air situation while operating in an IDLH environment. Apply the result to the model shown in figure 22–8 to build your own risk management pyramid. Without addressing the low-air and out-of-air situations that are occurring on a regular basis, departments are building the pyramid—brick by brick, day by day.

Throughout this book, the need for an air management mandate has been advocated, and a solution has been offered that prevents low-air and out-of-air events in the fire service. Air management is a firefighter emergency prevention activity; air management is a firefighter Mayday prevention activity; air management is a firefighter fatality prevention activity; air management is a lawsuit prevention activity; air management is a wellness/fitness activity; air management is a cancer prevention activity; and finally, air management is an emergency room prevention activity.

Hope is not a plan

I sincerely hope that fire departments will adopt air management to improve the safety and health of the firefighters under their charge. This book aims, in some small way, to initiate a change in the way fire departments operate, to provide a safer and more effective fireground. Hopefully, these changes will take place.

But hope is not a plan. We hope it happens, but it is you who makes it happen. Use the information, the drills, and the ideas in this book to make a plan. Plan to change your fire department, and plan to have an impact. Make a plan for the future, for a safe return from the fire, and for a long, cancer-free retirement. Plan to survive the fireground by operating safely. When your plan is in place, implement, implement, implement. This book, stemming from our own experience in planning and implementing these ideas in the Seattle Fire Department, is intended for that very purpose. You can do it! You can have an impact! You can make it happen!

STUDY QUESTIONS

1. What type of question is the most important when implementing change in an organization?

2. What organization produces national consensus standards that affect fire departments?

3. If an organization does not follow a standard and experiences an injury or fatality, who will hold the organization accountable?

4. What is one role of any training program as defined in this chapter?

5. How long does it take well-trained firefighters to check their air levels and report to their company officer?

6. True or false: It is acceptable to violate safe fireground practices if you are too busy to follow them.

7. When operating in times of high stress, what do firefighters fall back on?

8. Should firefighters rely solely on the IC or division/sector supervisor to monitor the air level for the team in the hazard area?

9. Will an increase in the time that it takes to activate the low-air alarm solve the problem of air management?

10. Does my fire department have an air management problem?

NOTES

[1]   American Academy of Orthopedic Surgeons. 2006. *Emergency Care and Transportation of the Sick and Injured.* 9th ed. Jones and Bartlett Publishers, Sudbury, MA.

[2]   NFPA 1404—Standard for Fire Service Respiratory Protection Training.

[3]   Klaene, B., and R. Sanders. 2000. *Structural Firefighting.* Quincy, MA: National Fire Protection Association.

[4]   Ibid.

[5]   Ibid.

[6]   NIOSH FACE program.

[7]   NIOSH FACE report 2004-05.

[8]   Bird and Germaine, Practical Loss Control Leadership, International Loss Control Institute, Loganville, GA . 1996.

ANSWER KEY

CHAPTER 1

1. They wet their mustaches and stayed low.

2. In 1863, by A. Lacour.

3. After brief stints in service, the sensitivity decreased so much that the PASS device was hard to reset, causing it to alarm frequently.

4. Firefighters would not turn the PASS device on.

5. The National Fire Protection Association (NFPA).

6. Resistance.

7. NFPA 1404.

8. 1. Exit from an IDLH atmosphere should be completed before consumption of reserve air supply begins. 2. Low-air alarm is notification that the individual is consuming his or her reserve air supply. 3. Activation of the reserve-air alarm is an immediate-action item for the individual and the team.

9. Since the alarm is as a regular part of operations, it is not considered an action item.

10. Bret Tarver (Phoenix, AZ FD).

CHAPTER 2

1. Volume.

2. There is no margin for error in this case.

3. LUNAR.

4. Air, location, and RIT.

5. 150 liters, or about 90–120 seconds.

6. Up to 45 minutes with proper training in air conservation techniques.

7. Altered gait; increased stress on hamstrings, lower back, and abdomen; and fatigue of the muscles of the shoulders and neck.

8. Increased load on the muscles in the chest wall adversely affect respiratory muscle function. This may result in suboptimal firefighting performance.

9. By increasing demand on the muscles of the legs, back, shoulders, abdomen, and chest, the 2400L cylinder causes the firefighter to produce more body heat, thereby increasing the thermal load on the firefighter.

10. An 1800L cylinder in conjunction with the ROAM provides almost the same work/rest cycle as the 1200L cylinder while providing 450 liters of air as an emergency reserve.

CHAPTER 3

1. 25%.

2. Air for normal operations and the emergency reserve.

3. A sense of urgency.

4. False-alarm mentality.

5. It increases time for RIT to perform rescue.

6. Catch-up mode.

7. Emergency reserve.

8. It is your air.

CHAPTER 4

1. Approximately 105 firefighters per year, on average.

2. True.

3. True.

4. Greater than 20,000 parts per million.

5. The air that they carry on their backs.

6. The energy crisis.

7. True.

8. One of the primary reasons for lightweight construction collapse is the gusset plate connector, found holding together the webs and chords of trusses used for roofs and floors.

9. Air.

10. Lost and disoriented in heavy-smoke conditions.

11. Smoke, thermal insult, structural collapse, getting lost, and running out of air.

CHAPTER 5

1. A comprehensive air management program.

2. Wood products.

3. Carbon, nitrogen, polycyclic aromatic hydrocarbons (PAHs), formaldehyde, acid gases (hydrochloric acid, sulfuric acid, and nitric acid), phosgene, benzene, and dioxin.

4. It is colorless and odorless.

5. In Hitler's death camps and by the Japanese religious cult Aum Shinrikyo.

6. Gases emitted during the decomposition stage of a fire are more toxic than those emitted during actual burning.

7. It is poisonous; it displaces oxygen; and it causes gradual disorientation, confusion, and fainting.

8. Human hemoglobin has 200–240 times the affinity for CO as it does for oxygen.

9. Large particles and soot.

10. No.

11. Dereliction of duty.

CHAPTER 6

1. They are based on federal, state, and local laws, past practices, and recognized standards.

2. Because firefighters were being injured or dying in training events.

3. Part 5.1.4 and Appendix 5.1.4(2).

4. Rotate members that are in positions of heavy work to light work, so that air consumption is equalized among team members.

5. The final 25%.

6. A Mayday, an activated PASS device, and a low-air alarm.

CHAPTER 7

1. Smoke inhalation.

2. Three—inhaling smoke, getting lost, and running out of air.

3. SCBA air is critically important to survival in structural firefighting; rapid intervention is not rapid; ICs must preload rescue efforts.

4. Calling the Mayday with one's dying breath is too late to prevent close calls from becoming fatalities.

CHAPTER 8

1. Literally, the Vigiles Urbani were the watchers of the city, who patrolled the streets of Rome and watched for fires. The Vigiles Urbani were organized in AD 6 by Augustus Caesar.

2. Jan van der Heiden invented the first fire hose in 1672.

3. CAFS stands for compressed-air foam systems, initially used by wildland firefighters in initial attack lines.

4. Desktop computers in the fire stations provide firefighters with access to communications and online training materials; touch-screen MDCs are included on the fire apparatus; building-inspection files have been converted to electronic databases; dispatch has been revolutionized by CAD; AED data are downloaded over the Internet; and computers are even found in the SCBA and air monitors.

5. To offset the inevitability that technology will eventually fail, firefighters must have a sound foundation in what? Answer: Basic skills such as search techniques, hose handling, pump operations, ladder operations, fire behavior, vertical ventilation skills, PPV techniques, building construction, fireground decision-making skills.

6. Training in the basics.

7. The SCBA, because it allows firefighters to breathe fresh air while inside a structure fire, instead of the toxic smoke and gases that these fires produce.

8. Air management is the skill of managing the air supply in your SCBA and keeping an emergency reserve in case you find yourself in trouble inside the IDLH atmosphere.

9. Most firefighters wait until the low-air alarm activates, which is dangerous because the emergency reserve of SCBA air will be consumed on exit, leaving no air in case things go wrong.

CHAPTER 9

1. The SCBA.

2. The behavior of a person who intentionally inflicts harm by strictly following the orders of management.

3. Develop, train, and implement a policy.

4. By reversing the peer pressure, turning the negative focus on those who are not in line with stated department goals.

5. When they were broken, firefighters would not have them repaired. All other tools are repaired or replaced on a regular basis

6. A good buddy check.

7. Yes, when the change may adversely affect the safety of firefighters.

8. Responding to and returning from alarms.

9. Firefighters should wear their seat belts.

10. The why questions are easy to overlook but are generally the most important.

CHAPTER 10

1. The ROAM.

2. Instruct firefighters to exit before their low-air alarms activate. This will remove the noise of routine activation of low-air alarms from the fireground; in turn, alarm activation becomes recognizable as indication of an actual emergency situation.

3. When firefighters work into their emergency reserve of air in these commercial structures, they are using a mind-set geared toward single-family dwellings.

4. Both breathe the air they bring with them on their backs, and both know that if they run out of air, there is a high likelihood that they will become seriously injured or die. For decades, the dive industry has mandated that divers check their air before they go underwater and at intervals while they are underwater; in addition they require that all divers return with an emergency reserve of 500 psi, only to be used if something bad happens to the diver while underwater.

5. Hazmat and confined-space entry teams are following the ROAM. This is because they understand that work time is limited by SCBA air supply, and they need to exit the IDHL environment—and go through decontamination in the case of hazmat technicians—with air remaining, to keep from breathing life-threatening gases.

6. Firefighters must know that they have a full air cylinder and that their cylinders are functioning properly before they enter the IDLH environment. Inside the fire building, they should check their air at intervals based on the demands of their workload.

7. The officer bases the decision on when to leave the IDLH environment of the fire building on training, as well as several specific factors affecting the duration of the air supply: distance from the entry point,

length of time it took to get where the company is now, whether conditions are worsening, fatigue levels, how fast the member with the lowest air level is consuming air, and stress levels.

8. Before the low-air alarm activates.

CHAPTER 11

1. False.

2. The point at which you start being part of the solution and stop being part of the problem.

3. Emotional preparation.

4. Smoke, heat, and structural stability.

5. True.

6. False.

7. Lack of accountability and poor crew rotation.

8. 12.

9. These operations are done in a methodical manner that is allowed by the type of response.

10. Running out of air.

CHAPTER 12

1. Tradition, fear of not being "one of the guys," and laziness.

2. True.

3. SCBA mastery, realistic operational drills, and communications.

4. False.

5. Situational awareness and the elimination of tunnel vision.

6. Egress and time-to-exit decisions.

7. 50%.

8. Appropriate supervisory ration.

9. Access.

10. False.

11. True.

CHAPTER 13

1. Training in basic firefighting skills.

2. A hands-on training environment.

3. Flowing water, throwing and extending ground ladders, rooftop ventilation, forcible entry with irons and rescue saws, pulling hoselines, interior searches in single-family dwellings and apartment houses, conducting rescue operations from the outside and inside, vehicle extrication, and CPR.

4. Training "refers to the repetitive performance of skills and application of knowledge in an effort to achieve and maintain mastery," whereas learning opportunities are the beginning of the process of becoming a competent firefighter.

5. Following the ROAM.

6. Fire prevention led, as intended, to a decrease in the number of fires; consequently, opportunities for on-the-job training in structural fire attack have all but disappeared, except in impoverished and dense areas.

7. The Seattle Fire Department designed a hands-on training program, called operational skills enhancement training (OSET). OSET takes a back-to-basics approach to structural firefighting and has as its philosophy that operations firefighters need to learn and continually practice the fundamentals of fire behavior, fire attack, rescue, ventilation, water supply, hose streams, ladders, and air management, among other basic skills, in acquired structures.

8. Preventing a firefighter from running out of air in a structure fire means that the RIT does not need to be launched.

9. The 75% solution dictates that 75% of a fire department's training budget, resources, and energy focus on preventing firefighter deaths, while the remaining 25% of a fire department's training budget, resources, and energy focus on firefighter rescue.

CHAPTER 14

1. Themselves and other teams.

2. Radio.

3. So that critical messages get to the appropriate place.

4. Equipment.

5. False.

6. Air.

7. On routine false alarms.

8. Duties.

9. Assignments going uncompleted, unnecessary turmoil in situations that are already confusing, and added risk for the team.

10. Yes, you are prepared to make entry.

CHAPTER 15

1. Congress mandated that the U.S. Forest Service design such a system.

2. The Phoenix Fire Department.

3. Command, operations, planning, logistics/finance, administration, and intelligence.

4. Strategic, tactical, and task levels.

5. NFPA 1561—Standard on Emergency Services Incident Management.

6. True.

7. Failure to do so will lead to ever-increasing injuries or worse, as firefighters are dying at an increasing rate specifically because they are running out of air.

CHAPTER 16

1. Communication is a cycle that requires continual completion; the sender encodes and transmits the message, and the receiver decodes it and feeds back to the sender.

2. Level 1 is acknowledgment of routine information. Level 2 is acknowledgment and feedback of key information. Level 3 is used to acknowledge more complex instructions. This level of acknowledgment also provides an opportunity to conduct dialogue on interpretation of the instructions or to request clarification. The receiver may repeat the order on where to go and what to do, or clarify what is expected if unclear.

3. Incident-related messages, which pertain directly to the incident at which companies are operating, and messages that are not related to the incident, including dispatching, routine radio traffic, and other incidents taking place simultaneously.

4. Radio discipline.

5. Radio discipline, incident management, chain of command, problem reporting, and progress reports.

6. Conditions, actions, resources, and air.

CHAPTER 17

1. Weapons of mass destruction, incorporating such hazardous materials as nuclear, biological, and chemical agents, are the weapons of choice for terrorist organizations.

2. A dirty bomb is a radiological dispersal device. Terrorist use of a dirty bomb is plausible because nuclear material stolen from a health care facility or a university could be incorporated into a pipe bomb, car bomb, suitcase bomb, or suicide bomb and detonated in the middle of a city center.

3. Their health and safety depends on it.

4. Bacteria, viruses, rickettsia, and toxins. Anthrax and cholera are examples of bacterial diseases. Ebola, Machupo, and smallpox are examples of deadly viruses. Q fever, typhus, and Rocky Mountain spotted fever are examples of rickettsial diseases. Ricin, botulinum toxin, and mycotoxins are examples of toxins.

5. Bacterial.

6. Because these weapons are easy to obtain or manufacture, are inexpensive, and have already been used by terrorists against civilian populations.

7. Nerve agents, blister agents, blood agents, choking agents, and irritating agents.

8. Nerve agents act on neurotransmitters to block or disrupt the pathways by which nerves transmit messages.

9. Cyanogen chloride (CK) or hydrogen cyanide (AC), also known as Zyklon B.

10. A choking agent.

11. They know that they must have enough SCBA air to get out of the hot zone and through decontamination; moreover, they are aware that, since something could still go wrong while they are in the warm zone, an emergency reserve is critical to life safety.

CHAPTER 18

1. OSHA.

2. One in five (20%) firefighters got into trouble during training.

3. It took more than 21 minutes.

4. 1200L, 1800L, and 2400L, respectively.

5. Working firefighters will consume between 90 and 105 liters of air per minute.

6. 63% of fireground fatalities are directly attributed to firefighters running out of air (asphyxiation).

7. Of the six fatalities, four were rescuers sent in as rapid intervention resources.

CHAPTER 19

1. 30–45 minutes.

2. Make better decisions and perform at higher levels.

3. Survival.

4. Landmarks.

5. One that was never sent.

6. Strength of construction and openings to the outside.

7. Skip-breathing, the count method, and the REBT.

8. False.

9. 50% of the donor's current air.

10. You are attached, and movement is limited.

CHAPTER 20

1. Heavy smoke in the hallway.

2. Firefighters were taught to leave the hazard area when the low-air alarm started to ring.

3. The sound of a low-air alarm was considered to be a normal noise on the fireground. Because of this false-alarm mentality, the other firefighters did nothing to assist.

4. They were forced to filter-breathe heavy smoke, which resulted in rapid deterioration of their skills.

5. The Washington State Department of Labor and Industries.

6. Air management provides a framework to assist firefighters, company officers, and the command staff in identifying the length and number of work cycles a firefighter has performed.

7. After using two 30-minute cylinders, with or without ROAM; after using 1800L cylinders and following the ROAM; after using one 1800L cylinder without following the ROAM; after using one 2400L cylinder; after using one 1800L cylinder, having used two during the previous work shift; and when the IC says so.

8. The union was concerned that the fire department would increase the work cycle of firefighters without addressing the rest cycle and the overall work/rest interval.

9. If firefighters use the 1800L cylinder and the ROAM, there is only a very slight or no increase in the work cycle. Therefore, the work/rest interval remains the same, and a huge increase in the margin for error is achieved.

10. Through regular and consistent training on the reasons and techniques for air management, rather than through policy implementation.

11. The Phoenix Fire Department, with its on-deck or three-deep system.

12. Specifically address the situation and find out how it happened and how it may be prevented in the future.

CHAPTER 21

1. At each entry point, a dedicated entry control officer is stationed, to maintain firefighter accountability and monitor each firefighter's air supply, with the goal of having all firefighters out of the structure before any low-air alarms activate.

2. Crew work cycles are kept short (to about 10 minutes), and on-deck crews are ready to rotate in when interior crews must leave the hazard area to get another cylinder or go to rehabilitation. Firefighters keep track of their own air supply while in the hazard area, and the company officer is also monitor crew members' air supply.

3. Radio transmissions from dispatch serve as a cue for firefighters breathing air from their SCBA to check how much air remains in their cylinders.

4. Firefighters work until their SCBA cylinders are half empty and then immediately begin exiting the fire building.

5. For someone to be at the command post actively monitoring the computer screen.

6. The entry control officer maintains firefighter accountability and monitor each firefighter's air supply, with the goal of having firefighters out of the structure before any low-air alarms activate.

7. The Phoenix work-cycle system; the UK, Swedish, and Austrian systems; technology-driven systems; and the time-on-scene system.

CHAPTER 22

1. The *why* questions.

2. The National Fire Protection Association (NFPA).

3. Juries, courts, and regulatory authorities like OSHA use the NFPA standards as best practices.

4. To generate acceptance of operational evolutions for coordination and skill.

5. Two to three seconds.

6. False.

7. Answer: Training and experience.

8. No, air management is an individual responsibility.

9. Increasing the low-air warning alarm to 50% of the cylinder size will increase, not decrease, the propensity to consider a low-air warning alarm as a "false alarm." Increasing the "false alarm" mentality will decrease firefighter safety.

10. Conduct the exercise described under "Myth 8" and find out.

INDEX

A

B

C

G

H

I

N

P

S

S

U

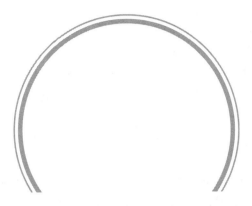

BIOGRAPHICAL SKETCHES
OF THE AUTHORS

Mike Gagliano has more than 20 years of fire, crash, and rescue experience with the Seattle Fire Department and the United States Air Force. He currently serves as the captain of training and is a member of the Seattle Fire Department Operational Skills Enhancement Development Team. Captain Gagliano teaches across the country on air management, firefighter safety and fireground strategy and tactics. He is also a member of the FDIC Advisory Board.

Casey Phillips is a 20-year veteran of the fire service, the last 13 years serving with the Seattle Fire Department. He currently holds the position of captain on Engine 28, and is a member of the Seattle Fire Department Operational Skills Enhancement Development Team. Captain Phillips teaches across the country on air management and firefighter safety.

Phillip Jose is a 20-year veteran of the Seattle Fire Department where he is the captain of Ladder 5 and Rapid Intervention Group 31. He has served as a Lieutenant and Captain in the Training Division and is a member of the Seattle Fire Department Operational Skills Enhancement Development Team. Captain Jose teaches across the country on Air Management and Firefighter Safety with the "Seattle Guys."

Steve Bernocco is a 17-year veteran of the Seattle Fire Department, where he is a lieutenant on Ladder 10. He has served as a training officer, and is currently a member of the SFD's Operational Skills Enhancement Development Team. Lt. Bernocco has written numerous articles and teaches across the country on the topics of air management, firefighter safety and survival, and fireground strategy and tactics with the "Seattle Guys."